Lecture Notes in Computer Science 11493

Commenced Publication in 1973
Founding and Former Series Editors:
Gerhard Goos, Juris Hartmanis, and Jan van Leeuwen

More information about this series at http://www.springer.com/series/7407

Ian McQuillan · Shinnosuke Seki (Eds.)

Unconventional Computation and Natural Computation

18th International Conference, UCNC 2019
Tokyo, Japan, June 3–7, 2019
Proceedings

Editors
Ian McQuillan
University of Saskatchewan
Saskatoon, SK, Canada

Shinnosuke Seki
University of Electro-Communications
Tokyo, Japan

ISSN 0302-9743 ISSN 1611-3349 (electronic)
Lecture Notes in Computer Science
ISBN 978-3-030-19310-2 ISBN 978-3-030-19311-9 (eBook)
https://doi.org/10.1007/978-3-030-19311-9

LNCS Sublibrary: SL1 – Theoretical Computer Science and General Issues

This Springer imprint is published by the registered company Springer Nature Switzerland AG
The registered company address is: Gewerbestrasse 11, 6330 Cham, Switzerland

Preface

This volume of *Lecture Notes in Computer Science* contains the papers presented at the 18th International Conference on Unconventional Computation and Natural Computation (UCNC 2019) organized by the Algorithmic "Oritatami" Self-Assembly Laboratory at the University of Electro-Communications (UEC) during June 3–7, 2019, in Tokyo, Japan as part of the 100th Anniversary Commemorative Event of the University of Electro-Communications (UEC).

The UCNC conference series is one of the major international conference series in unconventional computation, natural computation, and related areas.

It was established by Cristian S. Calude in Auckland, New Zealand in 1998 as Unconventional Models of Computation (UMC).

The second and third editions of UMC were held as a biannual conference in Brussels (2000) and in Kobe (2002).

Dramatic advancement in the fields of unconventional computation and natural computation made it critical to cover also experiments and applications as well as to hold the conference more often.

UMC therefore changed its name to UC (Unconventional Computation) and further to UCNC in 2012, and it has been held annually: in Seville (2005), York (2006), Kingston (2007), Vienna (2008), Ponta Delgada (2009), Tokyo (2010), Turku (2011), Orléans (2012) as UCNC, Milan (2013), London (2014), Auckland (2015), Manchester (2016), Fayetteville (2017), and Fontainebleau (2018).

The International Conference on Unconventional Computation and Natural Computation (UCNC) series provides a genuinely interdisciplinary forum for computation that goes beyond the Turing model to study human-designed computation inspired by nature, and computational nature of processes taking in place in nature.

Its scope includes, among others, the following topics and areas: hypercomputation; chaos and dynamical systems-based computing; granular, fuzzy, and rough computing; mechanical computing; cellular, evolutionary, molecular, neural, and quantum computing; membrane computing; amorphous computing, swarm intelligence; artificial immune systems; physics of computation; chemical computation; evolving hardware; the computational nature of self-assembly, developmental processes, bacterial communication, and brain processes.

There were 32 qualified submissions from 16 countries: China, Finland, France, Germany, Hungary, India, Japan, Latvia, Norway, Romania, Russia, South Korea, Spain, Taiwan, UK, and USA.

Each of the submissions was reviewed by at least three reviewers and thoroughly discussed by the Program Committee (PC).

The PC decided to accept 19 papers for oral presentation.

The volume also includes the abstracts or full papers of the five invited talks given by Noura Al Moubayed, Ho-Lin Chen, Akira Kakugo, Ion Petre, and Susan Stepney.

We warmly thank all the invited speakers and all authors of the submitted papers for making UCNC 2019 quite successful.

As the PC co-chairs, we, Ian McQuillan and Shinnosuke Seki, would like to express our cordial gratitude to the members of the PC and the external reviewers for reviewing the papers and participating in the selection process and helping to maintain the high standard of the UCNC conferences.

We appreciate the help of the EasyChair conference system for facilitating our work of organizing UCNC 2019 very much.

Furthermore, we would like to thank the editorial staff at Springer, and in particular Alfred Hofmann and Anna Kramer, for their guidance and help during the process of publishing this volume.

Last but not the least, we are grateful to the Organizing Committee members: Szilard Zsolt Fazekas, Kohei Maruyama, Reoto Morita, Kei Taneishi, and Fumie Watanabe.

UCNC 2019 was financially supported by JSPS-Kakenhi, Kubota Information Technology, Tateishi Science and Technology Foundation, and the University of Electro-Communications.

We would like to express our sincere gratitude for their support.

We are all looking forward to UCNC 2020 in Vienna.

June 2019

Ian McQuillan
Shinnosuke Seki

Organization

Program Committee

Martyn Amos	Northumbria University, UK
Cristian S. Calude	The University of Auckland, New Zealand
Da-Jung Cho	LRI, Université Paris-Sud, France
Erzsébet Csuhaj-Varjú	Eötvös Loránd University, Hungary
David Doty	University of California, Davis, USA
Jérôme Durand-Lose	Université d'Orléans, France
Giuditta Franco	University of Verona, Italy
Rudolf Freund	Vienna University of Technology, Austria
Masami Hagiya	The University of Tokyo, Japan
Yo-Sub Han	Yonsei University, Republic of Korea
Jacob Hendricks	University of Wisconsin River Falls, USA
Mika Hirvensalo	University of Turku, Finland
Sergiu Ivanov	Université d'Évry Val d'Essonne, France
Natasha Jonoska	University of South Florida, USA
Jarkko Kari	University of Turku, Finland
Lila Kari	University of Waterloo, Canada
Ian McQuillan	University of Saskatchewan, Canada
Makoto Naruse	The University of Tokyo, Japan
Turlough Neary	University of Zurich, Switzerland
Pekka Orponen	Aalto University, Finland
Matthew Patitz	University of Arkansas, USA
Kai Salomaa	Queen's University, Canada
Shinnosuke Seki	The University of Electro-Communications, Japan
Martin Trefzer	University of York, UK
Gunnart Tufte	Norwegian University of Science and Technology, Norway

Additional Reviewers

Aaser, Peter
Arrighi, Pablo
Bernard, Jason
Brijder, Robert
Hader, Daniel
Imai, Katsunobu
Isokawa, Teijiro
Jensen, Johannes
Jirasek, Jozef
Kawamata, Ibuki
Kim, Hwee
Ko, Sang-Ki
Laketic, Dragana
Lazar, Katalin A.
Lykkebø, Odd Rune
Mahalingam, Kalpana

Marquezino, Franklin
Mohammed, Abdulmelik
Nagy, Marius
Nakanishi, Masaki
Ng, Timothy
Orellana-Martín, David
Potapov, Igor
Prigioniero, Luca
Saarela, Aleksi
Winslow, Andrew
Yakaryilmaz, Abuzer

Abstracts of Invited Talks

Deep Learning: Fundamentals, Challenges, and Applications

Noura Al Moubayed

Department of Computer Science, Durham University, Durham, UK
Noura.al-moubayed@durham.ac.uk

Abstract. We live in an era of unprecedented success for machine learning with models used in a wide range of applications, e.g. self-driving cars, medical diagnostics, cyber-security, etc. This revolution is fueled by innovation in deep neural networks combined with a strong backing from tech giants and increased social and scientific interest. Building deep learning models that operate reliably in real-life scenarios comprises solving multiple challenging problems: - The quality and volume of the training data and the availability of labels influence the complexity and sophistication of the models. - Imbalance labels could lead to biased models reducing their reliability. - How to model complex relationships in the data with variable temporal and spatial dependencies? For machine learning models to be deployed in critical fields, other factors come to play including providing explanations of their decisions, ensure algorithmic fairness, identify potential bias, consistent performance, and safety against adversarial attacks. The talk will walk through the fundamental concepts behind deep learning, its current limitations and open questions, and will demonstrate through our work how to apply it in real-life scenarios. Our research focuses on the advancement of deep learning in three main tracks: (I) medical applications: diagnosis of type-2 diabetes, Brain-Computer Interfaces, mortality rate prediction. (II) cyber security in the cloud: early detection and prevention of cyber-attacks (III) natural language processing: sentiment analysis, multi-task learning, multi-modal learning.

On the Complexity of Self-assembly Tasks

Ho-Lin Chen*

National Taiwan University

Abstract. In the theory of computation, the Chomsky hierarchy provides a way to characterize the complexity of different formal languages. For each class in the hierarchy, there is a specific type of automaton which recognizes all languages in the class. There different types of automaton can be viewed as standard requirements in order to recognize languages in these classes. In self-assembly, the main task is to generate patterns and shapes within certain resource limitations. Is it possible to separate these tasks into different classes? If yes, can we find a standard set of self-assembly instructions capable of performing all tasks in each class? In this talk, I will summarize the works towards finding a particular boundary between different self-assembly classes and some trade-offs between different types of self-assembly instructions.

*National Taiwan University. Research supported in part by MOST grant number 107-2221-E-002-031-MY3. Email: holinchen@ntu.edu.tw.

Construction of Molecular Swarm Robot

Jakia Jannat Keya[1] and Akira Kakugo[1,2]

[1] Faculty of Science, Hokkaido University, Sapporo 060-0810, Japan
[2] Graduate School of Chemical Sciences and Engineering, Hokkaido University, Sapporo 060-0810, Japan

Abstract. In nature swarming behavior evolves repeatedly among motile organisms offering a variety of beneficial emergent properties. Switching between solitary and swarm behavior these can include improved information gathering, protection from predators, and resource utilization. Inspired by the aspects of swarming behavior, attempts have been made to construct swarm robots from motile supramolecular systems composed of biomolecular motor system where cross-linkers induce large scale organization. Here, we demonstrate that the swarming of DNA-functionalized microtubules (MTs) propelled by surface-adhered kinesin motors can be programmed and reversibly regulated by DNA signals. Emergent swarm behavior, such as translational and circular motion, can be selected by tuning the MT stiffness. Photoresponsive DNA containing azobenzene groups enables switching of swarm behavior in response to stimulation with visible or ultraviolet light. Such control of swarming offers design and development of molecular swarm robots based on natural molecular devices.

Keywords. Swarming · Biomolecular motors system · DNA · Swarm robot

1 Introduction

Swarm robot has appeared as the most attractive area in the field of robotics in which large number of robots are coordinated in a distributed and decentralized way. Compared to the complexity of the task to achieve, simple robots are programmed into swarm based on the local rules which is solely inspired by the nature. In nature swarming enables the living organisms to obtain several advantages including parallelism, robustness, and flexibility, which are unachievable by a single entity.[1,2] Such features of the living organisms have motivated the researchers for construction of swarm robots. Although utilization of a large number building blocks and realizing their programmable interactions among the building blocks, as the nature does, has remained the biggest challenge.[3] Hence, we took the advantage of natural supramolecular system MT/kinesin providing large number of molecular sized self-propelled objects which are powered by adenosine tri-phosphate (ATP). By utilizing the unique hybridization properties of DNA[4] together with MT/kinesin, swarming of a large number of self-propelled units was successfully controlled.[5] Swarming of such self-propelled molecular system can be reversibly regulated by DNA signals with translational and rotational motion depending on various stiffness of MTs.

Moreover, introducing a photosensor, swarming of the MTs is repeatedly regulated by photoirradiation. The highly programmable system comprising of three essential components; actuator, processor, and sensor has brought a breakthrough in the field of swarm robotics based on natural molecular systems.

2 Experimental

Swarm units were prepared by polymerizing MTs from azide (N_3)-modified tubulin using either guanosine-5'-triphosphate (GTP) or guanylyl-(α, β-methylene-triphosphate) (GMPCPP) to construct the swarm robots. N_3-modified MTs were modified with fluorescent dye labeled dibenzocyclooctyne (DBCO) DNA of different sequences to obtain DNA modified MTs by azide-alkyne cycloaddition (click reaction). In vitro motility of different DNA modified MTs was demonstrated on kinesin coated surface in presence of ATP.

3 Results and Discussion

About five million self-propelled MTs on a kinesin-coated surface was used as molecular actuators in presence of ATP which were modified by bimolecular processor DNA as receptors. DNA crosslinker (*l*-DNA) complementary to the receptors was added which can pair up DNA tethered MTs to form swarms with both translational and rotational motion. Single strand DNA complementary to *l*-DNA was introduced in

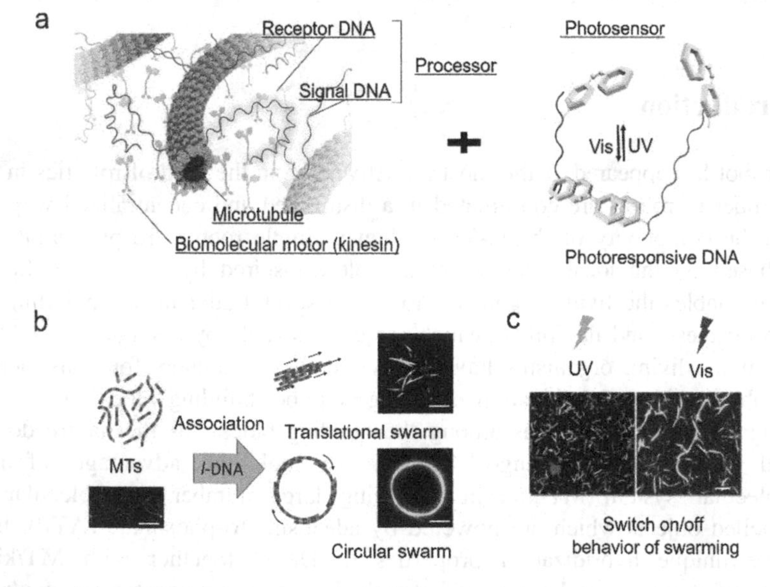

Fig. 1 Schematic representation of molecular swarm robots comprised of biomolecular actuators, DNA processors and azobenzene photosensors (a). Translational and circular swarm of MTs associated by DNA signal (b). Switch on/off control of swarming of MTs by photoirradiation. Scale bar: 20 μm.

the system to dissociate the assembled structures of DNA modified MTs. The swarming was able to reversibly control with programmable complex behavior. Incorporating a light sensitive azobenzene unit in DNA strands, MT swarms were further repeatedly regulated by photoirradiation up to several cycles utilizing cis-trans isomerization of azobenzene. Tuning the length of *l*-DNA sequences swarming can be controlled with thermodynamic and kinetic feasibility. Such regulation of swarming in molecular level interaction enables the construction of molecular swarm robots.

References

1. Whitesides, G.M., Grzybowski, B.: Science **295**, 2418–2421 (2002)
2. Bonabeau, E., Dorigo, M., Theraulaz, G.: Swarm Intelligence: From Natural to Artificial Systems. Oxford University Press, Oxford, New York (1999)
3. Rubenstein, M., Cornejo, A., Nagpal, R.: Science **345**, 795–799 (2014)
4. Qian, L., Winfree, E.: Science **332**, 1196–1201 (2011)
5. Keya, J.J., Suzuki, R., Kabir, A.M.R., Inoue, D., Asanuma, H., Sada, K., Hess, H., Kuzuya, A., Kakugo, A.: Nat. Commun. **9**, 453 (2018)

Network Controllability: Algorithmics for Precision Cancer Medicine

Ion Petre[1,2]

[1] Department of Mathematics and Statistics, University of Turku, Finland
[2] National Institute for Research and Development in Biological Science, Romania
ion.petre@utu.fi

Abstract. The intrinsic robustness of living systems against perturbations is a key factor that explains why many single-target drugs have been found to provide poor efficacy or to lead to significant side effects. Rather than trying to design selective ligands that target individual receptors only, network polypharmacology aims to modify multiple cellular targets to tackle the compensatory mechanisms and robustness of disease-associated cellular systems, as well as to control unwanted off-target side effects that often limit the clinical utility of many conventional drug treatments. However, the exponentially increasing number of potential drug target combinations makes the pure experimental approach quickly unfeasible, and translates into a need for algorithmic design principles to determine the most promising target combinations to effectively control complex disease systems, without causing drastic toxicity or other side-effects. Building on the increased availability of disease-specific essential genes, we concentrate on the target structural controllability problem, where the aim is to select a minimal set of driver/driven nodes which can control a given target within a network. That is, for every initial configuration of the system and any desired final configuration of the target nodes, there exists a finite sequence of input functions for the driver nodes such that the target nodes can be driven to the desired final configuration. We investigated this approach in some pilot studies linking FDA-approved drugs with cancer cell-line-specific essential genes, with some very promising results.

Co-Designing the Computational Model and the Computing Substrate (Invited Paper)

Susan Stepney[1,2]

[1] Department of Computer Science, University of York, York, UK
susan.stepney@york.ac.uk
[2] York Cross-disciplinary Centre for Systems Analysis, University of York, York, UK

Abstract. Given a proposed unconventional computing substrate, we can ask: Does it actually compute? If so, how well does it compute? Can it be made to compute better? Given a proposed unconventional computational model we can ask: How powerful is the model? Can it be implemented in a substrate? How faithfully or efficiently can it be implemented? Given complete freedom in the choice of model and substrate, we can ask: Can we co-design a model and substrate to work well together?

Here I propose an approach to posing and answering these questions, building on an existing definition of physical computing and framework for characterising the computing properties of given substrates.

Contents

On the Complexity of Self-assembly Tasks

Ho-Lin Chen(✉)

National Taiwan University, Taipei, Taiwan
holinchen@ntu.edu.tw

Abstract. In the theory of computation, the Chomsky hierarchy provides a way to characterize the complexity of different formal languages. For each class in the hierarchy, there is a specific type of automaton which recognizes all languages in the class. There different types of automaton can be viewed as standard requirements in order to recognize languages in these classes. In self-assembly, the main task is to generate patterns and shapes within certain resource limitations. Is it possible to separate these tasks into different classes? If yes, can we find a standard set of self-assembly instructions capable of performing all tasks in each class? In this talk, I will summarize the works towards finding a particular boundary between different self-assembly classes and some trade-offs between different types of self-assembly instructions.

Self-assembly is the process in which simple monomers autonomously assemble into large, complex structures. There are many spectacular examples in nature, ranging from crystal formation to embryo development. In nanotechnology, as the size of the system gets smaller, precise external control becomes prohibitively costly. It is widely believed that algorithmic self-assembly will eventually become an important tool in nanotechnology, enabling the fabrication of various nano-structures and nano-machines. DNA molecules have received a lot of attention due to its small size and programmable nature. Algorithmic self-assembly of DNA molecules has been used in many tasks including performing computation [3,24,29], constructing molecular patterns [12,16,22,23,26,34], and building nano-scale machines [4,13,17,25,27,33].

Many different theoretical models have been proposed to describe these DNA self-assembly processes. The first algorithmic DNA self-assembly model is the abstract Tile Assembly Model (aTAM) proposed by Rothemund and Winfree [21,29]. In the aTAM model, DNA molecules (called "tiles") attaches to an initial seed structure one by one and assemble into large two-dimensional structures. Many theoretical designs in aTAM have been successfully implemented using DNA molecules. [3,7,23,30] are some of the earliest examples. Self-assembly systems in aTAM can perform Turing-universal computation [29] and produce arbitrary shapes [20,28] using only a small number of different tile types (if arbitrary scaling is allowed). Many studies also focused on constructing structures of fixed sizes such as counters [1,9] and squares [14,19,21]. It is also known that aTAM is intrinsically universal [15].

Research supported in part by MOST grant number 107-2221-E-002-031-MY3.

I. McQuillan and S. Seki (Eds.): UCNC 2019, LNCS 11493, pp. 1–4, 2019.
https://doi.org/10.1007/978-3-030-19311-9_1

There are many different theoretical extensions to the aTAM model. The hierarchical (2-handed) self-assembly model [2] allows two large assemblies to attach to each other in one step. The nubot model [31] allows tiles to change states and make relative movement against each other (and push other tiles around) after attaching. One of the most significant difference between these two models and the original aTAM model is the ability to perform superlinear growth. In the aTAM model, assembling any structure of diameter n takes $\Omega(n)$ expected time. In hierarchical self-assembly model, there exists a tile system which assembles a structure of length n in time $O(n^{4/5} \log n)$ [5]. In the nubot model, there exists a system which assembles a line of length n in time $O(\log n)$ [31]. Furthermore, squares and arbitrary shapes of diameter n can also be assembled in poly-logarithmic time without any scaling [31].

A natural question to ask here is "what is the minimal extension to aTAM in order to achieve superlinear/exponential growth speed?" Several works focused on finding a minimal set of instructions from the nubot model which allows exponential growth to happen [6,10]. First, Chen et al. [6] proved that the exponential growth can be achieved without directional movements and flexible (soft) structures. Random agitation on rigid structures is sufficient. Second, Chin et al. [10] studied the problem of finding the minimum state change required for each tile. They proved that two state changes are required under very strict restriction on extra spaces, but one state change is sufficient if the restrictions are slightly relaxed. Third, Hou and Chen [18] constructed a nubot system which grows exponentially without any state changes. However, the construction requires several modifications to the nubot model. Furthermore, unlike the two works stated above, this construction do require soft structures which can be pushed around by the directional movements of other tiles.

Although the above works provide some basic understanding on the minimum requirement of exponential growth, many related questions remain open. First, it would be interesting to check whether the extra assumptions in [18] is necessary or not. Second, some other self-assembly models different from the nubot model also achieves exponential growth [11,32]. Finding the minimum set of instructions for exponential growth in these models is also a natural problem. Third, the minimum requirement for superlinear growth remains open. There might even be more complexity classes between linear growth and exponential growth. Last, Chen et al. [8] proved that the nubot model can perform parallel computation which provides an exponential speed up compared to computation using aTAM. Comparing the minimum requirement for this exponential speed up with the minimum requirement for exponential growth may provide more information about the complexity of self-assembly tasks.

References

1. Adleman, L., Cheng, Q., Goel, A., Huang, M.-D.: Running time and program size for self-assembled squares. In: Proceedings of the 33rd Annual ACM Symposium on Theory of Computing, pp. 740–748 (2001)
2. Aggarwal, G., Cheng, Q., Goldwasser, M.H., Kao, M.-Y., Espanés, P., Schweller, R.: Complexities for generalized models of self-assembly. SIAM J. Comput. **34**, 1493–1515 (2005)
3. Barish, R.D., Rothemund, P.W.K., Winfree, E.: Two computational primitives for algorithmic self-assembly: copying and counting. Nano Lett. **5**(12), 2586–2592 (2005)
4. Bishop, J., Klavins, E.: An improved autonomous DNA nanomotor. Nano Lett. **7**(9), 2574–2577 (2007)
5. Chen, H.-L., Doty, D.: Parallelism and time in hierarchical self-assembly. In: Proceedings of the Twenty-Third Annual ACM-SIAM Symposium on Discrete Algorithms, pp. 1163–1182 (2012)
6. Chen, H.-L., Doty, D., Holden, D., Thachuk, C., Woods, D., Yang, C.-T.: Fast algorithmic self-assembly of simple shapes using random agitation. In: Murata, S., Kobayashi, S. (eds.) DNA 2014. LNCS, vol. 8727, pp. 20–36. Springer, Cham (2014). https://doi.org/10.1007/978-3-319-11295-4_2
7. Chen, H.-L., Schulman, R., Goel, A., Winfree, E.: Error correction for DNA self-assembly: preventing facet nucleation. Nano Lett. **7**, 2913–2919 (2007)
8. Chen, M., Xin, D., Woods, D.: Parallel computation using active self-assembly. Nat. Comput. **14**(2), 225–250 (2015)
9. Cheng, Q., Goel, A., Moisset, P.: Optimal self-assembly of counters at temperature two. In: Proceedings of the 1st Conference on Foundations of Nanoscience: Self-Assembled Architectures and Devices, pp. 62–75 (2004)
10. Chin, Y.-R., Tsai, J.-T., Chen, H.-L.: A minimal requirement for self-assembly of lines in polylogarithmic time. In: Brijder, R., Qian, L. (eds.) DNA 2017. LNCS, vol. 10467, pp. 139–154. Springer, Cham (2017). https://doi.org/10.1007/978-3-319-66799-7_10
11. Dabby, N., Chen, H.-L.: Active self-assembly of simple units using an insertion primitive. In: Proceedings of the Twenty-Fourth Annual ACM-SIAM Symposium on Discrete Algorithms, pp. 1526–1536. SIAM (2013)
12. Dietz, H., Douglas, S., Shih, W.: Folding DNA into twisted and curved nanoscale shapes. Science **325**, 725–730 (2009)
13. Ding, B., Seeman, N.: Operation of a DNA robot arm inserted into a 2D DNA crystalline substrate. Science **384**, 1583–1585 (2006)
14. Doty, D.: Randomized self-assembly for exact shapes. SIAM J. Comput. **39**(8), 3521–3552 (2010)
15. Doty, D., Lutz, J.H., Patitz, M.J., Schweller, R.T., Summers, S.M., Woods, D.: The tile assembly model is intrinsically universal. In: Proceedings of the 53rd Annual IEEE Symposium on Foundations of Computer Science (2012)
16. Douglas, S., Dietz, H., Liedl, T., Hogberg, B., Graf, F., Shih, W.: Self-assembly of DNA into nanoscale three-dimensional shapes. Nature **459**, 414–418 (2009)
17. Green, S., Bath, J., Turberfield, A.: Coordinated chemomechanical cycles: a mechanism for autonomous molecular motion. Phys. Rev. Lett. **101**, 238101 (2008)
18. Hou, C.-Y., Chen, H.-L.: An exponentially growing nubot system without state changes. In: McQuillan, I., Seki, S. (eds.) UCNC 2019. LNCS, vol. 11493, pp. 122–135 (2019)

19. Kao, M.-Y., Schweller, R.: Reducing tile complexity for self-assembly through temperature programming. In: Proceedings of the 17th Annual ACM-SIAM Symposium on Discrete Algorithms, pp. 571–580 (2006)
20. Lagoudakis, M., LaBean, T.: 2D DNA self-assembly for satisfiability. In: Proceedings of the 5th DIMACS Workshop on DNA Based Computers in DIMACS Series in Discrete Mathematics and Theoretical Computer Science, vol. 54, pp. 141–154 (1999)
21. Rothemund, P., Winfree, E.: The program-size complexity of self-assembled squares (extended abstract). In: Proceedings of the 32nd Annual ACM Symposium on Theory of Computing, pp. 459–468 (2000)
22. Rothemund, P.W.K.: Folding DNA to create nanoscale shapes and patterns. Nature **440**, 297–302 (2006)
23. Rothemund, P.W.K., Papadakis, N., Winfree, E.: Algorithmic self-assembly of DNA Sierpinski triangles. PLoS Biol. **2**, 424–436 (2004)
24. Seelig, G., Soloveichik, D., Zhang, D., Winfree, E.: Enzyme-free nucleic acid logic circuits. Science **314**, 1585–1588 (2006)
25. Sherman, W.B., Seeman, N.C.: A precisely controlled DNA bipedal walking device. Nano Lett. **4**, 1203–1207 (2004)
26. Shih, W.M., Quispe, J.D., Joyce, G.F.A.: A 1.7-kilobase single-stranded DNA that folds into a nanoscale octahedron. Nature **427**, 618–621 (2004)
27. Shin, J.-S., Pierce, N.A.: A synthetic DNA walker for molecular transport. J. Am. Chem. Soc. **126**, 10834–10835 (2004)
28. Soloveichik, D., Winfree, E.: Complexity of self-assembled shapes. SIAM J. Comput. **36**, 1544–1569 (2007)
29. Winfree, E.: Algorithmic self-assembly of DNA. Ph.D. thesis, California Institute of Technology, Pasadena (1998)
30. Winfree, E., Liu, F., Wenzler, L., Seeman, N.: Design and self-assembly of two-dimensional DNA crystals. Nature **394**, 539–544 (1998)
31. Woods, D., Chen, H.-L., Goodfriend, S., Dabby, N., Winfree, E., Yin, P.: Active self-assembly of algorithmic shapes and patterns in polylogarithmic time. In: Proceedings of the 4th Conference on Innovations in Theoretical Computer Science, ITCS 2013, pp. 353–354 (2013)
32. Yin, P., Choi, H.M.T., Calvert, C.R., Pierce, N.A.: Programming biomolecular self-assembly pathways. Nature **451**(7176), 318–322 (2008)
33. Yurke, B., Turberfield, A., Mills Jr., A., Simmel, F., Neumann, J.: A DNA-fuelled molecular machine made of DNA. Nature **406**, 605–608 (2000)
34. Zhang, Y., Seeman, N.: Construction of a DNA-truncated octahedron. J. Am. Chem. Soc. **116**(5), 1661 (1994)

Co-Designing the Computational Model and the Computing Substrate
(Invited Paper)

Susan Stepney[1,2(✉)]

[1] Department of Computer Science, University of York, York, UK
susan.stepney@york.ac.uk
[2] York Cross-disciplinary Centre for Systems Analysis, University of York, York, UK

Abstract. Given a proposed unconventional computing substrate, we can ask: Does it actually compute? If so, how well does it compute? Can it be made to compute better? Given a proposed unconventional computational model we can ask: How powerful is the model? Can it be implemented in a substrate? How faithfully or efficiently can it be implemented? Given complete freedom in the choice of model and substrate, we can ask: Can we co-design a model and substrate to work well together?

Here I propose an approach to posing and answering these questions, building on an existing definition of physical computing and framework for characterising the computing properties of given substrates.

1 Introduction

There are many proposed unconventional computational models: reservoir computing, general purpose analogue computing, membrane computing, reaction-diffusion computing, quantum computing, morphogenetic computing, and more. There are just as many proposed unconventional computing substrates: slime moulds, carbon nanotubes, gene engineered bacteria, gold nanoparticle networks, memristors, optical systems, and more. But how to match model and substrate to get an effective unconventional computer?

In order to tackle this question, I work through several stages. I describe one definition of what is meant by *physical computing*, to distinguish a system that is computing from one which is just 'doing its thing' (Sect. 2). I describe a method for determining how well a given substrate performs as a physical computer implementing some computational model (Sect. 3). I then discuss how this method could be adapted to determine how well a given computational model captures the computing performed by some substrate (Sect. 4), and then how these approaches might be combined to co-design model and substrate (Sect. 5).

I. McQuillan and S. Seki (Eds.): UCNC 2019, LNCS 11493, pp. 5–14, 2019.
https://doi.org/10.1007/978-3-030-19311-9_2

2 When Does a Physical System Compute?

2.1 Abstraction/Representation Theory

It is necessary to be able to determine when a substrate is specifically computing, as opposed to merely undergoing the physical processes of that substrate, before we can determine how well it is doing so.

To address the question of when a physical system is computing, we use *abstraction/representation* theory (AR theory) [14–17,35], in which science, engineering, and computing are defined as a form of *representational activity*, requiring the use of a 'representation relation' to link a physical system and an abstract model in order to define their operation. We use AR theory to distinguish scientific experimentation on a novel substrate from the performance of computation by that substrate.

The compute cycle is shown in Fig. 1. An abstract problem A is encoded into the computational model as $m_{\mathbf{p}}$; abstractly the computation C produces $m'_{\mathbf{p}}$; this is decoded as the solution to the problem, A'. To implement this abstract computation, the encoded model is instantiated into the physical computer state $\mathbf{p}$; the computer calculates via $\mathbf{H}(\mathbf{p})$, evolving into physical state $\mathbf{p}'$; the final state is represented as the final abstract model $m_{\mathbf{p}'}$.

For the abstract computation to have been correctly physically computed, we require $m_{\mathbf{p}'} \simeq m'_{\mathbf{p}}$ (where how close the approximate equality needs to be is device- or problem-dependent). Ensuring this equality holds is a process of debugging the physical system, including how it is instantiated (engineered, programmed and provided with input data), and how its output is represented. Then we say that the initial instantiation, physical evolution, and final representation together implement the desired abstract computation $C(m_{\mathbf{p}})$.

From this model, Horsman et al. [14] define computing as *the use of a physical system to predict the outcome of an abstract evolution.*

2.2 Example: AR Theory Applied to Reservoir Computing

In order to demonstrate how AR theory can be applied to unconventional computing, we here outline its use applied to reservoir computing (RC) *in materio* [6]. RC is a popular choice of model for many unconventional substrates, because it treats the substrate as a black box. The stages of AR theory specialised to RC with a carbon nanotube substrate are:

- computational model: a formal model of reservoir computing, such as the Echo State Network model [18].
- instantiation:
  - the substrate is engineered to be an RC: here the physical RC is a blob of carbon nanotubes in polymer, deposited on an electrode array
  - the substrate is configured (programmed) to be an RC suitable for a particular task: here by applying voltages to a subset of the electrodes; which voltages to apply to which electrodes for a particular task are typically evolved during the programming phase
  - the substrate is provided input data, through another subset of electrodes

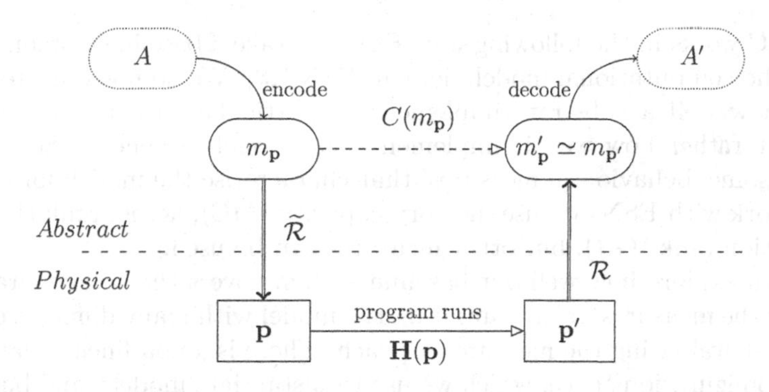

Fig. 1. Physical computing in AR theory. The full compute cycle, starting from initial abstract problem A, encoding in terms of an abstract computational model, instantiation into a physical computer $\mathbf{p}$, physical evolution of the device, followed by representation of the final physical state, then decoding to the abstract answer to the problem A'.

- physical substrate evolution: given the input configuration (program plus data) the physical system evolves under the laws of physics, producing an output
- representation: the output is extracted as voltages from a further subset of the electrodes, then passed through the RC output filter (trained during the programming phase) here implemented in a PC.

This description makes it clear that there is computation done in both the instantiation stage and the representation stage, in addition to the physical compute stage. Here the computing in the instantiation stage occurs during the 'programming' substage; this effort can be amortised over many runs on specific data, much as how classical computing ignores the cost of compilation and testing when analysing the computational resource costs of algorithms. The computing performed in the representation stage here, however, of processing with the output filter, occurs for each run, and so needs to be included in the overall cost of the computation.

Such instantiation and representation costs are one way to 'hide' the true cost of unconventional computation [3], and care must be taken to analyse these fully, in addition to the use of other unconventional computing resources [2].

3 How Well Does a Physical System Compute?

3.1 How Well Does a Carbon Nanotube Reservoir Compute?

AR theory defines *when* a substrate is computing with respect to a model. We have employed this to developed CHARC, a method to characterise *how well* some material substrate can instantiate and implement a given computational model [5,7]. We have applied CHARC to a reservoir computing model instantiated in carbon nanotubes, and also in simulations.

CHARC works in the following way. First, we take a baseline system: a simulation of the computational model, here an ESN [18]. We do not want to merely examine how well a substrate implements a particular task running under a model, but rather how well it implements the model in general. So next, we decide on some 'behavioural measures' that characterise the model: for our definitional work with ESNs we use memory capacity (MC), kernel rank (KR), and generalisation rank (GR), but other measures can be used.

We then explore how well our baseline system covers the behavioural space defined by the measures, by instantiating the model with many different configurations, and evaluating the measures for each. There is a non-linear relationship between configuration space, which we use to instantiate models, and behaviour space, where we measure their properties. This means that random selection in configuration space leads to a biased sampling of behaviour space. Instead we use Novelty Search [25–27] to explore as much of behaviour space as possible.

Once we have a baseline of the amount of behaviour space covered by the baseline system (Fig. 2, black), we repeat the process with the candidate substrate, and measure how much of the behaviour space it covers: how many different reservoirs it can instantiate (Fig. 2, grey). We see from the figure that our current carbon nanotube system can instantiate reservoirs, but only relatively small ones.

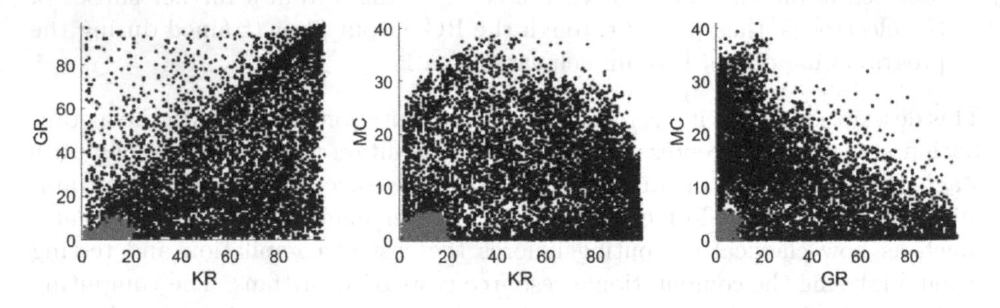

Fig. 2. Carbon nanotube in polymer behaviour space (grey) superimposed on a 100 node ESN behaviour space (black). From [7]

3.2 How Well Could a Carbon Nanotube Reservoir Compute?

We have demonstrated that our carbon nanotubes in polymer, deposited on a 64 electrode array, can be instantiated as small reservoir computers.

We could use the CHARC framework to engineer bigger and better *in materio* RCs, by using the amount of behaviour space covered as a fitness function in a search process, exploring carbon nanotube and polymer densities and types.

3.3 How Well Do Other Substrates Compute?

The current CHARC framework can be used directly to assess other proposed RC substrates, simulated or physical, in the same manner.

The CHARC approach is not limited to reservoir computing, however. It can be adapted to evaluate how well a substrate implements other computational models. A new set of measures that provide a behaviour space specific to the new computational model need to be defined. Then the process is the same: define the behaviour space; use a simulation of the model itself as a baseline; discover what region of behaviour space the proposed substrate can be instantiated to cover, relative to the baseline.

4 How Well Does a Computational Model Fit?

So far we have been focussing on evaluating, and eventually designing, substrates with respect to a given computational model: given the top abstract process of Fig. 1, evaluate the suitability of a given substrate to instantiate the bottom physical process. In this section I discuss the reverse problem: evaluating, and eventually designing, computational models with respect to a given substrate.

Most historical computational models were abstracted from existing substrates, for example: electronic analogue computing from the ability of circuits to implement differential equations; the Turing machine model from human 'computers' calculating by following precise instructions. Some unconventional computational models are inspired by the behaviours of specific physical, chemical and biological substrates: for example, reaction-diffusion models, and membrane models. However, there is often no explicit experimental validation of such theoretical models: how well do they capture the computing done by the substrate, or are they an abstraction too far?

Consider reversing the CHARC framework, to explore the space of computational model representations with respect to a given substrate. Performing such an exploration would need a language to express the computational models that provide the search landscape. A dynamical systems view of computation [34] could provide one such language. A dynamical system comprises a set of state variables, with equations over these variables defining the system dynamics. The relevant behaviour space would comprise dynamical properties: trajectories through state space, transients, attractors, bifurcation structure of parameterised systems, and so on.

Such a system can be visualised as graph, where nodes represent state variables, and contain the equations as state transition automata or time evolution definitions; the links show the explicit dependencies between variables. Such a visualisation is typically used for Cellular Automata, where the 'cells' correspond to the graph nodes, and the links are implicit connections to the neighbourhood nodes. Moving around the model space moves through the space of such graphs.

We are used to thinking of computational models in dynamical systems terms, even if not explicitly. Several examples include:

- Cellular Automata [1,36–38]: discrete time, discrete space, regular topology, node equations are boolean functions (Fig. 3a)
- Random Boolean Networks [9,20,21]: discrete time, discrete space, random topology, node equations are boolean functions (Fig. 3b)

- Coupled Map Lattices [19], threshold coupled maps [29–31]: discrete time, continuous variable, linear topology, each node equation is a (typically logistic) map with input (Fig. 3c)
- Reservoir computers: the ESN model is discrete time, continuous variable, random topology, node equations are some non-linear sum of inputs (Fig. 3d)

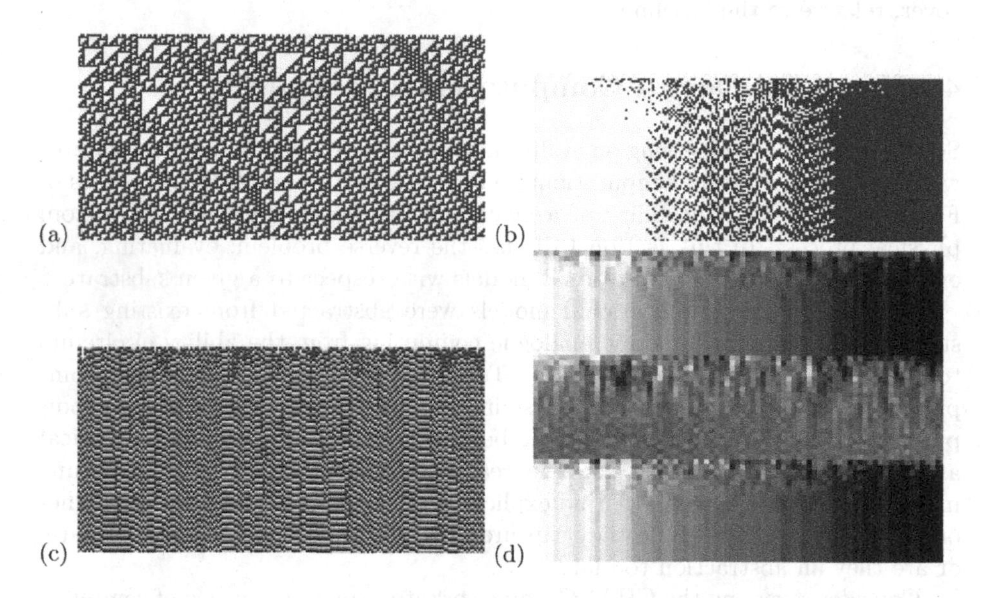

Fig. 3. The time evolution of a range of computational dynamical systems, showing their complex dynamics. The N components of the state vector run along the x axis; the T timesteps run down the y axis. (a) Elementary CA rule 110, $N = 400$, $T = 200$, random (50% 0s, 50% 1s) initial condition, periodic boundary conditions. (b) A $K = 2$ RBN with $N = 200$, $T = 80$; state components are ordered to expose the attractor and the 'frozen core'. (c) A threshold coupled lattice, threshold $x_* = 0.971$, $N = 200$, $T = 100$; each component's value $x_n \in [0, 1]$ is indicated by its shading from white $= 0$ to black $= 1$. (d) An ESN reservoir, with $N = 100$, $T = 100$; each node's value $x_n \in [-1, 1]$ is indicated by its shading from white $= -1$ to black $= 1$. Input is a square function, cycling through 1 for 20 timesteps then 0 for 20 timesteps. This ESN is in the chaotic dynamics regime.

A computational dynamical system needs to allow inputs and outputs. Inputs may be provided on initialisation, as the intial state, as is typically the case with CAs and RBNs. Alternatively, the dynamical system can be *open*, allowing inputs through time as the dynamics evolves, as is typical with RCs. Outputs may comprise the final attractor state, or may be read by observing (a projection of) the system state through time.

One thing all the systems in Fig. 3 have in common is a fixed size state space. However, classical computational models allow the state space to change as the computation progresses: in a Turing Machine, symbols can be written to a growing tape; in higher-level programming languages, variables can come in and go out of scope. This implies that the language of dynamical systems we wish to use should support developmental, or *constructive*, dynamical systems, where the state space can change in response to the progression of the computation. There is no need for a substrate to mirror such growth explicitly: it could instead allow computing processes to move into new areas of existing material (as in classical computing memory). However, more biological substrates with actual material growth might require such a growth model. Such a dynamical system has a higher level *meta-dynamics*. Approaches such as BIOMICS [8,28] and MGS [12,33] provide suggestive starting points for such languages.

Dynamical systems may be coupled in flat or hierarchical architectures. For example:

- Coupled Lattice Maps and threshold coupled maps are themselves a coupling of multiple instances of some discrete-time dynamical system, typically the logistic map
- Quasi-uniform cellular automata comprise coupled patches of CAs, each with different rules [13,32]
- Reservoirs can be coupled together, to form 'reservoirs of reservoirs' [4]
- RBN 'atoms' can be combined into 'molecular' boolean networks using an Artificial Chemistry [10,11,24]

One reason for considering unconventional substrates is that they may perform some tasks 'better' than conventional substrates: if not faster, then maybe with lower power, in hostile environments, in ways matched to what they are processing, or some other advantage. This leads to the idea of combining various disparate substrates, each doing what it does best, into a *heterotic* computing system [17,22,23]: a multi-substrate system that provides an advantage over a single-substrate system.

So, in summary, an appropriate language for capturing computational models that could be explored in a CHARC-like framework is one that supports discrete and continuous open dynamical systems, including constructive dynamical systems with a meta-dynamics, that can be combined and coupled in hierarchical architectures and in heterotic architectures.

Once such an approach is established, it could be extended to include stochastic systems and quantum systems in order to cover the full gamut of potential computing substrates.

5 Co-Designing Models and Substrates

I have described a demonstrated approach, CHARC, that can be used to evaluate substrates for how well they can instantiate the RC model. This approach can be extended to other computational models, and could be put in a design loop to engineer improved substrates.

I have also discussed reversing this process, to design computational models appropriate for given substrates.

The final suggested step is to combine both these processes into a co-design approach: design models to fit substrates, whilst engineering those substrates to better fit the models. This would allow other features to be considered in the overall system, from ease of manufacture of the substrate to expressivity and naturalness of the model.

Additionally, the vertical lines in Fig. 1 (instantiation and representation; including input and output) can be addressed here in a unified manner.

6 Conclusion

In order to take unconventional computing to the next level, from small demonstrators and simple devices to large scale systems, we need a systematic engineering approach.

The discussion here is meant to point the way to one potential such process: a framework for co-designing computational models and computing substrates, to ensure the benefits of unconventional computing can be achieved without sacrificing novelty in either side of the design.

Acknowledgements. Thanks to my colleagues Matt Dale, Dom Horsman, Viv Kendon, Julian Miller, Simon O'Keefe, Angelika Sebald, and Martin Trefzer for illuminating discussions, and collaboration on the work that has led to these ideas.

This work is part-funded by the SpInspired project, UK Engineering and Physical Sciences Research Council (EPSRC) grant EP/R032823/1.

References

1. Adamatzky, A. (ed.): Game of Life Cellular Automata. Springer, London (2010). https://doi.org/10.1007/978-1-84996-217-9
2. Blakey, E.: Unconventional computers and unconventional complexity measures. In: Adamatzky, A. (ed.) Advances in Unconventional Computing. ECC, vol. 22, pp. 165–182. Springer, Cham (2017). https://doi.org/10.1007/978-3-319-33924-5_7
3. Broersma, H., Stepney, S., Wendin, G.: Computability and complexity of unconventional computing devices. In: Stepney, S., Rasmussen, S., Amos, M. (eds.) Computational Matter. NCS, pp. 185–229. Springer, Cham (2018). https://doi.org/10.1007/978-3-319-65826-1_11
4. Dale, M.: Neuroevolution of hierarchical reservoir computers. In: GECCO 2018, Kyoto, Japan, pp. 410–417. ACM (2018)
5. Dale, M., Dewhirst, J., O'Keefe, S., Sebald, A., Stepney, S., Trefzer, M.A.: The role of structure and complexity on reservoir computing quality. In: McQuillan, I., Seki, S. (eds.) UCNC 2019. LNCS, vol. 11493, pp. 52–64. Springer, Heidelberg (2019)
6. Dale, M., Miller, J.F., Stepney, S.: Reservoir computing as a model for ***In-Materio*** computing. In: Adamatzky, A. (ed.) Advances in Unconventional Computing. ECC, vol. 22, pp. 533–571. Springer, Cham (2017). https://doi.org/10.1007/978-3-319-33924-5_22
7. Dale, M., Miller, J.F., Stepney, S., Trefzer, M.A.: A substrate-independent framework to characterise reservoir computers. arXiv:1810.07135 (2018)

8. Dini, P., Nehaniv, C.L., Rothstein, E., Schreckling, D., Horváth, G.: BIOMICS: a theory of interaction computing. Computational Matter. NCS, pp. 249–268. Springer, Cham (2018). https://doi.org/10.1007/978-3-319-65826-1_13
9. Drossel, B.: Random boolean networks. In: Schuster, H.G. (ed.) Reviews of Nonlinear Dynamics and Complexity, vol. 1. Wiley, Weinheim (2008)
10. Faulconbridge, A., Stepney, S., Miller, J.F., Caves, L.S.D.: RBN-World: a subsymbolic artificial chemistry. In: Kampis, G., Karsai, I., Szathmáry, E. (eds.) ECAL 2009. LNCS (LNAI), vol. 5777, pp. 377–384. Springer, Heidelberg (2011). https://doi.org/10.1007/978-3-642-21283-3_47
11. Faulkner, P., Krastev, M., Sebald, A., Stepney, S.: Sub-symbolic artificial chemistries. In: Stepney, S., Adamatzky, A. (eds.) Inspired by Nature. ECC, vol. 28, pp. 287–322. Springer, Cham (2018). https://doi.org/10.1007/978-3-319-67997-6_14
12. Giavitto, J.L., Michel, O.: MGS: a rule-based programming language for complex objects and collections. ENTCS **59**(4), 286–304 (2001)
13. Gundersen, M.S.: Reservoir Computing using Quasi-Uniform Cellular Automata. Master's thesis, NTNU, Norway (2017)
14. Horsman, C., Stepney, S., Wagner, R.C., Kendon, V.: When does a physical system compute? Proc. R. Soc. A **470**(2169), 20140182 (2014)
15. Horsman, D., Kendon, V., Stepney, S.: Abstraction/representation theory and the natural science of computation. In: Cuffaro, M.E., Fletcher, S.C. (eds.) Physical Perspectives on Computation, Computational Perspectives on Physics, pp. 127–149. Cambridge University Press, Cambridge (2018)
16. Horsman, D., Stepney, S., Kendon, V.: The natural science of computation. Commun. ACM **60**(8), 31–34 (2017)
17. Horsman, D.C.: Abstraction/representation theory for heterotic physical computing. Philos. Trans. R. Soc. A **373**, 20140224 (2015)
18. Jaeger, H.: The "echo state" approach to analysing and training recurrent neural networks - with an erratum note. GMD Report 148, German National Research Center for Information Technology (2010)
19. Kaneko, K.: Spatiotemporal intermittency in coupled map lattices. Progress Theoret. Phys. **74**(5), 1033–1044 (1985)
20. Kauffman, S.A.: Metabolic stability and epigenesis in randomly constructed genetic nets. J. Theor. Biol. **22**(3), 437–467 (1969)
21. Kauffman, S.A.: The Origins of Order. Oxford University Press, Oxford (1993)
22. Kendon, V., Sebald, A., Stepney, S.: Heterotic computing: past, present and future. Phil. Trans. R. Soc. A **373**, 20140225 (2015)
23. Kendon, V., Sebald, A., Stepney, S., Bechmann, M., Hines, P., Wagner, R.C.: Heterotic computing. In: Calude, C.S., Kari, J., Petre, I., Rozenberg, G. (eds.) UC 2011. LNCS, vol. 6714, pp. 113–124. Springer, Heidelberg (2011). https://doi.org/10.1007/978-3-642-21341-0_16
24. Krastev, M., Sebald, A., Stepney, S.: Emergent bonding properties in the Spiky RBN AChem. In: ALife 2016, Cancun, Mexico, July 2016, pp. 600–607. MIT Press (2016)
25. Lehman, J., Stanley, K.O.: Exploiting open-endedness to solve problems through the search for novelty. In: ALIFE XI. pp. 329–336. MIT Press (2008)
26. Lehman, J., Stanley, K.O.: Efficiently evolving programs through the search for novelty. In: GECCO 2010, pp. 837–844. ACM (2010)
27. Lehman, J., Stanley, K.O.: Abandoning objectives: evolution through the search for novelty alone. Evol. Comput. **19**(2), 189–223 (2011)

28. Nehaniv, C.L., Rhodes, J., Egri-Nagy, A., Dini, P., Morris, E.R., Horváth, G., Karimi, F., Schreckling, D., Schilstra, M.J.: Symmetry structure in discrete models of biochemical systems: natural subsystems and the weak control hierarchy in a new model of computation driven by interactions. Phil. Trans. R. Soc. A **373**, 20140223 (2015)
29. Sinha, S., Biswas, D.: Adaptive dynamics on a chaotic lattice. Phys. Rev. Lett. **71**(13), 2010–2013 (1993)
30. Sinha, S., Ditto, W.L.: Dynamics based computation. Phys. Rev. Lett. **81**(10), 2156–2159 (1998)
31. Sinha, S., Ditto, W.L.: Computing with distributed chaos. Phys. Rev. E **60**(1), 363–377 (1999)
32. Sipper, M.: Quasi-uniform computation-universal cellular automata. In: Morán, F., Moreno, A., Merelo, J.J., Chacón, P. (eds.) ECAL 1995. LNCS, vol. 929, pp. 544–554. Springer, Heidelberg (1995). https://doi.org/10.1007/3-540-59496-5_324
33. Spicher, A., Michel, O., Giavitto, J.-L.: A topological framework for the specification and the simulation of discrete dynamical systems. In: Sloot, P.M.A., Chopard, B., Hoekstra, A.G. (eds.) ACRI 2004. LNCS, vol. 3305, pp. 238–247. Springer, Heidelberg (2004). https://doi.org/10.1007/978-3-540-30479-1_25
34. Stepney, S.: Nonclassical computation: a dynamical systems perspective. In: Rozenberg, G., Bäck, T., Kok, J.N. (eds.) Handbook of Natural Computing, vol. 4, pp. 1979–2025. Springer, Heidelberg (2012). https://doi.org/10.1007/978-3-540-92910-9_59
35. Stepney, S., Kendon, V.: The role of structure and complexity on reservoir computing quality. In: McQuillan, I., Seki, S. (eds.) UCNC 2019. LNCS, vol. 11493, pp. 52–64. Springer, Heidelberg (2019)
36. von Neumann, J.: Theory of Self-Reproducing Automata (edited by A.W. Burks). University of Illinois Press, Urbana (1966)
37. Wolfram, S.: Statistical mechanics of cellular automata. Rev. Mod. Phys. **55**(3), 601–644 (1983)
38. Wolfram, S.: Universality and complexity in cellular automata. Physica D **10**(1), 1–35 (1984)

Generalized Membrane Systems with Dynamical Structure, Petri Nets, and Multiset Approximation Spaces

Péter Battyányi, Tamás Mihálydeák, and György Vaszil(✉)

Department of Computer Science, Faculty of Informatics,
University of Debrecen, Kassai út 26, Debrecen 4028, Hungary
{battyanyi.peter,mihalydeak.tamas,vaszil.gyorgy}@inf.unideb.hu

Abstract. We study generalized P systems with dynamically changing membrane structure by considering different ways to determine the existence of communication links between the compartments. We use multiset approximation spaces to define the dynamic notion of "closeness" of regions by relating the base multisets of the approximation space to the notion of chemical stability, and then use it to allow communication between those regions only which are close to each other, that is, which contain elements with a certain chemical "attraction" towards each other. As generalized P systems are computationally complete in general, we study the power of weaker variants. We show that without taking into consideration the boundaries of regions, unsynchronized systems do not gain much with such a dynamical structure: They can be simulated by ordinary place-transition Petri nets. On the other hand, when region boundaries also play a role in the determination of the communication structure, the computational power of generalized P systems is increased.

1 Introduction

Membrane systems, or P systems, were introduced in [14] as computing models inspired by the functioning of the living cell. The main component of such a system is a membrane structure with membranes enclosing regions as multisets of objects. Each membrane has an associated set of operators working on the objects contained by the region. These operators can be of different types, they can change the objects present in the regions or they can provide the possibility of transferring the objects from one region to another one. The evolution of the objects inside the membrane structure from an initial configuration to a final configuration (which can be specified in different ways) corresponds to a computation having a result which is derived from some properties of the final configuration.

Several variants of the basic notion have been introduced and studied proving the power of the framework, see the monograph [15] for a comprehensive introduction, the handbook [16] for a summary of notions and results of the area, and the volumes [2,17] for various applications.

I. McQuillan and S. Seki (Eds.): UCNC 2019, LNCS 11493, pp. 15–29, 2019.
https://doi.org/10.1007/978-3-030-19311-9_3

The structure of a membrane system can be represented by a graph where nodes correspond to the regions enclosed by the membranes and edges correspond to their "neighbourhood" relation, that is, to the communication links between the regions. So called cell-like membrane systems have a membrane structure which can be represented by a tree (corresponding to the arrangement where membranes and regions are physically embedded in each other), the structure of tissue-like systems, on the other hand, might correspond to arbitrary graphs. Here we study variants of tissue-like systems called generalized P systems (see [1]) which we equip with a dynamical communication structure, that is, with a communication structure that can change during the functioning of the system. Communication links will be established using a dynamic neighbourhood relation based on the notion of closeness of regions specified by a multiset approximation space associated with the system.

The notion of closeness of regions has already been studied by different methods. In [4], an abstract notion of closeness was defined using general topologies, and the approach was also extended to generalized membrane systems in [5]. Multiset approximation spaces provide the possibility to define the lower and upper approximations of multisets, and then consider the difference (informally speaking) of the two approximations as a notion of the border of a multiset. As the membrane regions are multisets, this framework can be used to define communication channels (as done in [6,7]) based on the idea that interaction between regions can only involve objects which are present in the borders. This approach also aims to model chemical stability which was the explicit motivation of the investigations in [8] where so-called symport/antiport P systems were studied. The idea was to connect the base multisets of the multiset approximation space with chemical compounds that are stable. (The base multisets of a multiset approximation space are the "building blocks" of the lower and upper approximations, they constitute collections of elements which, in some sense "belong together", thus, if we look at the objects in the membranes as some type of chemical ingredients, they provide a natural model for chemical stability.) In this context, the borders (those objects of the approximated multiset which cannot be "covered" by the base multisets) constitute those parts of the regions which contain the objects that are not part of any such stable compound, thus, which are ready to interact with other objects in other regions.

Here we use a similar approach to define a dynamic notion of closeness in the framework of generalized P systems, and study the effect of the resulting dynamical structure on their computational power. We investigate how possible variants of the neighbourhood relation (of the method for establishing the closeness of membranes) influence the functioning of these systems. We show that using a simple model of chemical stability based on the set of base sets of the approximation space is not sufficient to increase the computing capabilities, but on the other hand, with a refined notion involving also the idea of membrane borders, an increase of the computational power can be obtained.

In the following, we first recall the necessary definitions, then introduce and study two variants of generalized P systems with dynamically changing communication structure based on multiset approximation spaces. They are called generalized P systems with base multisets and generalized P systems with multiset approximation spaces. As maximal parallel rule application makes already the basic model of generalized P systems computationally complete, we study the weaker, unsynchronized variants. We first show that generalized P systems with base multisets have the same power as generalized P systems, and that both of these models can be simulated by simple place-transition Petri nets, then prove that generalized P systems with multiset approximation spaces are able to generate any recursively enumerable set, thus, they are computationally as powerful as Turing machines, even when using simple, so-called non-cooperative rules in an unsynchronized manner.

2 Preliminaries

Let $\mathbb{N}$ and $\mathbb{N}_{>0}$ be the set of non-negative integers and the set of positive integers, respectively, and let O be a finite nonempty set (the set of objects). A *multiset* M over O is a pair $M = (O, f)$, where $f : O \to \mathbb{N}$ is a mapping which gives the *multiplicity* of each object $a \in O$. The set $supp(M) = \{a \in O \mid f(a) > 0\}$ is called the *support* of M. If $supp(M) = \emptyset$, then M is the empty multiset. If $a \in supp(M)$, then $a \in M$, and $a \in^n M$ if $f(a) = n$.

Let $M_1 = (O, f_1), M_2 = (O, f_2)$. Then $(M_1 \sqcap M_2) = (O, f)$ where $f(a) = \min\{f_1(a), f_2(a)\}$; $(M_1 \sqcup M_2) = (O, f')$, where $f'(a) = \max\{f_1(a), f_2(a)\}$; $(M_1 \oplus M_2) = (O, f'')$, where $f''(a) = f_1(a) + f_2(a)$; $(M_1 \ominus M_2) = (O, f''')$ where $f'''(a) = \max\{f_1(a) - f_2(a), 0\}$; and $M_1 \sqsubseteq M_2$, if $f_1(a) \leq f_2(a)$ for all $a \in O$.

For any $n \in \mathbb{N}$, n-times addition of M, denoted by $\oplus_n M$, is given by the following inductive definition: (1) $\oplus_0 M = \emptyset$, (2) $\oplus_1 M = M$, and (3) $\oplus_{n+1} M = (\oplus_n M) \oplus M$. Let $M_1 \neq \emptyset, M_2$ be two multisets. For any $n \in \mathbb{N}$, $M_1 \sqsubseteq^n M_2$, if $\oplus_n M_1 \sqsubseteq M_2$ but $\oplus_{n+1} M_1 \not\sqsubseteq M_2$.

A tuple of multisets $(M_1, M_2, \ldots, M_l)$ is called a *partition* of the multiset M, if $M = M_1 \oplus \ldots \oplus M_l$ such that $M_i \sqcap M_j = \emptyset$ for any two different indices $1 \leq i < j \leq l$, and l is called the *order* of the partition.

The number of copies of objects in a finite multiset $M = (O, f)$ is its cardinality: $card(M) = \Sigma_{a \in supp(M)} f(a)$. Such an M can be represented by any string w over O for which $|w| = card(M)$, and $|w|_a = f(a)$ ($|w|$ denotes the length of the string w, and $|w|_a$ denotes the number of occurrences of symbol a in w).

We define the $\mathcal{MS}^n(O)$, $n \in \mathbb{N}$, to be the set of all multisets $M = (O, f)$ over O such that $f(a) \leq n$ for all $a \in O$, and we let $\mathcal{MS}^{<\infty}(O) = \bigcup_{n=0}^{\infty} \mathcal{MS}^n(O)$.

A *register machine* is a construct $W = (m, H, l_0, l_h, Inst)$, where m is the number of registers, H is the set of instruction labels, l_0 is the start label, l_h is the halting label, and $Inst$ is the set of instructions. Each label from H labels only one instruction from $Inst$. There are several types of instructions which can be used. For $l_i, l_j, l_k \in H$ and $r \in \{1, \ldots, m\}$ we have:

- $l_i : (\mathtt{nADD}(r), l_j, l_k)$ - *nondeterministic add*: Add 1 to register r and then go to one of the instructions with labels l_j or l_k, nondeterministically chosen.
- $l_i : (\mathtt{ADD}(r), l_j)$ - *deterministic add*: Add 1 to register r and then go to the instruction with label l_j.

- $l_i : (\mathtt{SUB}(r), l_j)$ - *subtract*: If register r is non-empty, then subtract one from it and go to the instruction with label l_j, otherwise check the register once more, and try again (thus, enter into an infinite loop, making it impossible to successfully finish the computation).
- $l_i : (\mathtt{CHECK(r)}, l_j, l_k)$ - *zero check*: If the value of register r is zero, go to instruction l_j, otherwise go to l_k.
- l_h : HALT - *halt*: Stop the machine.

Such a register machine W computes a set of numbers by starting with empty registers and proceeding by applying instructions as indicated by the labels (and made possible by the contents of the registers), beginning with the instruction l_0. If the halt instruction is reached, then the number stored at that time in register 1 is said to be computed by W. Note that because of the nondeterminism in choosing the continuation of the computation in the case of nondeterministic add instructions, $l_i : (\mathtt{nADD}(r), l_j, l_k)$ the computed set can also be infinite.

Note also, that register machines are usually defined as deterministic computing devices (without the nondeterministic add instructions) which compute functions of input values placed initially in the registers. It is known (see, for example [9]) that in this way they can compute all functions which are Turing computable. By adding the nondeterministic add instruction, we obtain a device which generates sets of numbers starting from a unique initial configuration. Since any recursively enumerable set can be obtained as the range of a Turing computable function on the set of non-negative integers, this way we can generate any recursively enumerable set of numbers.

A *place-transition Petri net* ([12]) is a quintuple $U = (P, T, F, V, m_0)$ such that P, T are finite sets with $P \cap T = \emptyset$, $P \cup T \neq \emptyset$, the sets of *places* and *transitions*, respectively. The set $F \subseteq (P \times T) \cup (T \times P)$, is a set of "arcs" connecting places and transitions, the *flow relation* of U. The function $V : F \to \mathbb{N}_{>0}$ determines the multiplicity (the *weight*) of the arcs, and $m_0 : P \to \mathbb{N}$ is a function called the initial marking. In general, a *marking* is a function $m : P \to \mathbb{N}$ associating nonnegative integers (the number of *tokens*) to the places of the net. Moreover, for every transition $t \in T$, there is a place $p \in P$ such that $f = (p, t) \in F$ and $V(f) \neq 0$.

Let $x \in P \cup T$. The *pre- and postsets* of x, denoted by ${}^\bullet x$ and $x^\bullet$, respectively, are defined as ${}^\bullet x = \{y \mid (y, x) \in F\}$ and $x^\bullet = \{y \mid (x, y) \in F\}$.

For each transition $t \in T$, we define two markings, $t^-, t^+ : P \to \mathbb{N}$ as follows:

$$t^-(p) = \begin{cases} V(p,t), & \text{if } (p,t) \in F, \\ 0 & \text{otherwise} , \end{cases} \quad t^+(p) = \begin{cases} V(t,p), & \text{if } (t,p) \in F, \\ 0 & \text{otherwise} . \end{cases}$$

A transition $t \in T$ is said to be *enabled* if $t^-(p) \leq m(p)$ for all $p \in {}^\bullet t$. Let $\Delta\, t(p) = t^+(p) - t^-(p)$ for $p \in P$, and let us define the *firing of a transition* as follows. A transition $t \in T$ can fire in m (notation: $m \longrightarrow^t$) if t is enabled in m. After the firing of t, the Petri net obtains a new marking $m' : P \to \mathbb{N}$, where $m'(p) = m(p) + \Delta\, t(p)$ for all $p \in P$ (notation: $m \longrightarrow^t m'$).

Petri nets can be considered as computing devices: Starting with the initial marking, going through a series of configuration changes by the firing of a series of transitions, we might obtain a marking where no transitions are enabled. This final marking is the result of the computation. Petri nets in this simple setting are strictly weaker in computational power than Turing machines, [12,13].

2.1 Multiset Approximation Spaces

Different ways of set approximations go back (at least) to rough set theory originated in the early 1980's, [10,11]. In this theory and its different generalizations, lower and upper approximations of sets appear which are based on different kinds of indiscernibility relations. An indiscernibility relation on a given set of objects provides the set of base sets by which any set can be approximated from lower and upper sides. Its generalization, the so-called partial approximation of sets (see [3]) gives a possibility to embed available knowledge into an approximation space. The lower and upper approximations of sets rely on base sets which represent the available knowledge. If we have concepts of lower and upper approximations, the concept of boundary can also be introduced (as the difference between the lower and upper approximations).

From a set-theoretical point of view, regions in membrane systems can be represented by multisets, therefore we have to generalize the theory of set approximations for multisets. The notion of multiset approximation spaces has been introduced in [6] (see also [7] for more details). Multiset approximations also rely on a beforehand given set of base multisets. Using the usual approximation technique (creating the lower and upper approximations), the boundaries of multisets (boundaries of membrane regions) can also be defined, and we will make use of this feature in subsequent parts of the paper.

The basic constituents of a multiset approximation space over a finite alphabet O are the following: (1) A *domain*: $\mathcal{MS}^{<\infty}(O)$, the elements of the domain are approximated using the approximation space. (2) A *base system*: $\mathfrak{B} \subseteq \mathcal{MS}^{<\infty}(O)$, a nonempty set of finite *base multisets* serving as the basis for the approximation process. (3) The *approximation functions*: $\mathsf{l}, \mathsf{u}, \mathsf{b} : \mathcal{MS}^{<\infty}(O) \to \mathcal{MS}^{<\infty}(O)$ determining the lower or upper approximations and the boundaries of multisets of the domain.

A quintuple $(O, \mathfrak{B}, \mathsf{l}, \mathsf{u}, \mathsf{b})$ is called a *multiset approximation space* where O is a finite set, $\mathfrak{B} \subseteq \mathcal{MS}^{<\infty}(O)$ is a base system, and $\mathsf{b}, \mathsf{u}, \mathsf{l} : \mathcal{MS}^{<\infty}(O) \to \mathcal{MS}^{<\infty}(O)$ are the approximation functions generated by $\mathfrak{B}$.

Let $M = (O, f)$ be a multiset over O, we define the *lower approximation function*:

$$\mathsf{l}(M) = \bigsqcup\{\oplus_n B \mid B \in \mathfrak{B},\ B \sqsubseteq M, \text{and } B \sqsubseteq^n M\};$$

the *boundary function*:

$$\mathsf{b}(M) = \bigsqcup\{\oplus_n B \mid B \in \mathfrak{B}, \text{ and } B \sqcap (M \ominus \mathsf{l}(M)) \sqsubseteq^n M \ominus \mathsf{l}(M)\};$$

with $\mathsf{b}^{\mathsf{e}}(M) = \mathsf{b}(M) \ominus M$ being the *external part* of the boundary of M, and $\mathsf{b}^{\mathsf{i}}(M) = \mathsf{b}(M) \sqcap M$ being the *internal part* of the boundary of M; and the *upper approximation function*:

$$\mathsf{u}(M) = \mathsf{l}(M) \sqcup \mathsf{b}(M).$$

Example 1. Let $O = \{a, b, c, d\}$, $\mathfrak{B} = \{ab, ac^2, cd\}$, and consider the multsiet approximation space $(O, \mathfrak{B}, \mathsf{l}, \mathsf{u}, \mathsf{b})$. Given the multiset $M = a^4b^2c^5$, we obtain the lower/upper approximations and the border of M as

- $\mathsf{l}(M) = (ab)^2 \sqcup (ac^2)^2$,
- $\mathsf{b}(M) = (ab)^2 \sqcup cd \sqcup ac^2$ where $\mathsf{b}^\mathsf{i}(M) = a^2c$ and $\mathsf{b}^\mathsf{e}(M) = b^2d$, and
- $\mathsf{u}(M) = (ab)^4 \sqcup (ac^2)^2 \sqcup cd$.

2.2 Generalized P Systems

Now we present the notion of generalized P systems, variants of tissue P systems introduced in [1].

An $n+3$-tuple $\Pi = (O, w_1, w_2, \ldots, w_n, R, i_o)$ is a *generalized P system* of degree $n \geq 1$, where

- O is a finite set called the alphabet of objects;
- $w_i \in \mathcal{MS}^{<\infty}$, $1 \leq i \leq n$, the initial contents of compartment i of Π;
- R is a finite set of rules of the form $(x_1, \alpha_1) \ldots (x_k, \alpha_k) \to (y_1, \beta_1) \ldots (y_l, \beta_l)$, where $x_i, y_j \in \mathcal{MS}^{<\infty}(O)$, and $1 \leq \alpha_i, \beta_j \leq n$ for all $1 \leq i, j \leq n$;
- $1 \leq i_o \leq n$ is the label of the output compartment.

For a generalized P system Π as above, an n-tuple $c = (u_1, u_2, \ldots, u_n)$ with $u_i \in MS^{<\infty}(O)$, $1 \leq i \leq n$, is called a *configuration* of Π, and $c_0 = (w_1, w_2, \ldots, w_n)$ is called its *initial configuration.* The multisets $u_1, u_2, \ldots, u_n$ are also called the *contents* of the corresponding compartments $1, 2, \ldots, n$, in configuration c.

A generalized P system changes its configurations by applying its rules. A rule $r \in R$, as given above, is *applicable* to a configuration c, if and only if x_i is a submultiset of u_{α_i} for all $1 \leq i \leq k$. As a result of applying r to c, each multiset x_i is removed from the region u_{α_i}, $1 \leq i \leq k$, and each multiset y_j is added to the region u_{β_j}, $1 \leq j \leq l$.

The rules of a generalized P system are *non-cooperative*, if they have the special form $(x, \alpha) \to (y_1, \beta_1) \ldots (y_l, \beta_l)$ where $x \in O, 1 \leq \alpha \leq n$.

We say that the configuration $c' = (v_1, \ldots, v_n)$ of Π is obtained directly from $c = (u_1, \ldots, u_n)$ by applying the rules in a *maximally parallel* manner, if there is a multiset R' of rules from R, such that all of them are simultaneously applicable to different copies of objects in configuration c, and the configuration c' is the result of the application of the rules in R'. Moreover, the set R' is maximal: for any $r \in R$, the rule multiset $\{r\} \oplus R$ is not applicable simultaneously to c. If R' is not necessarily maximal, then the configuration c' is obtained from c by applying the rules in the *unsynchronized* manner.

A sequence of configurations $c_0, c_1, \ldots$ of Π, where each element is obtained directly from the previous one starting from the initial configuration, is called a *computation.* The computation halts if no rule can be applied. The result of a *halting computation* is the number of objects that are present in the output compartment (compartment i_o) in the halting configuration.

3 Membrane Systems with Dynamically Changing Structure

In this section we study generalized P systems with dynamical communication structure induced by associated multiset approximation spaces. We study variants of the notion of membrane closeness based on the model of chemical stability provided by mutliset approximation spaces.

The underlying idea is to use the base multisets of the multiset approximation space as the model for those collections of objects which together form a coherent, chemically stable unit, that is, to think of the system of base multisets as the representation of compounds whose potential energy is lower than the potential energy of their parts. Similarly to natural systems attempting to reach the configuration with the lowest possible energy, the objects of base multisets that are split by the membranes' borders are ready for moving towards the stable states, they have a certain attraction towards each other. In other words, the membranes containing submultisets that form a partition of a base multiset together are more likely to engage into interactions, or in other words, closer to each other.

Our first notion of closeness is based on the set of base multisets, the second one is more refined, it also takes into account the boundaries defined by the multiset approximation space.

3.1 Dynamical Structure Induced by the System of Base Multisets

Now we present a definition of closeness using the base multisets of the multiset approximation space. A set of regions is close to each other in this sense, if there is a base multiset which is "covered" by the contents of the regions.

Definition 1. Let $M_1, M_2, \ldots, M_k \in \mathcal{MS}^{<\infty}(O)$, and let $\mathfrak{B}$ be a system of base multisets over O. We say that $M_1, M_2, \ldots, M_k$ are *close to each other*, if there is a $B \in \mathfrak{B}$ such that $B \sqsubseteq M_1 \sqcup \ldots \sqcup M_k$ and, for every $1 \leq i \leq k$, $B \sqcap M_i \neq \emptyset$.

Using closeness, we restrict the applicability of rules to those configurations where they move objects between regions that are close to each other (based on their currents contents).

Definition 2. The P system $\Pi = (O, \mathfrak{B}, w_1, w_2, \ldots, w_n, R, i_0)$ is called a *generalized P system with base multisets* (of degree n) if $(O, w_1, w_2, \ldots, w_n, R, i_0)$ is a generalized P system and $\mathfrak{B} \subseteq \mathcal{MS}^{<\infty}(O)$ is a (nonempty) set of finite multisets over O.

A rule $r = (x_1, \alpha_1) \ldots (x_k, \alpha_k) \to (y_1, \beta_1) \ldots (y_l, \beta_l) \in R$ is applicable to a configuration $c = (u_1, u_2, \ldots u_n)$, if

- x_i is a submultiset of u_{α_i}, $1 \leq i \leq k$, and
- the regions $u_{\alpha_1}, \ldots, u_{\alpha_k}$, $u_{\beta_1} \ldots, u_{\beta_l}$, $1 \leq i \leq k$, $1 \leq j \leq l$, are close to each other according to the notion of closeness induced by the base system $\mathfrak{B}$.

Note how the closeness relation induced by $\mathfrak{B}$ determines the communication structure: A rule $r \in R$ is applicable only if all regions appearing on the left or the right sides of r are close to each other.

As generalized P systems with maximal parallel rule application are computationally complete (see, for example [1]), we study the variant working with unsynchronized rule application. Unfortunately, as the next theorem shows, using the notion of closeness determined by a system of base multisets does not add to the computational power of generalized P systems if they use the unsynchronized manner of rule application.

Theorem 1. *For any generalized P system with base multisets, Π, there is a generalized P system, Π', such that Π and Π' compute the same set of numbers using the unsynchronized manner of rule application.*

Proof. Let $\Pi = (O, \mathfrak{B}, w_1, w_2, \ldots, w_n, R, i_0)$ be a generalized P system with base multisets. We construct $\Pi' = (O, w_1, w_2, \ldots, w_n, R', i_o)$ with the rules being defined as follows.

Let $r = (x_1, \alpha_1) \ldots (x_k, \alpha_k) \to (y_1, \beta_1) \ldots (y_l, \beta_l)$ be a rule in R, and let $(B_1, \ldots, B_s)$ be a partition of $B \in \mathfrak{B}$, such that $s = k + l$. Observe that the number of partitions is a fixed value depending on $Card(B)$, the cardinality of B and s. Let $par(B, s)$ denote the set of all partitions of B with order s.

We define for all

$$r = (x_1, \alpha_1) \ldots (x_k, \alpha_k) \to (y_1, \beta_1) \ldots (y_l, \beta_l) \in R \text{ and } B \in \mathfrak{B}$$

the set of rules

$$\begin{aligned} R_{r,B} = \{&(B_1 \sqcup x_1, \alpha_1) \ldots (B_k \sqcup x_k, \alpha_k)(C_1, \beta_1) \ldots (C_l, \beta_l) \\ &\to (B_1 \ominus x_1, \alpha_1) \ldots (B_k \ominus x_k, \alpha_k)(y_1 \oplus C_1, \beta_1) \ldots (y_l \oplus C_l, \beta_l) \mid \text{for all} \\ &(B_1, \ldots, B_k, C_1, \ldots, C_l) \in par(B, k + l)\}, \end{aligned}$$

and we define the rule set of Π' as

$$R' = \bigcup_{r \in R, B \in \mathfrak{B}} R_{r,B}.$$

To see how Π' simulates Π, notice that a rule in $R_{r,B}$ corresponding to a rule $r \in R$ with $r = (x_1, \alpha_1) \ldots (x_k, \alpha_k) \to (y_1, \beta_1) \ldots (y_l, \beta_l)$ and a partition $(B_1, \ldots, B_k, C_1, \ldots, C_l)$ of B can be applied if not only all x_i, but also all B_i, C_j are submultisets of the respective regions α_i, β_j, $1 \leq i \leq k$, $1 \leq j \leq l$, thus if and only if the rule $r \in R$ is applicable in Π.

Note also that applying such a rule in Π' produces the same result as the corresponding rule $r \in R$ does, when applied in Π.

Moreover, there exists a rule $r \in R$ which is applicable in Π, if and only if, there exist a $B \in \mathfrak{B}$, such that there is an applicable rule in $R_{r,B}$ in Π'. □

Next, we also show that generalized P systems with base multisets working in the unsynchronized manner of rule application can be simulated by simple place-transition Petri nets.

Theorem 2. *For any generalized P system with base multisets, Π, there is a place-transition Petri net N, such that N generates the same set of numbers as Π in the unsynchronized manner of rule application.*

Proof. Let $\Pi = (O, \mathfrak{B}, w_1^0, w_2^0, \ldots, w_n^0, R, i_o)$ be a generalized P system with base multisets. We use the same idea, as in the proof of the previous theorem. For a given rule $r = (x_1, \alpha_1)\ldots(x_k, \alpha_k) \to (y_1, \alpha_{k+1})\ldots(y_l, \alpha_s) \in R$, we can check the applicability of r by examining whether there exist a base multiset $B \in \mathfrak{B}$, a partition $\pi = (B_1, \ldots, B_s)$ of B such that $B_i \sqsubseteq u_{\alpha_i}$, $1 \le i \le s$, and whether there are enough objects in the corresponding membranes such that r can be executed. To this aim, being in configuration $(u_1, u_2, \ldots, u_n)$, it is sufficient to check that, for all $a \in O$, there is at least $max(x_i(a), B_i(a))$ number of as in u_{α_i}, $1 \le i \le k$, and that the number of as is at least $B_i(a)$ in u_{α_i}, $k+1 \le i \le s$.

For a rule $r = (x_1, \alpha_1)\ldots(x_k, \alpha_k) \to (y_1, \alpha_{k+1})\ldots(y_l, \alpha_s)$, let us denote s, the number of multisets on the left- and on the right-hand side of r, by $deg(r) = s$ (degree of r), and let us consider the set of all triples $RP = \{(r, B, \pi) \mid r \in R,\ B \in \mathfrak{B},\ deg(r) = s,\ \pi = (B_1, \ldots, B_s) \text{ is a partition of } B\}$.

Let us define the Petri net $N = (P, T, F, V, m_0)$ with $P = O \times \{1, \ldots, n\} \cup \{p_{ini}\}$, A place $(a, j) \in P$ represents the number of objects $a \in O$ inside the jth membrane at every step of the computational sequence, so let us set $m_0(p_{ini}) = 1$, and $m_0(p) = w_j^0(a)$ for every place $p = (a, j) \in O \times \{1, \ldots, n\}$.

The net N consists of subnets for each triple $(r, B, \pi) \in RP$ responsible for the simulation of r, and the place p_{ini} makes sure that only one of the subnets can operate at a time, thus, that the simulation of the rule executions are mutually exclusive.

To define these subnets, let $T = \{t_\gamma \mid \gamma \in RP\}$. Let $\gamma = (r, B, B_1, \ldots, B_s) \in RP$ with r denoted as $r = (x_1, \alpha_1)\ldots(x_k, \alpha_k) \to (x_{k+1}, \alpha_{k+1})\ldots(x_s, \alpha_s)$, and let $p = (a, \alpha_j) \in {}^\bullet t_\gamma \cap t_\gamma^\bullet$ if and only if $a \in x_{\alpha_j}$, $1 \le j \le s$. In addition, $p_{ini} \in {}^\bullet t_\gamma \cap t_\gamma^\bullet$.

For each $p = (a, \alpha_j)$ with $1 \le j \le k$, and for all $t_\gamma \in T$, $\gamma \in RP$, the weights of the arcs are computed as $V(p, t_\gamma) = max\{x_{\alpha_j}(a), B_j(a)\}$, that is, the maximum of the number of occurrences of $a \in O$ in x_{α_j} or in B_j. These weights indicate that the transition corresponding to $\gamma = (r, B, B_1, \ldots, B_s) \in RP$ can only fire if the number of objects in the regions appearing on the left side of r is at least as many as required by B_j of the partition in γ, and also sufficient for the rule r to be executed. In addition, we have $V(t_\gamma, p) = max\{B_j(a) - x_{\alpha_j}(a), 0\}$ in order to return the necessary amount of tokens to $p = (a, \alpha_j)$ so that the simulation of the rule execution decreases the number of tokens in p by $x_{\alpha_j}(a)$ only.

For $p = (a, \alpha_j)$ with $k+1 \le j \le s$, we have $V(p, t_\gamma) = B_j(a)$ and $V(t_\gamma, p) = B_j(a) + x_{\alpha_j}(a)$. That is, we remove as many as $B_j(a)$ tokens from $p = (a, \alpha_j)$ necessary for checking if there are enough objects for B_j in the partition) and put them back together with the tokens corresponding to the objects introduced by the right side of the rule that is simulated. Furthermore, $V(p_{ini}, t_\gamma) = V(t_\gamma, p_{ini}) = 1$.

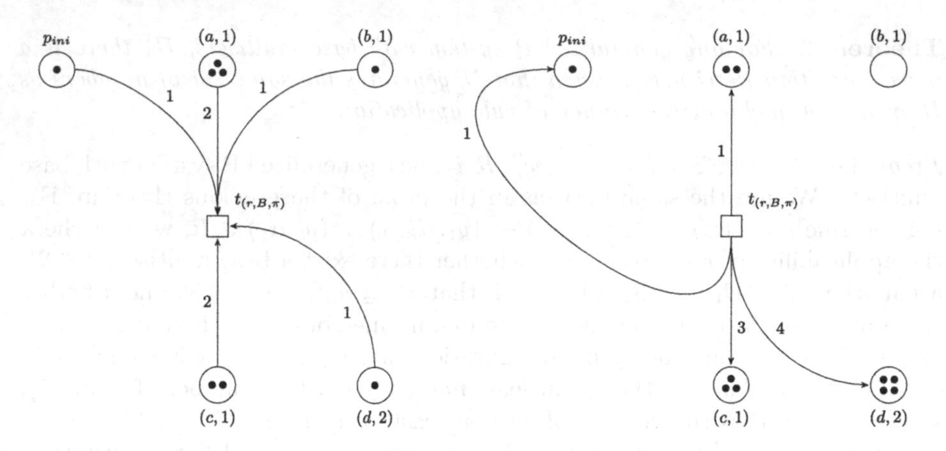

Fig. 1. The incoming and outgoing edges of the transition $t_{(r,B,\pi)}$ are shown in the left and right part of the figure, respectively. The transition corresponds to the rule $r = (ab, 1) \to (c, 1)(d, 2)^3$, the base set $B =^2 bc^2d$, and the partition $\pi = (a^2bc^2, d)$ of B. The number of tokens at the places $(a, 1)$, $(b, 1)$, $(c, 1)$, and $(d, 2)$ on the left corresponds to a configuration of the membrane system with $w_1 = a^3bc^2d$ and $w_2 = d$, the number of tokens on the right show the number of tokens after the firing of $t_{(r,B,\pi)}$.

See Fig. 1 for an example demonstrating the structure of the subnets constituting the Petri net N corresponding to each $(r, B, \pi) \in RP$.

To sum up the above construction, the places of the Petri net represent the objects in the different compartments of the P system. For every triple $\gamma = (r, B, \pi)$ where r is a rule, B is a base multiset and π is a partition of B (of appropriate order), we define a subnet consisting of all the places and a transition t_γ together with the corresponding arcs. This subnet simulates an application of r when B is the base multiset that ensures that the membranes containing the multisets on the left hand side and right hand side of r are close to each other. The whole process is governed by the place p_{ini}. Each of the subnets is connected with p_{ini} in such a manner that only one subnet is able to operate at a time.

Thus, unsynchronized rule execution of the generalized membrane system is simulated in a sequential manner. The Petri net halts if and only if the membrane system halts, and the number of objects in the output membrane is indicated by the number of tokens in the corresponding places. □

As the expressive power of place-transition Petri nets are less than that of Turing machines (see, for example [12,13]), we obtain the following corollary.

Corollary 1. *Generalized membrane systems with base multisets using the unsynchronized manner of rule application are strictly weaker in computational power than Turing machines, that is, they are not computationally complete.*

3.2 Dynamical Structure Based on Membrane Boundaries

Now we present how the dynamical communication structure of a generalized P system can be induced by a notion of closeness based on membrane boundaries provided by the multiset approximation space.

In the next definition we formalize the idea that changes in the configurations can only be the result of manipulating the objects in the membrane boundaries: We only allow the development of objects that are parts of the membrane boundaries, and the notion of closeness is defined in such a way that it only depends on the boundaries of the respective regions. This variant of closeness can also related to the idea of chemical stability, since membrane boundaries are collections of base multisets which are split into two or more parts by the membrane borders. If we consider the base multisets as the model of collections of chemical ingredients which together form stable compounds, then the boundaries of regions can be thought of corresponding to the unstable parts of membrane contents where chemical reactions can occur.

Definition 3. Let M_1, $M_2, \ldots, M_k \in MS^{\leq\infty}(O)$ be multisets over O, and let $(O, \mathfrak{B}, \mathsf{l}, \mathsf{u}, \mathsf{b})$ be a multiset approximation space. We say that M_1, $M_2, \ldots, M_k$ are *close to each other by boundaries*, if there is a $B \in \mathfrak{B}$ such that $B \sqsubseteq \mathsf{b}^{\mathsf{i}}(M_1) \sqcup \ldots \sqcup \mathsf{b}^{\mathsf{i}}(M_k)$, and for every $1 \leq i \leq k$, $B \sqcap M_i \neq \emptyset$.

Definition 4. The system $\Pi = (O, \mathfrak{B}, w_1, w_2, \ldots, w_n, R, i_o)$ is a *generalized P system with an associated multiset approximation space* (of degree n), if the $n+3$-tuple $(O, w_1, w_2, \ldots, w_n, R, i_o)$ is a generalized P system, and $\mathfrak{B}$ is a nonempty set of finite multisets over O.

A rule $r = (x_1, \alpha_1) \ldots (x_k, \alpha_k) \to (y_1, \beta_1) \ldots (y_l, \beta_l) \in R$ is applicable to a configuration $c = (u_1, u_2, \ldots u_n)$, if

- x_i is a submultiset of $\mathsf{b}^{\mathsf{i}}(u_{\alpha_i})$, the boundary of the respective region, $1 \leq i \leq k$, and
- the regions $u_{\alpha_1}, \ldots u_{\alpha_k}$, $u_{\beta_1} \ldots u_{\beta_l}$, $1 \leq i \leq k$, $1 \leq j \leq l$, are close to each other by boundaries.

Theorem 3. *Generalized P systems with associated multiset approximation spaces generate any recursively enumerable set of numbers in the unsynchronized manner of rule application, even if the rules are non-cooperative.*

Proof. Let L be a recursively enumerable set of numbers, and consider the register machine $W = (m, H, l_0, l_h, Inst)$ generating L. We construct a generalized P system with an associated multiset approximation space, such that it also generates L in the sense that the generated numbers correspond to the multiplicity of a certain object in the output region when the computation halts. Let $\Pi = (O, \mathfrak{B}, w_1, w_2, w_3, R, 2)$ with

$$O = \{l, l', l'', l''' \mid l \in H\} \cup \{a_r \mid 1 \leq r \leq m\} \cup \{b, c, d_1, d_2, e\},$$

$$\begin{aligned}
\mathfrak{B} = \{&l_i b,\ l'_j c,\ l'_k c \mid l_i : (\texttt{nADD}(r), l_j, l_k) \in \mathit{Inst}\} \\
&\cup \{l_i b,\ l'_j c \mid l_i : (\texttt{ADD}(r), l_j) \in \mathit{Inst}\} \\
&\cup \{l_i b,\ l'_j a_r,\ l'_j c,\ l'_k c,\ l''_k a_r,\ l'''_k c \mid l_i : (\texttt{CHECK}(r), l_j, l_k) \in \mathit{Inst}\} \\
&\cup \{l_i a_r,\ l_i l_j \mid l_i : (\texttt{SUB}(r), l_j) \in \mathit{Inst}\} \\
&\cup \{d_2 c, d_1 e\} \cup \{l_h c d_1 \mid l_h : \texttt{HALT}\},
\end{aligned}$$

$$w_1 = l_0 c d_1,\ w_2 = b,\ w_3 = e,$$

$$R = R_{Add} \cup R_{Check} \cup R_{Sub} \cup R_{Ex},$$

where

$$\begin{aligned}
R_{Add} = \{&(l_i, 1) \to (l'_j a_r, 2), (l_i, 1) \to (l'_k a_r, 2), (l'_j, 2) \to (l_j, 1), \\
&(l'_k, 2) \to (l_k, 1) \mid \text{for all } l_i : (\texttt{nADD}(r), l_j, l_k) \in \mathit{Inst}\} \cup \\
\{&(l_i, 1) \to (l'_j a_r, 2), (l'_j, 2) \to (l_j, 1) \mid \text{for all} \\
&l_i : (\texttt{ADD}(r), l_j) \in \mathit{Inst}\}, \\
R_{Check} = \{&(l_i, 1) \to (l'_j, 2), (l_i, 1) \to (l'_k, 2), (l'_j, 2) \to (l_j, 1), \\
&(l'_k, 2) \to (l''_k, 1), (l''_k, 1) \to (l'''_k, 2), (l'''_k, 2) \to (l_k, 1) \mid \text{for all} \\
&l_i : (\texttt{CHECK}(r), l_j, l_k) \in \mathit{Inst}\},
\end{aligned}$$

$$\begin{aligned}
R_{Sub} &= \{(a_r, 2) \to (l_j l_j, 2) \mid \text{for all } l_i : (\texttt{SUB}(r), l_j) \in \mathit{Inst}\}, \\
R_{Ex} &= \{d_1, 1) \to (d_2, 3), (d_2, 3) \to (d_1, 1)\}.
\end{aligned}$$

To see how Π simulates the computations of W, consider its initial configuration $(l_0 c d_1, b, e)$: it corresponds to the initial configuration of W, as the first region contains l_0, the label of the instruction that has to be executed next, and the number of occurrences of a_r, $1 \leq r \leq m$, in the second region are 0, corresponding to the fact that all registers are initially empty.

Notice that as long as l_h is not present, it is possible to exchange d_1 in the first region with d_2 in the third (and back), since both $d_1 e$ and $d_2 c$ are base multisets of $\mathfrak{B}$, and the two symbols are also in the boundary of the respective regions, so one of the rules of R_{Ex} is always applicable. When l_h appears in the first region, then after d_1 also appears there, they are "removed" from the boundary of the region, as $l_h c d_1 \in \mathfrak{B}$ is a base multiset of the multiset approximation space, and after this happens, no rule of R is applicable. From these considerations we can see that Π reaches a halting configuration only if the label of the halting instruction, l_h appears.

Let us consider the more general case when the generalized P system Π is in the configuration $(l_i c \delta_1 w_{1,in}, w b w_{2,in}, e \delta_2)$ with $w(a_r) = k_r$, $1 \leq r \leq m$, corresponding to a situation when W is going to execute instruction l_i, and the contents of register r is $k_r \geq 0$, $1 \leq r \leq m$. The symbols δ_1, δ_2 are used to denote either d_1 or d_2, their exact meaning is not important, as they do not interfere with the simulation process until l_h appears. The submultisets $w_{1,in}$ and $w_{2,in}$ denote those elements of the first two regions that are not on the region boundary (as rules can only be applied to objects in the boundary, they are not important from the point of view of the computation).

If l_i is the label of an add, or nondeterministic add instruction, then since $l_i b = B \in \mathfrak{B}$ and $l_i c$ is not a base multiset, the two regions are close to each other by boundaries, and l_i is part of the boundary of the first region, so the rule simulating the instruction $l_i : (\texttt{nADD}(r), l_j, l_k)$ is applicable, yielding the configuration $(c\delta_1 w_{1,in}, wl' a_r b w_{2,in}, e\delta_2)$ with $l' \in \{l'_j, l'_k\}$ (or the configuration $(c\delta_1, wl'_j a_r b, e\delta_2)$ if $l_i : (\texttt{ADD}(r), l_j)$ is simulated). In any of these cases, since $l'_j c$ and $l'_k c$ are both base multisets, and none of these labels are elements of other base multisets in $\mathfrak{B}$, the two regions are again close to each other, and the primed label objects are in the boundary of the second region, so the corresponding rules are applicable, and we get a configuration $(lc\delta_1, wa_r b, e\delta_2)$ where l corresponds to the instruction that has to be executed next, while the second region contains one more object a_r, that is, the number stored in register r was incremented, as required by the simulated add instructions.

Suppose now, that Π is in a configuration $(l_i c\delta_1 w_{1,in}, wbw_{2,in}, e\delta_2)$ and the instruction to be executed is $l_i : (\texttt{CHECK}(r), l_j, l_k) \in Inst$. By applying the rules in R_{Check} we either get $(c\delta_1 w_{1,in}, wbl'_j w_{2,in}, e\delta_2)$, or $(c\delta_1 w_{1,in}, wbl'_k w_{2,in}, e\delta_2)$. In the first case, if $w(a_r) > 0$, then as $l'_j a_r$ is a base multiset, l'_j is removed from the region boundary and the computation continues "forever" by the rules of R_{Ex}. On the other hand, if $w(a_r) = 0$, then l_j can appear in the first region, and the simulation can continue in order. If we have $(c\delta_1 w_{1,in}, wbl'_k w_{1,in}, e\delta_2)$, then we get $(l''_k c\delta_1 w_{1,in}, wbw_{2,in}, e\delta_2)$. If $w(a_r) = 0$, then the computation can only continue with the rules of R_{Ex}, but if $w(a_r) > 0$, then k'' can be removed from the first region, as $k'' a_r$ is a base multiset, and we first obtain $(c\delta_1 w_{1,in}, wbl'''_k w_{2,in}, e\delta_2)$, and then $(l_k c\delta_1 w_{1,in}, wbw_{2,in}, e\delta_2)$, as required by the correct simulation of W.

Now, if we have a configuration $(l_i c\delta_1 w_{1,in}, wbw_{2,in}, e\delta_2)$ and l_i is the label of a subtract instruction $l_i : (\texttt{SUB}(r), l_j)$, then the computation can only continue if $w(a_r) > 0$, as $l_i a_r$ is a base multiset, but $l_i b$ is not. If $w(a_r) > 0$, we get $(l_i l_j l_j c\delta_1 w_{1,in}, w' bw_{2,in}, e\delta_2)$ where $w = w' a_r$. Since $l_i l_j$ is a base multiset, its elements are not part of the boundary, we get $(l_j c\delta_1 w'_{1,in}, w' bw_{2,in}, e\delta_2)$, and the simulation of W can continue.

The simulation is finished when the object l_h appears in the first region. The only rules that are applicable are the rules of R_{ex}, but when d_1 also appears in the first region, the computation halts, because $l_h c d_1$ is a base multiset, so all these objects disappear from the region boundary.

After halting, the result of the computation is the number of a_1 objects in the second region, as they correspond to the contents of the first register (the output register) of the register machine W. □

4 Concluding Remarks

We have used multiset approximation spaces to obtain dynamically changing communication structures for generalized P systems. It turned out that closeness defined using the base multisets only has no effect on the power of unsynchronized systems. On the other hand, using the multiset approximation space in a more sophisticated way, that is, taking membrane borders into account, the resulting systems are as powerful as Turing machines, even with non-cooperative rules and unsynchronized rule application.

Acknowledgments. T. Mihálydeák and G. Vaszil was supported by the project TÉT_16-1-2016-0193 of the National Research, Development and Innovation Office of Hungary (NKFIH). G. Vaszil was also supported by grant K 120558 of the National Research, Development and Innovation Office of Hungary (NKFIH), financed under the K 16 funding scheme.

References

1. Bernardini, F., Gheorgue, M., Margenstern, M., Verlan, S.: Networks of cells and Petri nets. In: Díaz-Pernil, D., Graciani, C., Gutiérrez-Naranjo, M.A., Păun, G., Pérez-Hurtado, I., Riscos-Núñez, A. (eds.) Proceedings of the Fifth Brainstorming Week on Membrane Computing, pp. 33–62. Fénix Editora, Sevilla (2007)
2. Ciobanu, G., Pérez-Jiménez, M.J., Păun, G. (eds.): Applications of Membrane Computing. Natural Computing Series. Springer, Heidelberg (2006). https://doi.org/10.1007/3-540-29937-8
3. Csajbók, Z., Mihálydeák, T.: Partial approximative set theory: a generalization of the rough set theory. Int. J. Comput. Inf. Syst. Ind. Manag. Appl. **4**, 437–444 (2012)
4. Csuhaj-Varjú, E., Gheorghe, M., Stannett, M.: P systems controlled by general topologies. In: Durand-Lose, J., Jonoska, N. (eds.) UCNC 2012. LNCS, vol. 7445, pp. 70–81. Springer, Heidelberg (2012). https://doi.org/10.1007/978-3-642-32894-7_8
5. Csuhaj-Varjú, E., Gheorghe, M., Stannett, M., Vaszil, G.: Spatially localised membrane systems. Fundam. Inform. **138**(1–2), 193–205 (2015)
6. Mihálydeák, T., Csajbók, Z.E.: Membranes with boundaries. In: Csuhaj-Varjú, E., Gheorghe, M., Rozenberg, G., Salomaa, A., Vaszil, G. (eds.) CMC 2012. LNCS, vol. 7762, pp. 277–294. Springer, Heidelberg (2013). https://doi.org/10.1007/978-3-642-36751-9_19
7. Mihálydeák, T., Csajbók, Z.E.: On the membrane computations in the presence of membrane boundaries. J. Autom. Lang. Comb. **19**(1), 227–238 (2014)
8. Mihálydeák, T., Vaszil, G.: Regulating rule application with membrane boundaries in P systems. In: Rozenberg, G., Salomaa, A., Sempere, J.M., Zandron, C. (eds.) CMC 2015. LNCS, vol. 9504, pp. 304–320. Springer, Cham (2015). https://doi.org/10.1007/978-3-319-28475-0_21
9. Minsky, M.L.: Computation: Finite and Infinite Machines. Prentice-Hall, Inc., Upper Saddle River (1967)
10. Pawlak, Z.: Rough sets. Int. J. Comput. Inf. Sci. **11**(5), 341–356 (1982)
11. Pawlak, Z.: Rough Sets: Theoretical Aspects of Reasoning about Data. Kluwer Academic Publishers, Dordrecht (1991)
12. Peterson, J.L.: Petri Net Theory and the Modeling of Systems. Prentice Hall PTR, Upper Saddle River (1981)
13. Popova-Zeugmann, L.: Time and Petri Nets. Springer, Heidelberg (2013). https://doi.org/10.1007/978-3-642-41115-1
14. Păun, G.: Computing with membranes. J. Comput. Syst. Sci. **61**(1), 108–143 (2000)

15. Păun, G.: Membrane Computing: An Introduction. Springer, Heidelberg (2002). https://doi.org/10.1007/978-3-642-56196-2
16. Păun, G., Rozenberg, G., Salomaa, A.: The Oxford Handbook of Membrane Computing. Oxford University Press, Inc., New York (2010)
17. Zhang, G., Pérez-Jiménez, M.J., Gheorghe, M.: Real-life Applications with Membrane Computing. ECC, vol. 25. Springer, Cham (2017). https://doi.org/10.1007/978-3-319-55989-6

Quantum Dual Adversary for Hidden Subgroups and Beyond

Aleksandrs Belovs(✉)

Faculty of Computing, University of Latvia, Raiņa bulvāris 19, Riga LV1050, Latvia
aleksandrs.belovs@lu.lv

Abstract. An explicit quantum dual adversary for the S-isomorphism problem is constructed. As a consequence, this gives an alternative proof that the query complexity of the dihedral hidden subgroup problem is polynomial.

Keywords: Quantum algorithms · Hidden subgroup problem · Property testing · Isomorphism testing · Quantum aversary bound

1 Introduction

The Hidden Subgroup Problem (HSP) is one of the most famous problems in the field of quantum algorithms. It was proposed by Kitaev [13] as a common framework for both Simon's algorithm [24] and Shor's algorithms for factoring and discrete logarithm [23]. Time-efficient quantum algorithms are known for the HSP in Abelian groups and few other cases. Most of them are constructed using Fourier sampling [7,8,18]. However, the general case and the cases relevant for applications remain open.

On a positive note, quantum query complexity of the general HSP is known to be polynomial (in the logarithm of the size of the group). There are two different ways of showing this. The first one is by Ettinger *et al.* [6]. In this approach, a tensor product of a polynomial number of copies of the HSP state are prepared. Then, the algorithm proceeds by sequentially probing whether each element of the group is in the hidden subgroup. The idea is that each probe only disturbs the state slightly, so it is possible to perform exponentially many different probes without reconstructing the state. The running time of this algorithm is exponential by design.

The second approach is by Bacon *et al.* [3]. This approach also starts with a tensor product of polynomially many copies of the HSP state. The difference with the first approach is that the pretty good measurement is used [10]. A priori, there is nothing preventing polynomial-time implementation of this approach, and Bacon *et al.* indeed succeeded in constructing polynomial-time quantum algorithms for some new classes of groups. However, for the majority of the groups, it is not clear how to implement the pretty good measurement efficiently.

This research is partly supported by the ERDF grant number 1.1.1.2/VIAA/1/16/113.

I. McQuillan and S. Seki (Eds.): UCNC 2019, LNCS 11493, pp. 30–36, 2019.
https://doi.org/10.1007/978-3-030-19311-9_4

We give an alternative construction of a polynomial-query algorithm. It is based on the dual adversary bound. The latter is a characterisation of the quantum query complexity [12,21,22] using a semi-definite optimisation problem. In principle, this characterisation is known to be tight for every problem, but it is usually not easy to come up with an explicit solution. In this paper, we give a short explicit solution to the dual adversary optimisation problem corresponding to the HSP problem.

Unlike previous solutions based on measurements, the dual adversary is implemented using quantum walks. Sometimes it is possible to implement the corresponding walk time-efficiently, resulting in a time-efficient algorithm [1,4,5]. Most of the time, the solution to the dual HSP is only a guideline for the time-efficient implementation which deviates significantly from the general-case implementation of the dual adversary bound.

Although similar ideas can be used for other groups, for simplicity and concreteness, in this paper we focus on the Dihedral HSP as one of the most important special cases. The dihedral HSP is related to the lattice problems [19], and a sub-exponential (but super-polynomial) quantum algorithm is known [15,16,20]. See [14] for a survey.

It is known that the dihedral HSP is equivalent to the following hidden shift problem. Two strings x_1 and x_2 of length n over some alphabet $[q]$ are given. It is known that all the symbols in x_1 are distinct and all the symbols in x_2 are distinct. The task is to distinguish two cases: x_2 is a cyclic shift of x_1, or x_1 and x_2 do not have a common symbol.

Our algorithm solves a more general problem of testing S-isomorphism [2,9].

Definition 1. *The S-isomorphism problem is specified by a subset S of the set of permutations on $[n]$, and a real number $0 < \varepsilon < 1$. The problem is to evaluate the following partial function on $2n$ input variables over some alphabet $[q]$:*

$$f(x) = f\Big(x_{1,1}, x_{1,2}, \ldots, x_{1,n};\ x_{2,1}, x_{2,2}, \ldots, x_{2,n}\Big),$$

which is defined by

- $f(x) = 1$ *if there exists* $\sigma \in S$ *such that* $x_{2,i} = x_{1,\sigma i}$ *for all* $i \in [n]$*;*
- $f(y) = 0$ *if for all* $\sigma \in S$ *we have* $\Big|\big\{i \in [n] \mid y_{2,i} \neq y_{1,\sigma i}\big\}\Big| \geq \varepsilon n.$

For other inputs the function is not defined.

In other words, one has to detect whether x_2 can be obtained from x_1 using a permutation from S, or it is ε-far from that. Note that S is not required to be a group. The dihedral HSP is a special case of this problem, where S is the group of cyclical shifts, $\varepsilon = 1$, and it is additionally promised that there are no equal elements inside x_1 and no equal elements inside x_2.

The pretty-good-measurement approach does not directly apply to this problem because of the equal elements inside x_1 or x_2. Harrow *et al.* [9] showed how to modify the sequential measurement approach and proved the following theorem:

Theorem 2. *The S-isomorphism problem can be solved in $O\left(\frac{\log|S|}{\varepsilon}\right)$ quantum queries.*

As observed in the same paper, this leads to quantum algorithms for testing Boolean function isomorphism, graph isomorphism, and other problems.

In this paper, we give an alternative proof of Theorem 2 using the adversary bound.

2 Preliminaries

We use the following notation. For an integer n, $[n]$ denotes the set $\{1, 2, \dots, n\}$. For two strings $x, y \in [q]^n$, we use $d(x, y)$ to denote the Hamming distance between them.

We use standard linear-algebraic notation. For more detail on this the reader may refer to e.g. [11]. All matrices have real entries. For two vectors $u, v \in \mathbb{R}^d$, $\langle u, v\rangle$ denotes their inner product. For a matrix A, $A[\![x, y]\!]$ denotes the entry of the matrix at the intersection of the row labelled by x and the column labelled by y. For two matrices A and B of the same dimensions, $A \circ B$ denotes their Hadamard (entry-wise) product. If k is a non-negative integer, $A^{\circ k}$ stands for the k-th Hadamard power of A, that is $A^{\circ k} = A \circ A \circ \dots \circ A$, k times.

We use $X \succeq 0$ to denote that X is a positive semi-definite (PSD) matrix. We use the following well-known properties of PSD matrices. If $v = \{v_z\}_{z\in\mathcal{D}}$ is a collection of vectors in some inner-product space, then the Gram matrix $G(v)$ of this collection, defined by $G(v)[\![x, y]\!] = \langle v_x, v_y\rangle$ for $x, y \in \mathcal{D}$, is positive semi-definite. A sum (even an infinite one) of PSD matrices is a PSD matrix. Finally, a Hadamard product of two PSD matrices is a PSD matrix.

Our main technical tool is the dual adversary bound.

Definition 3. *Let $f\colon \mathcal{D} \to \{0,1\}$ with $\mathcal{D} \subseteq [q]^m$ be a function. The adversary bound $\mathrm{Adv}^{\pm}(f)$ is defined as the optimal value of the following optimisation problem:*

$$\text{minimise} \quad \max_{x\in\mathcal{D}} \sum_{j\in[m]} X_j[\![x, x]\!] \tag{1a}$$

$$\text{subject to} \quad \sum_{j\in[m]:\, x_j\neq y_j} X_j[\![x, y]\!] = 1 \qquad \textit{whenever } f(x) \neq f(y); \tag{1b}$$

$$X_j \succeq 0 \qquad \textit{for all } j \in [m]; \tag{1c}$$

where the optimisation is over positive semi-definite matrices X_j with rows and columns labeled by the elements of $\mathcal{D}$.

The general adversary bound characterises quantum query complexity. Let $Q(f)$ denote the query complexity of the best quantum algorithm evaluating f with a bounded error.

Theorem 4 ([12,17,22]). *Let f be as in Definition 3. Then, we have that $Q(f) = \Theta(\mathrm{Adv}^{\pm}(f))$.*

3 One-Matrix Problem

In this section, we describe our approach towards proving Theorem 2, and in the next section we finish the proof.

In our feasible solution to (1), all the matrices X_j will be equal. Let us denote their common value by X. It is easy to see that the necessary and sufficient condition on X is

$$X[\![x,y]\!] = \frac{1}{d(x,y)} \qquad \text{for all } x \in f^{-1}(1) \text{ and } y \in f^{-1}(0),$$

where $d(x, y)$ is the Hamming distance between x and y. Our goal is to embed this rectangular matrix into a full semi-definite matrix with small diagonal entries.

For brevity, denote $\mu(X) = \max_z X[\![z,z]\!]$. The objective value (1a) equals $\mu(Y)$ for $Y = 2nX$. For z in the domain, define unit vectors

$$\varphi_z = \frac{1}{\sqrt{2n}} \sum_j |j, z_j\rangle. \tag{2}$$

Note that

$$\langle \varphi_x, \varphi_y \rangle = 1 - \frac{d(x,y)}{2n},$$

Hence,

$$Y[\![x,y]\!] = \frac{1}{1 - \langle \varphi_x, \varphi_y \rangle} = 1 + \langle \varphi_x, \varphi_y \rangle + \langle \varphi_x, \varphi_y \rangle^2 + \cdots \tag{3}$$

Let

$$\Phi = G(\varphi),$$

where G denotes the Gram matrix of the collection φ_x. We have $\Phi \succeq 0$ and $\mu(\Phi) = 1$. Inspired by (3), we could define $Y = \sum_{k=0}^{\infty} \Phi^{\circ k}$, where $\Phi^{\circ k}$ is the Hadamard power. The problem is that this sum diverges. It is not surprising: we have not used anything about the problem yet. The solution is to find a substitute for Φ that works well for large k.

4 Substitute for Φ

For each positive input x, fix a permutation $\sigma \in S$ that witnesses that x is positive, that is, $x_{1,i} = x_{2,\sigma i}$ for all i. If there are several potential witnesses σ, fix any one of them, so that every $x \in f^{-1}(1)$ is witnessed by exactly one σ.

Fix a permutation $\sigma \in S$. Let x be a positive input witnessed by σ, and y be a negative input. Construct (non-normalised) vectors u_x and u_y as follows

$$u_x = \frac{1}{\sqrt{2n}} \sum_{i=1}^{n} |i\rangle|x_{1,i}\rangle \qquad \text{and} \qquad u_y = \frac{1}{\sqrt{2n}} \sum_{i=1}^{n} |i\rangle \big(|y_{1,i}\rangle + |y_{2,\sigma i}\rangle\big).$$

These vectors satisfy the following properties. First, $\langle u_x, u_y\rangle = \langle \varphi_x, \varphi_y\rangle$, where φ_x are as in (2). Next, $\|u_x\|^2 = 1/2$. Finally,

$$\|u_y\|^2 = \frac{1}{2n}\left(4\cdot\left|\{i\in[n] \mid y_{1,i} = y_{2,\sigma i}\}\right| + 2\cdot\left|\{i\in[n] \mid y_{1,i}\neq y_{2,\sigma i}\}\right|\right) \le 2-\varepsilon.$$

Let $c = \sqrt[4]{2(2-\varepsilon)}$. Define $u'_x = cu_x$ for $x\in f^{-1}(1)$ witnessed by σ, $u'_y = u_y/c$ for $y\in f^{-1}(0)$, and $u'_z = 0$ for $z\in f^{-1}(1)$ not witnessed by σ. The Gram matrix of this collection $u' = (u'_z)_{z\in\mathcal{D}}$ of vectors

$$\Psi_\sigma = G(u')$$

satisfies $\Psi_\sigma[\![x,y]\!] = \langle\varphi_x,\varphi_y\rangle$ for x and y as above, and $\mu(\Psi_\sigma)\le\sqrt{1-\varepsilon/2}$.

Now let us define

$$\Psi^{(k)} = \sum_{\sigma\in S}\Psi_\sigma^{\circ k}.$$

For this matrix, we have $\Psi^{(k)}\succeq 0$, and $\Psi^{(k)}[\![x,y]\!] = \langle\varphi_x,\varphi_y\rangle^k$ for all $x\in f^{-1}(1)$ and $y\in f^{-1}(0)$. Also,

$$\mu(\Psi^{(k)}) \le |S|\left(1-\frac{\varepsilon}{2}\right)^{k/2}.$$

Let ℓ be the smallest positive integer such that $|S|(1-\frac{\varepsilon}{2})^{\ell/2} < 1$. In particular, $\ell = O\left(\frac{\log|S|}{\varepsilon}\right)$. We define

$$Y = \sum_{k=0}^{\ell-1}\Phi^{\circ k} + \sum_{k=\ell}^{\infty}\Psi^{(k)}.$$

First, this matrix satisfies (3) for $x\in f^{-1}(1)$ and $y\in f^{-1}(1)$. Also,

$$\mu(Y) \le \ell + \sum_{k=\ell}^{\infty}|S|\left(1-\frac{\varepsilon}{2}\right)^{k/2} = O\left(\frac{\log|S|}{\varepsilon}\right) + O\left(\frac{1}{\varepsilon}\right) = O\left(\frac{\log|S|}{\varepsilon}\right).$$

Hence, if we define $X_j = Y/2n$ for all j, we get a feasible solution to (1) for the function in Theorem 2 with objective value $O\left(\frac{\log|S|}{\varepsilon}\right)$.

5 Discussion

We have described a dual adversary for the S-isomorphism testing problem, which implies a quantum query algorithm for the dihedral HSP problem. It is not clear whether these ideas can be used for a time-efficient implementation of any of these problems. It is interesting to identify other problems for which these ideas can give query-efficient quantum algorithms.

Acknowledgements. I am grateful to all the persons with whom I have discussed this problem. Especially, I would like to thank Martin Roetteler, Dmitry Gavinsky and Tsuyoshi Ito.

References

1. Ambainis, A., Belovs, A., Regev, O., de Wolf, R.: Efficient quantum algorithms for (gapped) group testing and junta testing. In: Proceedings of 27th ACM-SIAM SODA, pp. 903–922 (2016)
2. Babai, L., Chakraborty, S.: Property testing of equivalence under a permutation group action (2008)
3. Bacon, D., Childs, A.M., van Dam, W.: From optimal measurement to efficient quantum algorithms for the hidden subgroup problem over semidirect product groups. In: Proceedings of 46th IEEE FOCS, pp. 469–478 (2005)
4. Belovs, A., Childs, A.M., Jeffery, S., Kothari, R., Magniez, F.: Time-efficient quantum walks for 3-distinctness. In: Fomin, F.V., Freivalds, R., Kwiatkowska, M., Peleg, D. (eds.) ICALP 2013, Part I. LNCS, vol. 7965, pp. 105–122. Springer, Heidelberg (2013). https://doi.org/10.1007/978-3-642-39206-1_10
5. Belovs, A., Reichardt, B.W.: Span programs and quantum algorithms for *st*-connectivity and claw detection. In: Epstein, L., Ferragina, P. (eds.) ESA 2012. LNCS, vol. 7501, pp. 193–204. Springer, Heidelberg (2012). https://doi.org/10.1007/978-3-642-33090-2_18
6. Ettinger, M., Høyer, P., Knill, E.: The quantum query complexity of the hidden subgroup problem is polynomial. Inf. Process. Lett. **91**(1), 43–48 (2004)
7. Friedl, K., Ivanyos, G., Magniez, F., Santha, M., Sen, P.: Hidden translation and orbit coset in quantum computing. In: Proceedings of 35th ACM STOC, pp. 1–9 (2003)
8. Grigni, M., Schulman, L., Vazirani, M., Vazirani, U.: Quantum mechanical algorithms for the nonabelian hidden subgroup problem. Combinatorica **24**(1), 137–154 (2004)
9. Harrow, A.W., Lin, C.Y.Y., Montanaro, A.: Sequential measurements, disturbance and property testing. In: Proceedings of 28th ACM-SIAM SODA, pp. 1598–1611 (2017)
10. Hausladen, P., Wootters, W.K.: A "pretty good" measurement for distinguishing quantum states. J. Mod. Opt. **41**(12), 2385–2390 (1994)
11. Horn, R.A., Johnson, C.R.: Matrix Analysis. Cambridge University Press, Cambridge (1985)
12. Høyer, P., Lee, T., Špalek, R.: Negative weights make adversaries stronger. In: Proceedings of 39th ACM STOC, pp. 526–535 (2007)
13. Kitaev, A.: Quantum measurements and the Abelian stabilizer problem (1995)
14. Kobayashi, H., Le Gall, F.: Dihedral hidden subgroup problem: a survey. Inf. Media Technol. **1**(1), 178–185 (2006)
15. Kuperberg, G.: A subexponential-time quantum algorithm for the dihedral hidden subgroup problem. SIAM J. Comput. **35**, 170–188 (2005)
16. Kuperberg, G.: Another subexponential-time quantum algorithm for the dihedral hidden subgroup problem (2011)
17. Lee, T., Mittal, R., Reichardt, B.W., Špalek, R., Szegedy, M.: Quantum query complexity of state conversion. In: Proceedings of 52nd IEEE FOCS, pp. 344–353 (2011)
18. Moore, C., Rockmore, D., Russell, A., Schulman, L.J.: The power of basis selection in Fourier sampling: hidden subgroup problems in affine groups. In: Proceedings of 15th ACM-SIAM SODA, pp. 1113–1122 (2004)
19. Regev, O.: Quantum computation and lattice problems. SIAM J. Comput. **33**(3), 738–760 (2004)

20. Regev, O.: A subexponential time algorithm for the dihedral hidden subgroup problem with polynomial space (2004)
21. Reichardt, B.W.: Span programs and quantum query complexity: the general adversary bound is nearly tight for every boolean function. In: Proceedings of 50th IEEE FOCS, pp. 544–551 (2009)
22. Reichardt, B.W.: Reflections for quantum query algorithms. In: Proceedings of 22nd ACM-SIAM SODA, pp. 560–569 (2011)
23. Shor, P.W.: Polynomial-time algorithms for prime factorization and discrete logarithms on a quantum computer. SIAM J. Comput. **26**, 1484–1509 (1997)
24. Simon, D.: On the power of quantum computation. SIAM J. Comput. **26**, 1474–1483 (1997)

Further Properties of Self-assembly by Hairpin Formation

Henning Bordihn[1], Victor Mitrana[2,3,4(✉)], Andrei Păun[3,4], and Mihaela Păun[4,5]

[1] Department of Computer Science, University of Potsdam, Potsdam, Germany
henning@cs.unipotsdam.de

[2] Department of Information Systems, Polytechnic University of Madrid, Madrid, Spain
victor.mitrana@upm.es

[3] Faculty of Mathematics and Computer Science, University of Bucharest, Bucharest, Romania
apaun@fmi.unibuc.ro

[4] National Institute for Research and Development of Biological Science, Bucharest, Romania

[5] Faculty of Administration and Business, University of Bucharest, Bucharest, Romania
mihaela.paun@gmail.com

Abstract. We continue the investigation of three operations on words and languages with motivations coming from DNA biochemistry, namely unbounded and bounded hairpin completion and hairpin lengthening. We first show that each of these operations can be used for replacing the third step, the most laborious one, of the solution to the CNF-SAT reported in [28]. As not all the bounded/unbounded hairpin completion or lengthening of semilinear languages remain semilinear, we study sufficient conditions for semilinear languages to preserve their semilinearity property after applying once either the bounded or unbounded hairpin completion, or lengthening. A similar approach is then started for the iterated variants of the three operations. A few open problems are finally discussed.

Keywords: DNA hairpin formation · Hairpin completion · Bounded hairpin completion · Hairpin lengthening · Semilinearity property

1 Introduction

A *DNA strand* can be abstracted, if its spatial structure is ignored, as a word over the four-letter alphabet {A, C, G, T} where the letters represent the nucleotides

This work was supported by a grant of the Romanian National Authority for Scientific Research and Innovation, project number POC P-37-257. Victor Mitrana has also been supported by the Alexander von Humboldt Foundation.

I. McQuillan and S. Seki (Eds.): UCNC 2019, LNCS 11493, pp. 37–51, 2019.
https://doi.org/10.1007/978-3-030-19311-9_5

Adenine, Cytosine, Guanine, and Thymine, respectively. In the double helix of DNA, the two strands are end-to-end chemically oriented in opposite directions, namely from 5′ to 3′ and from 3′ to 5′, respectively, hence they are anti-parallel, which permits base pairing by the Watson-Crick complementarity, where A is complementary to T and C to G. This base pairing, by the hydrogen bonds between complementary nucleotides under some specific environment conditions, is one of the main properties of DNA which the process of DNA replication is based on.

Throughout this note, we use a bar-notation for the Watson-Crick complement; thus $\overline{\mathsf{A}} = \mathsf{T}$ and $\overline{\mathsf{T}} = \mathsf{A}$ as well as $\overline{\mathsf{C}} = \mathsf{G}$ and $\overline{\mathsf{G}} = \mathsf{C}$. We extend this bar-notation to sequences of nucleotides (words) by $\overline{s_1 \cdots s_n} = \overline{s_n} \cdots \overline{s_1}$.

Due to the base pairing discussed above, a single stranded DNA molecule may produce a hairpin structure as shown in Fig. 1.

```
GATCTGATGCATGAGATCGCATCAG

             ⇃↾

                     T  G
      GATCTGATGCA          A
         ||||||||
         GACTACGC          G
                     T  A
```

Fig. 1. Hairpin structure.

In many DNA-based algorithms, the single stranded DNA molecules that formed already or might form a hairpin structure cannot be used in the subsequent computations. Hairpin or hairpin-free DNA structures have numerous applications in DNA computing and molecular genetics. In a series of papers, see, e.g., [7,9,10], such structures are discussed in the context of finding sets of DNA sequences which are unlikely to lead to "bad" hybridizations, that is DNA fragments do not anneal to complementary segments in undesired ways. It has been claimed in different places that the potential of DNA molecules to form self-assembly structures might be employed for designing solutions to hard problems, see, e.g., [30]. Thus a first DNA-based solution to the CNF-SAT problem, where one of the main steps was implemented on the basis of hairpin formation by single-stranded DNA molecules was reported in [28]. In this DNA-based solution, the third phase is mainly based on the elimination of hairpin structured molecules. A rather long and complicated lab methodology is discussed for implementing this phase. We propose a modification of this algorithm that may use any of the three operations considered here and could be easily implemented. Different types of hairpin and hairpin-free languages were defined in [3,25], and [15], where they were studied from a language theoretical point of view.

A common lab technique to lengthen DNA or to make copies of regions of DNA is called *polymerase chain reaction* (PCR). This technique which is rather cheap, easy and reliable is widely used in molecular biology but also in

DNA computing. This technique is used to produce a complete double stranded DNA molecule starting from a single stranded molecule as informally follows: the starting single strand (usually called *primer*) is bonded to its 3′ end with another shorter strand (usually called *template*) by Watson-Crick complementarity. Then a *polymerization buffer* with many copies of the four nucleotides, and a DNA polymerase (an enzyme) will concatenate nucleotides to the primer by complementing the template. A very closely related intramolecular reaction, called whiplash PCR, which employs polymerization stop is also used here.

These principles discussed above have been the source of inspiration for introducing in [5] a new formal operation on words, namely *hairpin completion*. We now informally explain the hairpin completion operation and how it can be related to the aforementioned biological concepts. Let us consider the following hypothetical biological situation: we are given one single stranded DNA molecule z such that either a prefix or a suffix of z is Watson-Crick complementary to a subword of z. Then the prefix or suffix of z and the corresponding subword of z get annealed by complementary base pairing and then z is lengthened by DNA polymerases up to a complete hairpin structure. Finally, the linear structure of the molecule is restored by DNA denaturation such that the whole process described above can be resumed. This is illustrated in Fig. 2, where $z = \gamma\alpha\beta\overline{\alpha}$.

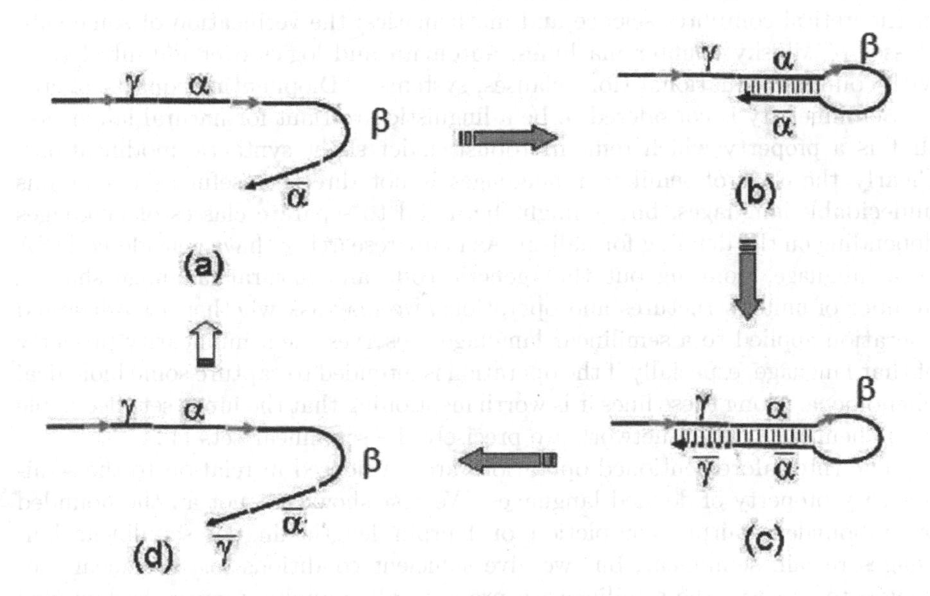

Fig. 2. (a) A single stranded DNA molecule; (b) hairpin formation with part α; (c) polymerization extension of γ; (d) restoring the linear structure.

The mathematical expression of this hypothetical situation defines the hairpin completion operation. This operation is considered in [5] as an abstract operation on formal languages. A series of subsequent works extended the investigation of this operation and proposed a few variants. Some algorithmic problems regarding the hairpin completion are investigated in [18]. In the aforementioned papers, no restriction is imposed on the length of prefix or suffix added by the hairpin completion. In [14] one considers a restricted variant of the hairpin completion, called bounded hairpin completion. Another operation derived from the biological phenomenon described above, namely hairpin lengthening has been introduced in [21] and further investigated in [22]. Two other operations inspired by these biological phenomena are: *WK-superposition* [2] and [20], *overlap assembly* [6] and [8].

We propose a modification of the solution to the CNF-SAT based on the DNA hairpin formation reported in [28]. This modification, which may use any of the operations considered here, eliminates the third step of that algorithm which seems to be the most laborious one. Our algorithm appears, at least theoretically, to be more easily implemented in a laboratory because our step might be implemented by gel electrophoresis.

A set of tuples of integers constructed as a multi-dimensional arithmetic progression is called linear. A finite union of linear sets is called semilinear. The semilinear sets are exactly the sets definable in Presburger arithmetic [12]. They also can be viewed as a generalization of ultimately periodic sets of natural numbers to any given dimension. Semilinear sets have numerous applications in theoretical computer science and mathematics: the verification of some subclasses of Minsky counter machines, automata and logics over unranked trees with counting, equational Horn clauses, systems of Diophantine equations, etc.

Semilinearity is considered to be a linguistic invariant for natural languages, that is a property which remains robust under slight syntactic modifications. Clearly the class of semilinear languages is not directly useful as it contains undecidable languages, but it might be useful to separate classes of languages depending on the defining formalism. As many researchers have considered DNA as a language, pointing out that genetic code and natural language share a number of units, structures and operations, we may ask whether a bio-inspired operation applied to a semilinear language preserves the semilinearity property of that language, especially if the operation is intended to capture some biological phenomena. Along these lines it is worth mentioning that the input sets decidable by a chemical reaction network are precisely the semilinear sets [4].

The three aforementioned operations are considered in relation to the semilinearity property of defined languages. We first show that not all the bounded or unbounded hairpin completion or hairpin lengthening of semilinear languages remain semilinear, but we give sufficient conditions for semilinear languages to preserve the semilinearity property after applying once the bounded or unbounded hairpin completion or hairpin lengthening. A similar investigation is then done for the iterated variants of the three operations. Finally, we discuss a few directions for further investigation.

2 Basic Definitions

We assume the reader to be familiar with the fundamental concepts of formal language theory and automata theory, see, e.g., [27].

An alphabet is a finite set of letters. For a finite set A we denote by $card(A)$ the cardinality of A. The set of all words over an alphabet V is denoted by V^*. The empty word is denoted by ε; moreover, $V^+ = V^* \setminus \{\varepsilon\}$. Given a word w over an alphabet V, we denote by $|w|$ its length, while $|w|_a$ denotes the number of occurrences of the letter a in w. If $w = xyz$ for some $x, y, z \in V^*$, then x, y, z are called prefix, subword, suffix, respectively, of w. If $0 < |x|, |z| < |w|$, then x and z are called proper prefix and proper suffix, respectively. If x or z is non-empty, then y is called a proper subword of w.

An *involution* over a set S is a bijective mapping $\sigma : S \longrightarrow S$ such that $\sigma = \sigma^{-1}$. In this paper's context, any involution σ over some set S such that $\sigma(a) \neq a$ for all $a \in S$ is called a *Watson-Crick involution.* Despite that this is nothing more than a fixed point-free involution, we prefer this terminology since the hairpin completion defined later is inspired by the DNA biochemistry, where the Watson-Crick base-pair complementarity plays an important role. Let $\bar{\cdot}$ be a Watson-Crick involution over some alphabet V; we extend this involution to an anti-morphism from V^* to V^* in the usual way, namely $\overline{a_1 a_2 \ldots a_n} = \overline{a_n} \ldots \overline{a_2}\, \overline{a_1}$. We say that the letters a and $\overline{a}$ are complementary to each other. If $\bar{\cdot}$ is a Watson-Crick involution over some alphabet V, then clearly $\overline{V} = \{\overline{a} \mid a \in V\} = V$; such an alphabet is called a Watson-Crick alphabet. If not otherwise stated, all the alphabets in this note are Watson-Crick alphabets and $\bar{\cdot}$ is a fixed involution such that $\overline{a} \neq a$ for any letter a in the alphabet. Remember that the DNA alphabet consists of four letters, $V_{DNA} = \{A, C, G, T\}$, which are abbreviations for the four nucleotides and we have set $\overline{A} = T$, $\overline{C} = G$.

Let V be an alphabet, for any $w \in V^+$ we define the *k-hairpin lengthening* of w, denoted by $HL_k(w)$, for some $k \geq 1$, as follows [21]:

$$\begin{aligned} HLP_k(w) &= \{\overline{\delta} w \mid w = \alpha\beta\overline{\alpha}\gamma, |\alpha| = k,\ \alpha, \beta, \gamma \in V^+, \\ &\qquad \text{and } \delta \text{ is a proper prefix of } \gamma\} \\ HLS_k(w) &= \{w\overline{\delta} \mid w = \gamma\alpha\beta\overline{\alpha}, |\alpha| = k,\ \alpha, \beta, \gamma \in V^+, \\ &\qquad \text{and } \delta \text{ is a proper suffix of } \gamma\}, \\ HL_k(w) &= HLP_k(w) \cup HLS_k(w) \end{aligned}$$

If in the above definitions, one replaces "δ is a proper prefix/suffix of γ" by "$\delta = \gamma$", then the operation is called *k-hairpin completion* [5] and it is denoted by HC_k. Furthermore, if $\delta = \gamma$ and $|\gamma| \leq p$, for some positive integer p, then the operation is called *p-bounded k-hairpin completion* [14] and it is denoted by pHC_k. The *hairpin lengthening/completion* and *bounded hairpin completion* of w is defined by

$$H(w) = \bigcup_{k \geq 1} H_k(w),$$

where $H \in \{HL, HC, pHC\}$, respectively. The hairpin lengthening/completion and bounded hairpin completion is naturally extended to languages by

$$H_k(L) = \bigcup_{w \in L} H_k(w) \qquad H(L) = \bigcup_{w \in L} H(w),$$

where $H \in \{HL, HC, pHC\}$, respectively. We want to stress that the biological phenomenon is just a source of inspiration for introducing the operations defined above. First, it is known that a "stable" hairpin structure as above is possible if the subword α is sufficiently long. Second, it is known that DNA polymerase can act continuously only in the $5' \longrightarrow 3'$ due to the greater stability of $3'$ when attaching new nucleotides. As one can see in our definitions, we have allowed polymerase to extend continuously in either end; however, polymerase can also act in the opposite direction, but in short "spurts" (Okazaki fragments).

For a class of languages $\mathcal{F}$ and an integer $k \geq 1$ we denote the class of the hairpin lengthening/completion and bounded hairpin completion of languages in $\mathcal{F}$ by $H_k(\mathcal{F}) = \{H_k(L) \mid L \in \mathcal{F}\}$.

3 A New Solution to the CNF-SAT Using Hairpin Operations

The Conjunctive Normal Form (CNF) - Satisfiability (SAT) problem asks to find, if any, Boolean assignments that satisfy a given formula in conjunctive normal form, that is a formula of the form $C_1 \wedge C_2 \wedge \cdots \wedge C_m$, where each C_i is a clause and $\wedge$ is the logical AND operator. Each clause C_i is of the form $C_i = l_{i_1} \vee l_{i_2} \vee \cdots \vee l_{i_{k_i}}$, where each l_{i_j} is a literal (a variable or its negation) and $\vee$ is the logical OR operation. A DNA-based algorithm for solving CNF-SAT having three main steps was proposed in [28]. We refer to [28] for all unexplained notions.

S1. Generate all the literal strings with respect to the given formula. The literal strings encode conjunctions of literals one per clause. Such a conjunction is

$$l_{1_{r_1}} \wedge l_{2_{r_2}} \wedge \cdots \wedge l_{m_{r_m}},$$

where $1 \leq r_j \leq k_j$ for all $1 \leq j \leq m$. This step is implemented by a routine ligation reaction which concatenates DNA segments encoding literal units to larger strings according to the input formula.

S2. Allow all literal strings formed in step S1 to form hairpins. Note that the hairpin structures formed in this step are not necessarily as that of (b) in Figure 2. More precisely, it is not necessary that one of the segments that get annealed by complementary base pairing be a prefix or a suffix. This step may be implemented by regulating the temperature.

S3. Remove all the molecules that formed a hairpin in step S2. This has been done by two techniques: (i) remove DNA molecules that formed hairpin by enzymatic digestion, and (ii) amplify the population of DNA molecules without hairpin by a variant of PCR, called "exclusive PCR". This is the reason why this step is very laborious.

The correctness of the algorithm follows from the fact that a molecule forms a hairpin in the second step, if and only if it encodes an inconsistent assignment, that is when both a variable and its negation are true. A long and rather complicated procedure based on the two techniques described above is proposed for implementing the third step. We propose the following modification of this algorithm. First, instead of literal units as in the previous algorithm, we use extended literal units, that is literal units as above linked to a unique segment, say α which is newly designed and cannot form hairpin structure with any other segment. As in the first step of the previous algorithm, literal strings are obtained by extending the already produced strings with extended literal units associated with a new clause which was not considered yet. However, before extending the current literal strings with the new extended literal unit, we propose to allow all current literal strings obtained in the previous step to form hairpins and be extended by one of the three hairpin operations. All those literal strings that contain the complement of α will be removed such that they will not be extended with new extended literal units. The algorithm is informally described below. This situation corresponds to the case of using hairpin completion.

S1. Generate all the exteded literal units associated with each clause.
S2. For each clause C that has not been considered yet, do the following:
S2.1. Extend by ligation the current literal strings with the extended literal units associated with C, say α.
S2.2. Allow all current literal strings obtained in the previous step to form hairpins and be extended by one of the three hairpin operations. Let us choose the hairpin completion.
S2.3. Remove all literal strings that contain the complement of α.
S3. All the literal strings at this step encode solutions to the input formula.

As one can see, this modification seems to simplify the implementation of the algorithm because the laborious step 3 in the initial algorithm, which is based on a rather complicated la procedures, namely either enzymatic digestion of the hairpin DNA molecules or exclusive PCR, is avoided and replaced by a sequence of a rather standard lab procedure which removes all single stranded DNA molecules that contain a given sequence. In the case of hairpin lengthening, this lab method could be gel electrophoresis.

4 Non-iterated Hairpin Formation and Semilinearity

A language is semilinear if its Parikh map is a semilinear set [24]. More formally, a subset S of $\mathbb{N}^k$ (k-tuples of natural numbers) is a linear set if there exist vectors $v_0, v_1, \ldots, v_n \in \mathbb{N}^k$ such that $S = \{v_0 + \sum_{i=1}^{n} x_i v_i | x_i \in \mathbb{N}, 1 \le i \le n\}$. A finite union of linear sets is called a semilinear set. Let $V = \{a_1, \ldots, a_k\}$ be an alphabet. The Parikh mapping of w is the vector $\psi(w) = (|w|_{a_1}, \ldots, |w|_{a_k})$, which is extended to languages, by $\psi(L) = \{\psi(w) | w \in L\}$. A language is semilinear if its Parikh mapping is a semilinear set. Two languages are said to be letter-equivalent if for every word in one language there exist at least one anagram of that word in the other language. Clearly, two letter-equivalent languages have the same Parikh mapping. It is known that a language L is semilinear if and only if it is letter-equivalent to a regular language [24]. Furthermore, a language family is called semilinear if all the languages in the family are semilinear. We denote by **SLin** the class of semilinear languages.

Theorem 1

1. **SLin** *is closed neither under* HC_k *nor under* HL_k*, for any* $k \geq 1$.

2. **SLin** *is not closed under* pHC_k*, for any* $k \geq 1$*,* $p \geq 2$.

Proof. Since our main source of inspiration is the DNA biochemistry, we give counterexamples that use at most 4-letter alphabets.

1. Let $k \geq 1$ and take the semilinear language

$$L_1 = \{ba^{2^n}b^k a^m \overline{b^k} \mid n, m \geq 1\},$$

where both a and $\overline{a}$ are different from b. It is easy to note that $HC_k(L_1) = \{ba^{2^n}b^k a^m \overline{b^k}\overline{a^{2^n}b} \mid n, m \geq 1\}$, which is not semilinear. This follows from the fact that semilinear sets are closed under projection, and the projection of $\psi(HC_k(L_1))$ on $\overline{a}$ is not semilinear.
Furthermore,

$$\begin{aligned} HL_k(L_1) = &\{ba^{2^n}b^k a^m \overline{b^k}\overline{a^r} \mid n, m \geq 1, 1 \leq r \leq 2^n\} \cup \\ &\{ba^{2^n}b^k a^m \overline{b^k}\overline{a^{2^k}b} \mid n, m \geq 1\}. \end{aligned}$$

By a similar reason as above, $HL_k(L_1)$ cannot be semilinear. We prove that the projection of $\psi(HL_k(L_1))$ on the letters $\overline{a}$ and $\overline{b}$ is not semilinear. This projection is the set

$$X = \{(t, k) \mid t \geq 1\} \cup \{(2^t, k + 1) \mid t \geq 1\}.$$

As the class of semilinear sets is closed under intersection [12], if X were semilinear, then the set $X \cap \{(t, k + 1) \mid t \geq 1\}$ would be semilinear and this is not the case.

2. We take the semilinear language $L_2 = \{ba^t b^k a^{2^n} \overline{b^k} \mid n, t \geq 1\}$. For every $p \geq 2$, $pHC_k(L_2) = \{ba^r b^k a^{2^n} \overline{b^k}\overline{a^r}\overline{b} \mid n \geq 1, 1 \leq r < p\}$, which is not semilinear. This statement is supported by the fact that the projection of $\psi(pHC_k(L_2))$ on a is $X = \{r + 2^n \mid 1 \leq r < p, n \geq 1\}$, which is not semilinear.

Assume that X is semilinear, hence X is a finite union of linear sets. At least one set in this union, say Y must be infinite. Assume that Y is a linear set with q periods, that is there are the natural numbers $c_1, c_2, \ldots, c_q$, for some $q \geq 1$, such that

$$Y = \{c_o + \sum_{j=1}^{q} x_j c_j \mid x_j \in \mathbb{N}\}.$$

It is known that any linear set is the union of a finite set and a linear set with one period [26]. Therefore, $Y = Z \cup \{c_0 + xy \mid x \in \mathbb{N}\}$. Now, let n_1 be a large number and $1 \leq r_1 < p$ such that $r_1 + 2^{n_1} = c_0 + xy$, for some x. Clearly, there exist $1 \leq r_2 < p$ and n_2 such that $r_2 + 2^{n_2} = c_0 + (x+p)y$. It follows that $2^{n_1}(2^{n_2 - n_1} - 1) + r_2 - r_1 = py$, which is a contradiction by the choice of n_1. □

Although the whole class **SLin** is closed under none of the hairpin completion, bounded hairpin completion, and hairpin lengthening, subclasses of **SLin** with further closure properties are closed.

A shuffle of two words is an arbitrary interleaving of subwords of these words such that it contains all letters of both words, like shuffling two decks of cards. This is a well-known language-theoretic operation with a long history in theoretical computer science, in particular within formal languages. Formally, the shuffle operation denoted by $\|\|$ is defined recursively for two words as follows:

$$x \,\|\!\|\, \varepsilon = \varepsilon \,\|\!\|\, x = \{x\},\ x \in V^*, \text{ and}$$
$$ax \,\|\!\|\, by = a(x \,\|\!\|\, by) \cup b(ax \,\|\!\|\, y),\ a, b \in V,\ x, y \in V^*.$$

The shuffle of two languages $L_1, L_2 \subseteq V^*$ is denoted by $L_1 \,\|\!\|\, L_2$ and is defined as the language consisting of all words that are a shuffle of a word from L_1 and a word from L_2. Thus

$$L_1 \,\|\!\|\, L_2 = \{w \in u \,\|\!\|\, v \mid u \in L_1,\ v \in L_2\}.$$

A restrictive variant of this operation that can be applied to equal length words only, called literal shuffle, is defined by $x \wr\!\wr\!\wr y = a_1 b_1 a_2 b_2 \ldots a_n b_n$, provided that $x = a_1 a_2 \ldots a_n$, $y = b_1 b_2 \ldots b_n$, for some $n \geq 1$, and $a_i, b_i \in V, 1 \leq i \leq n$.

Theorem 2. *Let $\mathcal{F}$ be a family of semilinear languages.*

1. *If $\mathcal{F}$ is closed under intersection and shuffle with regular languages, then both families $HC_k(\mathcal{F})$ and $HL_k(\mathcal{F})$ contain semilinear languages only, for any $k \geq 1$.*
2. *If $\mathcal{F}$ is closed under intersection with regular languages, then all languages in $pHC_k(\mathcal{F})$ are semilinear for all $p, k \geq 1$.*

Proof. 1. The proof of the first two statements is based on a similar idea to that used in [19]. We take a semilinear language $L \subseteq V^*$ in $\mathcal{F}$ and define the next two languages:

$$L_1 = \{(\gamma \wr\!\wr\!\wr \overline{\gamma})\alpha\beta\overline{\alpha}, \alpha, \beta, \gamma \in V^+, |\alpha| = k, \gamma\alpha\beta\overline{\alpha} \in L\},$$
$$L_2 = \{\mu(\delta \wr\!\wr\!\wr \overline{\delta})\alpha\beta\overline{\alpha}, \alpha, \beta \in V^+, |\alpha| = k, \mu\delta\alpha\beta\overline{\alpha} \in L\}.$$

□

Claim 1: $HC_k(L) \in \mathbf{SLin}$.

It is known that the class of semilinear languages is closed under union, hence it suffices to prove that $HCS_k(L)$ is semilinear. Clearly, this language is letter-equivalent to the language L_1. We consider a new alphabet U, disjoint of V, and define a one-to-one mapping $\sigma : V \longrightarrow U$; moreover let $\theta : (V \cup U)^* \longrightarrow V$ be a morphism such that $\theta(a) = \begin{cases} a, & \text{if } a \in V, \\ \overline{b}, & \text{if } a = \sigma(b). \end{cases}$

Now, L_1 can be written as $L_1 = \bigcup_{|\alpha|=k} \theta((L \between U^*) \cap \{a\sigma(a) \mid a \in V\}^+\{\alpha\}V^+\{\overline{\alpha}\})$.

By the closure properties of $\mathcal{F}$, the language $L \between U^*$ is in $\mathcal{F}$, moreover, $(L \between U^*) \cap \{a\sigma(a) \mid a \in V\}^+\{\alpha\}V^+\{\overline{\alpha}\}$ still belongs to $\mathcal{F}$ as each language $\{a\sigma(a) \mid a \in V\}^+\{\alpha\}V^+\{\overline{\alpha}\}$ is regular for any $|\alpha| = k$. For the class of semilinear languages is closed under non-erasing morphisms, it follows that L_1 is a finite union of semilinear languages, which verifies the first claim of the proof.

Claim 2: $HL_k(L) \in \mathbf{SLin}$.

The language $HLS_k(L)$, which is defined analogously to $HCS_k(L)$, is now letter-equivalent to the language L_2. Further on,

$$L_2 = \bigcup_{|\alpha|=k} \theta((L \between U^*) \cap V^*\{a\sigma(a) \mid a \in V\}^+\{\alpha\}V^+\{\overline{\alpha}\}),$$

which concludes the proof of the second claim.

2. Let $L \subseteq V^*$ be a semilinear language in $\mathcal{F}$, and $p, k \geq 1$. As in the previous case, it suffices to prove that $pHCP_k(L)$ is semilinear. To this aim, for every x with $1 \leq |x| \leq p$, we consider the language $L(x) = L \cap (\bigcup_{|\alpha|=k} \{\alpha\}V^+\{\overline{\alpha}x\})$. Since $\mathcal{F}$ is closed under intersection with regular sets, it follows $L(x) \in \mathcal{F}$ holds. Now, $pHCP_k(L) = \bigcup_{1 \leq |x| \leq p} \{\overline{x}\}L(x)$, which implies that $pHCP_k(L)$ is semilinear.

As a direct consequence of this theorem and of Parikh Theorem [24] we have:

Corollary 1. *The hairpin completion, hairpin lengthening, and bounded hairpin completion of every context-free language is semilinear.*

In the proof of the previous theorem, we have used the new alphabet U, hence the construction does not work if one wants to stay within the DNA alphabet with four letters. However, a similar result holds under related requirements. A *generalized sequential machine* (GSM for short) is a device that has finitely many states and each transition is defined as follows: it reads one input letter and outputs 0 or more letters, depending on the current state and the read letter. Every GSM defines a GSM mapping which is a function that associates a finite set of words with each input word. The GSM mapping of a language is defined in the standard way. A family of languages $\mathcal{F}$ is closed under GSM mapping if the GSM mapping of each language in $\mathcal{F}$ lies in $\mathcal{F}$.

Theorem 3. *If $\mathcal{F}$, a family of semilinear languages, is closed under GSM mappings, then all families $HC_k(\mathcal{F})$, $HL_k(\mathcal{F})$, and $pHC_k(\mathcal{F})$ contain semilinear languages only, for any $k \geq 1$.*

Proof. We give an informal explanation for $HC_k(\mathcal{F})$; the small differences for the other two cases can be easily set by the reader. We define a GSM M that follows the next stages:

Stage 1. For each read letter a, M outputs $a\bar{a}$. Nondeterministically, M passes to the next stage. Note that M reads at least one letter in Stage 1.

Stage 2. This stage is divided into two substages. In the first one, M reads k letters, outputs them and store them as a word into its internal memory. In the second substage, M outputs each read letter. At least one letter is read during the second substage. Nondeterministically, M passes to Stage 3.

Stage 3. M reads k letters, outputs them and checks whether these letters form a word that is the complement of the stored word. If this is the case, M enters a final state otherwise, it enters an error state and halts in both cases.

It is not difficult to note that, given a language L in $\mathcal{F}$, the GSM mapping of L is letter-equivalent to $HCS_k(L)$, which concludes the proof. □

5 Iterated Hairpin Formation and Semilinearity

We consider that a similar investigation of the iterated hairpin completion, hairpin lengthening, and bounded hairpin completion could be of interest for the reader. The iterated version of the hairpin completion is defined as usual by:

$$HC_k^0(w) = \{w\},\ HC_k^{n+1}(w) = HC_k(HC_k^n(w)),\ HC_k^*(w) = \bigcup_{n\geq 0} HC_k^n(w)$$
$$HC^0(w) = \{w\},\ HC^{n+1}(w) = HC(HC^n(w)),\ HC^*(w) = \bigcup_{n\geq 0} HC^n(w)$$

$$HC_k^*(L) = \bigcup_{w\in L} HC_k^*(w) \qquad HC^*(L) = \bigcup_{w\in L} HC^*(w).$$

In a similar way one can define the iterated hairpin lengthening as well as the bounded hairpin completion. The examples presented in Theorem 1 can be modified to prove the following statement.

Theorem 4. **SLin** *is closed neither under HC_k^* nor under pHC_k^*, for any $k \geq 3$ and $p \geq 2$.*

Proof. The modifications are as follows:

$$L_1 = \{ba^{2^n}b^k\bar{b}a^m b\overline{b^k} \mid n, m \geq 1\}, \qquad L_2 = \{ba^t b^s (ab)^{2^n}\overline{b^k} \mid n, t \geq 1, s \geq k\}.$$

The proof is based on the fact that the iteration is blocked after the first step. More precisely, $HC_k^*(L_1) = L_1 \cup HC_k(L_1)$ and $pHC_k^*(L_2) = L_2 \cup pHC_k(L_2)$. As $HC_k(L_1) = \{ba^{2^n}b^k\overline{b}a^m b\overline{b^k a^{2^n} b} \mid n, m \geq 1\}$, it follows that $HC_k(HC_k(L_1)) = \emptyset$. The projection of $HC_k^*(L_1)$ on $\overline{a}$ is not semilinear.

On the other hand, as $pHC_k(L_2) = \{ba^t b^s (ab)^{2^n} \overline{b^{k+q} a^t b} \mid n \geq 1, 0 \leq q \leq s-k, t+q+1 \leq p\}$. Again, $pHC_k(pHC_k(L_2)) = \emptyset$. If the alphabet of $pHC_k^*(L_2)$ is $\{a, \overline{a}, b, \overline{b}\}$, then the set

$$\psi(pHC_k(pHC_k(L_2))) \cap \{(n_1, n_2, n_3, n_4 \mid n_i \geq 1, 1 \leq i \leq 4\}$$

is not semilinear as its projection on the first component is not semilinear. □

Clearly, the statement is valid for any $k \geq 1$, as soon as alphabets with more than four letters are used. We do not know whether **SLin** is closed under HL_k^*.

We now consider the iterated hairpin completion of singletons. Two out of the three cases have been completely solved. More precisely,

Theorem 5. *Let w be a word and $k, p \geq 1$.*

1. *The iterated k-hairpin lengthening of w is regular, hence semilinear* [21].
2. *The iterated p-bounded k-hairpin completion of w is regular, hence semilinear* [14].

The case of iterated unbounded hairpin completion is open. It is known that the iterated unbounded hairpin completion of a single word can be even non-context-free [17], but we are not aware of any result about the semilinearity of this language. It is worth noting that the iterated unbounded hairpin completion of the word used in [17], namely

$$w = a^k b a^k \overline{a^k} a^k \overline{c a^k},$$

is a semilinear language, though it is not context-free. We strongly conjecture that the iterated unbounded hairpin completion of a single word is always semilinear. It is worth noting that there have been reported necessary and sufficient conditions such that the iterated unbounded hairpin completion of a single word with a special structure, namely a crossing $(2, 2)$-word, is a regular language [29]. The fact that the iterated p-bounded k-hairpin completion of a singleton is semilinear turns out to be useful for proving a more general result.

Theorem 6. *Let $p, k \geq 1$ and $\mathcal{F}$ be a class of semilinear languages closed under intersection with regular languages. Then $pHC_k^*(\mathcal{F}) \in$* **SLin**.

Proof. We take the language $L \in \mathcal{F}$ over the alphabet V. The language L can be written as the union of two languages L_1 and L_2, where

- L_1 contains all words of L shorter than $2(k+p)+1$.
- L_2 contains all words of L of a length at least $2(k+p)+1$.

The relation $pHC_k^*(L) = pHC_k^*(L_1) \cup pHC_k^*(L_2)$ is immediate. As the language L_1 is finite and the iterated bounded hairpin completion of every singleton is semilinear, it follows that $pHC_k^*(L_1)$ is semilinear. It remains to prove that $pHC_k^*(L_2)$ is also semilinear.

We first note that $L_2 \in \mathcal{F}$ as it can be obtained from L by intersection with a regular language. Second, let u, v be arbitrary words over V of length $k+p$. We set $L_2(u,v) = L_2 \cap \{u\}V^+\{v\}$. By the closure properties of $\mathcal{F}$, each language $L_2(u,v)$ belongs to $\mathcal{F}$ as well. As $L_2 = \bigcup_{|u|=|v|=k+p} L_2(u,v)$ it follows that $pHC_k^*(L_2) = \bigcup_{|u|=|v|=k+p} pHC_k^*(L_2(u,v))$. Every language $pHC_k^*(L_2(u,v))$ can be expressed as $pHC^*(L_2(u,v)) = \sigma(pHC_k^*(uZv))$, where Z is a letter that does not belong to V and σ is a substitution $\sigma : (V \cup \{Z\})^* \longrightarrow 2^{V^*}$ defined by:

$$\sigma(a) = \begin{cases} \{a\}, \text{ if } a \in V, \\ \{z \in V^+ \mid uzv \in L_2(u,v)\}, \text{ if } a = Z. \end{cases}$$

As one can easily see, each substitution σ as above is a substitution with semilinear languages which preserves the semiliniarity [11], therefore each language $pHC_k^*(L_2(u,v))$ is semilinear. In conclusion, $pHC_k^*(L_2)$ is semilinear as well, which concludes the proof. □

As we have mentioned above, the iterated bounded hairpin completion and the iterated hairpin lengthening of every singleton language are semilinear sets. Based on the constructive proofs of these results we may have the following brief discussion.

Let $p, k \geq 1$, w be a word over some alphabet V of cardinality n, and $X \subseteq \mathbb{N}^n$ be a semilinear set. Then the problems:

(i) Does $\psi(HL_k^*(w)) = X$ hold?
(ii) Does $\psi(pHC_k^*(w)) = X$ hold?

are decidable. Both are odd consequences of the following facts. From the proof in [21], it is possible to effectively construct a finite automaton that recognizes $HL_k^*(w)$, hence a regular expression for this language. From the regular expression one can construct the semilinear set $\psi(HL_k^*(w))$ following the original Parikh's proof [24] as well as some later variants [1,13]. A more recent work dealing with the complexity of this transformation is [16]. Thus, if $HL_k^*(w)$ is recognized by a nondeterministic finite automaton with n states, $\psi(HL_k^*(w))$ can be constructed in polynomial time in n (but exponential in $card(V)$). As the equivalence of semilinear sets is decidable, see, e.g., [23], the statement (i) follows. An analogous reasoning works for the second statement as well. Obviously, the algorithm described here has a huge complexity. Is it possible to decrease this complexity by considering another approach?

6 Concluding Remarks

Starting from the observation that the class of semilinear languages is not closed under any of the non-iterated or iterated hairpin operations, we have given some sufficient conditions for subclasses of semilinear languages to preserve this property after applying the hairpin operations. It would be of interest to find other sufficient and/or necessary conditions. Another possible and attractive way to continue this investigation might be to extend the investigation from Sect. 4 to other classes of languages.

References

1. Blattner, M., Latteux, M.: Parikh-bounded languages. In: Even, S., Kariv, O. (eds.) ICALP 1981. LNCS, vol. 115, pp. 316–323. Springer, Heidelberg (1981). https://doi.org/10.1007/3-540-10843-2_26
2. Bottoni, P., Labella, A., Manca, V., Mitrana, V.: Superposition based on Watson-Crick-like complementarity. Theory Comput. Syst. **39**, 503–524 (2006)
3. Castellanos, J., Mitrana, V.: Some Remarks on Hairpin and Loop Languages, Words, Semigroups, and Translations, pp. 47–59. World Scientific, Singapore (2001)
4. Chen, H.-L., Doty, D., Soloveichik, D.: Deterministic function computation with chemical reaction networks. Nat. Comput. **13**, 517–534 (2014)
5. Cheptea, D., Martin-Vide, C., Mitrana, V.: A new operation on words suggested by DNA biochemistry: hairpin completion. In: Proceedings of Transgressive Computing, pp. 216–228 (2006)
6. Csuhaj-Varjú, E., Petre, I., Vaszil, G.: Self-assembly of strings and languages. Theor. Comput. Sci. **374**, 74–81 (2007)
7. Deaton, R., Murphy, R., Garzon, M., Franceschetti, D.R., Stevens, S.E.: Good encodings for DNA-based solutions to combinatorial problems. In: Proceedings of DNA-Based Computers II, DIMACS Series, vol. 44, pp. 247–258 (1998)
8. Enaganti, S.K., Ibarra, O.H., Kari, L., Kopecki, S.: On the overlap assembly of strings and languages. Nat. Comput. **16**, 175–185 (2017)
9. Garzon, M., Deaton, R., Neathery, P., Murphy, R.C., Franceschetti, D.R., Stevens, E.: On the encoding problem for DNA computing. In: Proceedings of Third DIMACS Workshop on DNA-Based Computing, University of Pennsylvania, pp. 230–237 (1997)
10. Garzon, M., Deaton, R., Nino, L.F., Stevens Jr., S.E., Wittner, M.: Genome encoding for DNA computing. In: Proceedings of Third Genetic Programming Conference, Madison, MI, 1998, pp. 684–690 (1998)
11. Ginsburg, S.: AFL with the semilinear property. J. Comput. Syst. Sci. **5**, 365–396 (1971)
12. Ginsburg, S., Spanier, E.H., Henry, E.: Semigroups, Presburger formulas, and languages. Pac. J. Math. **16**, 285–296 (1966)
13. Goldstine, J.: A simplified proof of Parikh's theorem. Discrete Math. **19**, 235–239 (1977)
14. Ito, M., Leupold, P., Manea, F., Mitrana, V.: Bounded hairpin completion. Inf. Comput. **209**, 471–485 (2011)

15. Kari, L., Konstantinidis, S., Sosík, P., Thierrin, G.: On hairpin-free words and languages. In: De Felice, C., Restivo, A. (eds.) DLT 2005. LNCS, vol. 3572, pp. 296–307. Springer, Heidelberg (2005). https://doi.org/10.1007/11505877_26
16. Kopczyński, E., To, A.W. : Parikh images of grammars: complexity and applications. In: Proceedings of 25th Annual IEEE Symposium on Logic in Computer Science (LICS), 2010, pp. 80–89 (2010)
17. Kopecki, S.: On the iterated hairpin completion. Theor. Comput. Sci. **412**, 3629–3638 (2011)
18. Manea, F., Martín-Vide, C., Mitrana, V.: On some algorithmic problems regarding the hairpin completion. Discrete Appl. Math. **157**, 2143–2152 (2009)
19. Manea, F., Mitrana, V., Yokomori, T.: Two complementary operations inspired by the DNA hairpin formation: completion and reduction. Theor. Comput. Sci. **410**, 417–425 (2009)
20. Manea, F., Mitrana, V., Sempere, J.M.: Some remarks on superposition based on Watson-Crick-like complementarity. In: Diekert, V., Nowotka, D. (eds.) DLT 2009. LNCS, vol. 5583, pp. 372–383. Springer, Heidelberg (2009). https://doi.org/10.1007/978-3-642-02737-6_30
21. Manea, F., Mercas, R., Mitrana, V.: Hairpin lengthening and shortening of regular languages. In: Bordihn, H., Kutrib, M., Truthe, B. (eds.) Languages Alive. LNCS, vol. 7300, pp. 145–159. Springer, Heidelberg (2012). https://doi.org/10.1007/978-3-642-31644-9_10
22. Manea, F., Martín-Vide, C., Mitrana, V.: Hairpin lengthening: language theoretic and algorithmic results. J. Log. Comput. **25**, 987–1009 (2015)
23. Oppen, D.: A $2^{2^{2^{pn}}}$ upper bound on the complexity of Presburger arithmetic. J. Comput. Syst. Sci. **16**, 323–332 (1978)
24. Parikh, R.: On context-free languages. J. ACM **13**, 570–581 (1966)
25. Păun, G., Rozenberg, G., Yokomori, T.: Hairpin languages. Intern. J. Found. Comp. Sci. **12**, 837–847 (2001)
26. Rosales, J.C., García-Sánchez, P.A.: Numerical Semigroups. Springer-Verlag, New-York (2009). https://doi.org/10.1007/978-1-4419-0160-6
27. Rozenberg, G., Salomaa, A. (eds.): Handbook of Formal Languages. Springer-Verlag, Berlin (1997). https://doi.org/10.1007/978-3-642-59136-5
28. Sakamoto, K., Gouzu, H., Komiya, K., Kiga, D., Yokoyama, S., Yokomori, T., Hagiya, M.: Molecular computation by DNA hairpin formation. Science **288**, 1223–1226 (2000)
29. Shikishima-Tsuji, K.: Regularity of iterative hairpin completions of crossing (2, 2)-words. Int. J. Found. Comput. Sci. **27**, 375–390 (2016)
30. Winfree, E., Yang, X., Seeman, N.C.: Universal computation via self-assembly of DNA: some theory and experiments. In: DNA Bsed Computers II, vol. 44 of DIMACS (1999), pp. 191–213 (1999)

The Role of Structure and Complexity on Reservoir Computing Quality

Matthew Dale[1,4(✉)], Jack Dewhirst[1,4], Simon O'Keefe[1,4], Angelika Sebald[2,4], Susan Stepney[1,4], and Martin A. Trefzer[3,4]

[1] Department of Computer Science, University of York, York, UK
matt.dale@york.ac.uk
[2] Department of Chemistry, University of York, York, UK
[3] Department of Electronic Engineering, University of York, York, UK
[4] York Cross-disciplinary Centre for Systems Analysis, York, UK

Abstract. We explore the effect of structure and connection complexity on the dynamical behaviour of Reservoir Computers (RC). At present, considerable effort is taken to design and hand-craft physical reservoir computers. Both structure and physical complexity are often pivotal to task performance, however, assessing their overall importance is challenging. Using a recently proposed framework, we evaluate and compare the dynamical freedom (referring to quality) of neural network structures, as an analogy for physical systems. The results quantify how structure affects the range of behaviours exhibited by these networks. It highlights that high quality reached by more complex structures is often also achievable in simpler structures with greater network size. Alternatively, quality is often improved in smaller networks by adding greater connection complexity. This work demonstrates the benefits of using abstract behaviour representation, rather than evaluation through benchmark tasks, to assess the quality of computing substrates, as the latter typically has biases, and often provides little insight into the complete computing quality of physical systems.

Keywords: Reservoir computing · Unconventional computing · Echo state networks · Structure · Complexity

1 Introduction

Reservoir Computing (RC) [26,29] is a computational model used to train and exploit an increasingly rich variety of dynamical systems, ranging from virtual neural networks to novel physical systems, such as quantum, electrical, chemical, optical and mechanical (see reviews [20,28]). Every reservoir system is designed to harness the underlying dynamics of the *substrate* it is implemented with, whether that be a physical device or material, a simulated network, or a set of system equations. However, finding a suitable substrate, or designing one, with sufficient dynamics to compute specific tasks is challenging.

I. McQuillan and S. Seki (Eds.): UCNC 2019, LNCS 11493, pp. 52–64, 2019.
https://doi.org/10.1007/978-3-030-19311-9_6

Methods to assess the complete range of dynamics a substrate can exhibit are still undeveloped. Therefore, matching substrates to tasks is typically done through trial and error. This makes for a long and laborious exercise to determine how best to configure, perturb and alter the low-level design of substrates to improve performance.

In [10] we present an alternative method to assess and compare computing systems based on abstract behaviours of dynamical properties. The idea is to map and explore the full dynamical range of substrates and use this to determine the effects of configuration, perturbation and design alteration on substrate "quality".

Here we use that method to investigate how topology and structural complexity in simulated recurrent networks affect dynamical range and quality. Our hypothesis is that, when implementing these networks, compromises in size and structure lead to similar qualities, e.g., larger networks with simple structure are equivalent to smaller networks with complex structures; therefore, simple structures still exhibit complex behaviours, but require larger network size to do so.

In the reservoir computing literature, simple network topologies and regular structures produce competitive performances to complex networks [24,25,30]. Simple structures made from multiple processing units are easy to implement and control in hardware. However, complex structures are more abundant in physical systems, but harder to manipulate. Knowing how structure and size affect quality will provide a useful guide to future substrate designers.

2 Reservoir Computing

Embracing the intrinsic properties of physical systems has the potential to offer improvements in performance, efficiency and/or computational power compared with conventional computing systems [5,18]. However, to do so requires a model of computation that naturally fits the substrate rather than imposing an inappropriate model that fights its implementation [27].

The reservoir computing model's simplicity means that it naturally aligns with many physical systems. However, its simplicity also has drawbacks, for example, it struggles to solve complex tasks requiring high-order abstractions [11,20].

An input-driven reservoir computer is typically represented and divided into three layers: the input, the *reservoir*, and the readout. The reservoir is the dynamical system, and perturbed and observed as a black-box. The input and readout depend on the chosen method of encoding and decoding of information to and from the reservoir, and the material instantiation of the encoded information. Depending on the implementation, this could be discrete or continuous values encoded in electrical, optical, chemical or other signals.

As with many systems, each reservoir is configured, controlled and tuned to perform a desired function. This often requires the careful tuning of parameters in order to produce working and optimal reservoirs. Most reservoirs are hand-crafted to a task, often requiring expert domain knowledge to design an optimal system. However, the separation between layers allows the reservoir to be optimised and configured independently of the input and readout layer. Many techniques have been used to optimise virtual reservoirs [1], and more recently, physical reservoirs [7–9].

To interpret a substrate as a reservoir, we define that the observed reservoir states $x(n)$ form a combination of the substrate's implicit function and its *discrete* observation:

$$x(n) = \Omega(\mathcal{E}(W_{in}u(t), u_{config}(t))) \quad (1)$$

where $\Omega(n)$ is the observation of the substrate's macroscopic behaviour and $\mathcal{E}(t)$ the continuous microscopic substrate function, when driven by the input $u(t)$. Here, W_{in} symbolises a set of input weights common to all reservoir systems; this is typically random (see [19], a guide for creating reservoir weights). The variable $u_{config}(t)$ represents the substrate's configuration, whether that be through external control, an input-output mapping, or other method of configuration. Typically, $u_{config}(t)$ is not a function of time, but only of a given problem. However, switching between different configurations in time could add additional dynamical complexity.

This formalisation of the reservoir states separates the system into contributing parts, including the observation and configuration method, which as a whole represents the overall reservoir system.

The final output signal $y(n)$ is determined by the readout function g, on the observation $x(n)$:

$$y(n) = g(x(n)) \quad (2)$$

In Eq. 1, we give a simple case where no feedback is applied. To add feedback, the input variables y and W_{fb} are added to $\mathcal{E}(.)$, where W_{fb} represents feedback weights.

Note that $\mathcal{E}$, the intrinsic substrate function, is described as being fixed, clamped, or set by u_{config}; only g is adapted. However, depending on the system, $\mathcal{E}$ may change when interacted with or observed, and therefore be non-deterministic.

3 CHARC Framework

The CHARC (CHAracterisation of Reservoir Computers) framework [10] is used to map and compare the computational expressiveness of computing substrates. In this process, the applied computational model, e.g., encoding, decoding and abstract representation of the system, is also assessed, suggesting whether the chosen model is a suitable fit to the physical implementation.

To characterise the computing *quality* of substrates, the framework searches over metrics measuring dynamical properties. The framework shows that carbon nanotube composites possess a lower quality than small recurrent networks when configured and stimulated using current techniques [10]. This supports previous findings [7–9].

In addition to the quality measure, the framework can also be used to model relationships between dynamical behaviour and task performance, which can be used to predict performance across the different substrates, without the need to assess directly [10].

The basic framework is defined in terms of various levels to define and measure quality. The output of lower levels are used by higher levels to model relationships between parameters, dynamics and task performance.

At the base level, the computational model is chosen. The reservoir computing model is applied here: input-driven dynamics recorded as system states and extracted through a trained weighted readout layer. However, other models are possible. In the context of a model, an abstract *behaviour* space is defined. This space represents the dynamical behaviour of the substrate when configured. This space is typically different from the configuration space, where small changes in parameters result in large changes in behaviour, and vice versa.

To define the n-dimensional behaviour space, n independent property measures are used. Here, we define the same three-dimensional space used in [10], using three metrics measuring basic properties required for reservoir computing: Kernel Rank (*separation* property), Generalisation Rank (*generalisation* property), and Memory Capacity (*echo state* property). An example 3-dimensional space is shown in step 1 of the CHARC workflow, Fig. 1.

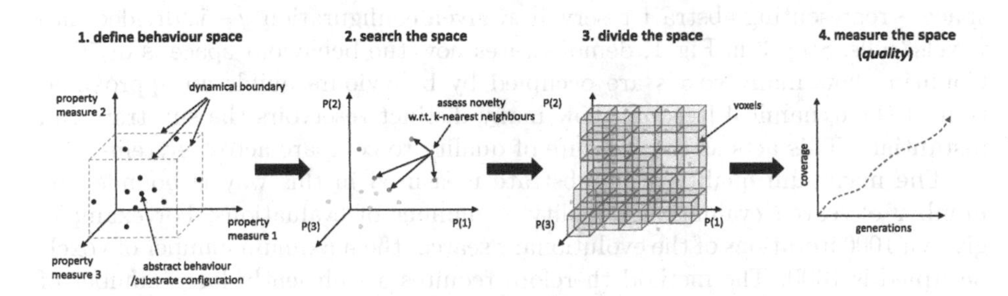

Fig. 1. CHARC framework basic workflow.

Kernel rank (KR) measures the reservoir's ability to produce a rich non-linear representation of the input u and its history $u(t-1), u(t-2), \ldots$. This measures the *linear separation property*, outlined by Legenstein & Maass [16]. The generalisation rank (GR), proposed at the same time, is a measure of the reservoir's capability to generalise given similar input streams. Reservoirs in ordered regimes typically have low ranking values in both measures, and in chaotic regimes both are high. In general, a good reservoir should possess a high kernel quality rank and a low generalisation rank [4]. However, in terms of matching reservoir dynamics to tasks, the right balance is less clear.

The measure for memory capacity (MC) captures the linear short-term memory capacity of a reservoir. This measure was first outlined in [14] to quantify the *echo state* property. For the echo state property to hold, the dynamics of the input driven reservoir must asymptotically wash out any information resulting from initial conditions, i.e., produce a fading memory.

For an outline of how to implement these measures consult [10]. As mentioned there, these three measures by themselves do not capture all the information about the reservoir's dynamical properties. To improve the accuracy of the model and the quality measure, more independent property measures are required. However, these three measures suffice to demonstrate the use of the framework to evaluate various substrates here.

The *Exploration & Mapping* level defines the search method used to construct the mapping between abstract reservoir and substrate configuration. This is shown as step 2 in Fig. 1.

Exploration of the space is done using an adapted implementation of novelty search [17]. Novelty search is an open-ended genetic algorithm that navigates the behaviour space searching for novel solutions, until some user-defined termination criterion is met. Novelty search discovers a wider range of behaviours than does random search; it explores the behaviour space until it reaches the dynamical boundaries of the system. The mapping process can therefore determine the practical use of the substrate, or whether the computational model and configuration method is appropriate.

The *Evaluation* level defines the mechanisms to evaluate quality. This constitutes the final level for measuring quality. To assess quality, the behaviour space – representing abstract reservoir x, given configuration y – is divided into voxels/cells. Step 3 in Fig. 1, demonstrates how the behaviour space is divided. Counting how many voxels are occupied by behaviours builds an approximation of the dynamical freedom: how many distinct reservoirs the substrate can instantiate. This acts as the measure of quality to compare across systems.

The maximum quality of a substrate measured in this way is bounded by number of search evaluations: quality $\leq$ number of evaluations. For example, given a 1000 iterations of the evolutionary search, the maximum number of voxels occupied is 1000. The method therefore requires a sufficiently large number of evaluations to get a good measure of quality.

4 Simulated Network Topologies

In [23,24], it was shown that simple and deterministic connection topologies tend to perform as well as, or better than, standard (fully-connected) randomly-generated reservoir networks on a number of benchmark tasks.

In the experiments described below, we use the CHARC framework to investigate the effect of network topology, by evaluating three simulated recurrent network topologies: *ring*, *lattice*, and *fully-connected* networks.

The ring topology (Fig. 2a) has the least complexity. Each node has a single self-connection and one connection to each of its neighbours to its left and right, resulting in every node having three connections.

A basic ring topology is the simplest network to implement in physical hardware as the number of connections required is small. Ring structures with various

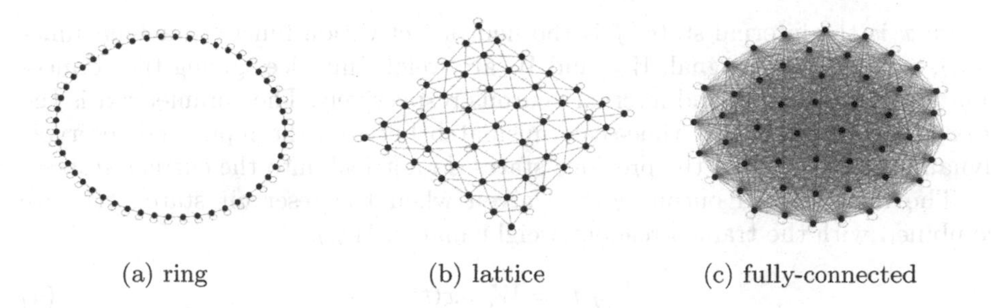

Fig. 2. Network structures investigated here.

connection types have been applied to reservoir computing systems, including minimum complexity echo state networks (ESN) [24], DNA reservoirs (deoxyribozyme oscillators) [12], Cycle reservoirs with regular jumps (CRJ) [23], and delay-based reservoirs using a single non-linear node with a delay line [3, 21].

The lattice topology (Fig. 2b) has greater connection complexity. Here, we define a square grid of neurons with each connected to its Moore neighbourhood (as commonly used in cellular automata like Conway's Game of Life [2]). So each node (except for the perimeter nodes) has eight connections to neighbours and one self-connection, resulting in each node having a maximum of nine connections.

Lattice networks/models are common in computational physics, condensed matter physics and beyond, modelling physical interactions, phase transitions and structure [15]. Examples include: discrete lattices like the Ising model with variables representing magnetic dipole moments of atomic spins, and the Gray-Scott reaction-diffusion model to simulate chemical systems [22]. Also, physical substrates often have a regular grid of connections. Lattice networks are therefore more realistic representations of many physical systems that would be considered for reservoir computing.

The fully-connected topology (Fig. 2c) has the most connections and is considered the most complex. This type of network is challenging to implement in physical hardware. It is typically used in recurrent neural network models, such as echo state networks [13]; however, its biological plausibility is debatable.

In early work on echo state networks (ESN) it was believed these networks often work best when sparsely connected [13, 19], decoupling dynamics into smaller subsystems. However, the reservoir community is still undecided on the actual benefits of sparsity on performance.

Fully-connected networks possess the most parameters (weights) and degrees-of-freedom in the configuration space, but how this translates to dynamical behaviour is undeveloped.

In the following experiments, each network's dynamics is given by the state update equation:

$$x(t) = (1 - \alpha)x(t - 1) + \alpha f(W_{in}u(t) + Wx(t - 1)) \quad (3)$$

where x is the internal state, f is the neuron activation function (a *tanh* function), u is the input signal, W_{in} and W are weight matrices giving the connection weights to inputs and internal neurons respectively. The parameter α is the *leak rate*, controlling the time-scale mismatch between the input and reservoir dynamics; when $\alpha = 1$, the previous states do not leak into the current states.

The final trained output $y(t)$ is given when the reservoir states $x(t)$ are combined with the trained readout weight matrix W_{out}:

$$y(t) = W_{out}x(t) \tag{4}$$

5 Experiment Parameter Settings

To quantify how network structure affects dynamical behaviour, that is, reservoir *quality*, we investigate the three topologies over multiple network sizes and two internal connection types.

These two connection types define whether a connection in the reservoir is *undirected* – the weights on a link are the same in both directions, and so the weight matrix W is symmetric – or *directed* – the weight w_{ij} from node x_i to x_j may be different from the weight w_{ji} from node x_j to x_i, and so the weight matrix W is not symmetric. Considering these two connection types provides additional limits on the complexity of each network. For the fully-connected topology, we investigate only the directed type, making them equivalent to echo state networks.

The network sizes assessed for each topology are: 25, 50, 100, 200 nodes. For the lattice, these are networks with the nearest square values: 25 (5×5), 49 (7×7), 100 (10×10) and 196 (14×14). By comparing different sizes, we can determine what relationships exist between network structure and quality independent of size, and therefore whether relationships scale with network size.

Each network has many local (weight) and global (scaling) parameters under manipulation in the novelty search process. Individual weights in the input layer matrix W_{in} and the reservoir matrix W are mutated between $[-1, 1]$. Global weight scaling parameters of both matrices are also evolved. For W scaling this is a value between $[0, 2]$, and for W_{in} scaling between $[-1, 1]$. The leak rate parameter α, controlling the "leakiness" of past states into current states, is restricted between $[0, 1]$. Input and internal connectivity, the weight distribution and sparseness of W_{in} and W, are evolved by mutating between zero-value weights and non-zero weights.

The output weight matrix W_{out} for each network is used only for the memory capacity measure, as both KR and GR are calculated using only the reservoir states. When the readout layer W_{out} is in use, training is carried out using ridge regression (see [19] for training details) to minimise the difference between the network output y and the target signal y_t.

The parameters used for the novelty search algorithm are those used in [10]: *population size* $= 200$, *deme* $= 40$, *recombination rate* $= 1$, *mutation rate* $= 0.2$, $\rho_{min} = 3$, and ρ_{min} *update* $= 200$ generations.

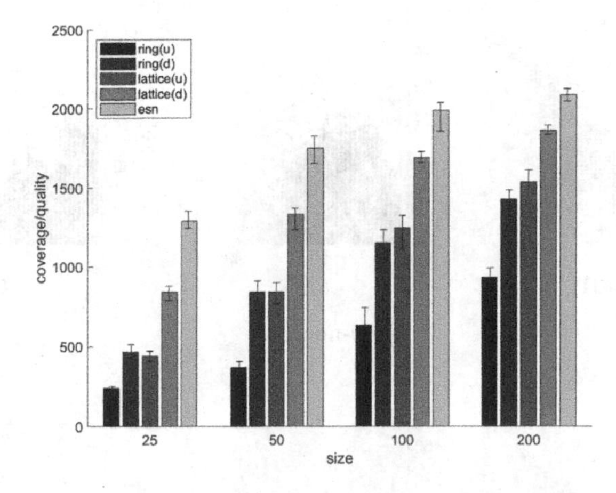

Fig. 3. Behaviour space coverage of all networks.

10 runs are conducted for each network topology, size and connection type, with 2000 generations for each run. In our implementation of novelty search (see [10]), only one new behaviour is possible every generation.

6 Results

6.1 Size and Structure

Here we compare total coverage of the behaviour space, how many voxels are occupied, by each topology. The argument in [10] is that this number represents a measure of substrate quality. The maximum size of the behaviour space in this work is bound by the largest network size to the power of total number of measures, i.e., total behaviours = 200^3. However, a maximum of 2200 behaviours are possible in a single run here, set by the number of generations and initial population size. Hence the maximum possibly quality is 2200.

The coverage results (Fig. 3) show that the fully-connected (directed) topology, the ESN network, occupies a greater area of the behaviour space, possessing a higher quality than the others, independent of size. This suggests that access to more adjustable parameters typically leads to more dynamical behaviours.

The larger ESN networks reach close to the maximum possible coverage in every run, suggesting more generations are required to better outline their limits. The others, however, struggle to reach the maximum coverage. This suggests that either new behaviours are more difficult to find, requiring additional search time, or the behavioural limits of the networks are nearly reached. A common pattern across all types is each network tends to improve in quality when increased in size. How many new behaviours are discovered by increasing in size, depends on topology and connection type.

Figure 3 shows that the undirected lattice topology is statistically similar to the directed ring topology, across all network sizes. As implementing the

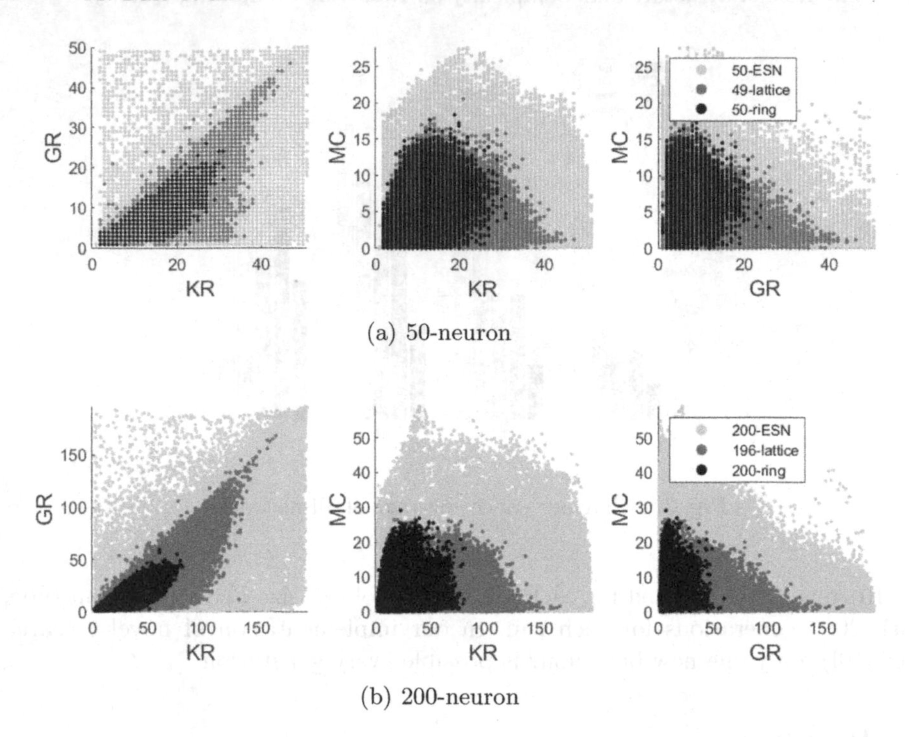

(a) 50-neuron

(b) 200-neuron

Fig. 4. Superimposed behaviour space of: (a) 50-neuron network, (b) 200 neuron network. Directed topologies only. Showing all 10 runs of 2000 generations each.

lattice connectivity is more demanding, requiring greater complexity (up to 9 connections per node versus 3 per node), this result suggests a way to get similar dynamical behaviour to an undirected lattice with less connection complexity.

A visualisation of how two network sizes (50 and 200-neuron) cover the behaviour space using each network topology is given in Fig. 4. Here, only the directed topologies are shown; for undirected results, see next section. The plot shows that ESNs tend to occupy regions the others cannot, such as chaotic regions, e.g., with high KR and high GR, and regions with larger memory capacities.

The ring and the lattice topologies have similar maximum memory capacities as each other; however, lattices typically exhibit greater non-linearity and chaotic behaviour (higher KR and GR values) than rings.

6.2 Directed vs. Undirected Networks

Here we compare the difference in quality between directed and undirected connection types. In the previous section, we show that network topology significantly affects quality. Here we show that connection type is equally as important.

To visualise this, the behaviour space coverage of a 100-neuron ring and lattice is shown in Fig. 5. Each plot shows the directed connection type (grey) and

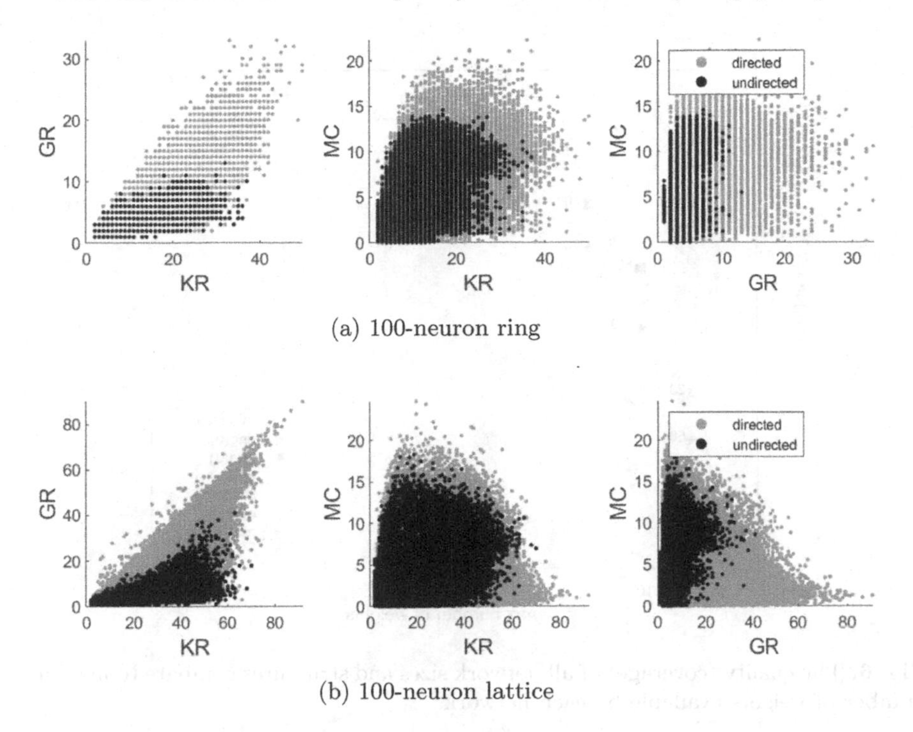

Fig. 5. Directed vs. Undirected: Superimposed behaviour space of: 100-neuron ring and lattice network. Showing all 10 runs of 2000 generations each.

undirected type (black). Here, directed connections typically result in broader dynamical behaviour, producing more "challenging" behaviours (high KR and high MC). The difficulty in producing such behaviours exists because non-linearity (KR) and ordered dynamics (MC) are often conflicting. The additional freedom of a non-symmetric weight matrix allows a broader set of behaviours to be realised.

6.3 Parameters vs. Quality

The results show the quality of each topology and connection type tends to scale linearly with the maximum number of weights available in each network configuration (Fig. 6). Here, the maximum number of weights is used to represent the total freedom in the network's configuration space. In reality, the actual number of connections – with non-zero weights – required to occupy different regions of the behaviour space may differ significantly.

In Fig. 6, several groups exist where networks with the same maximum number of weights produce different qualities. For example, the 25-neuron lattice (undirected), 100-neuron ring (undirected) and 50-neuron ring (directed) are each limited to roughly 100 weights, yet differ significantly in quality/coverage. The 50-neuron ring (directed), with simpler structure, produces many more

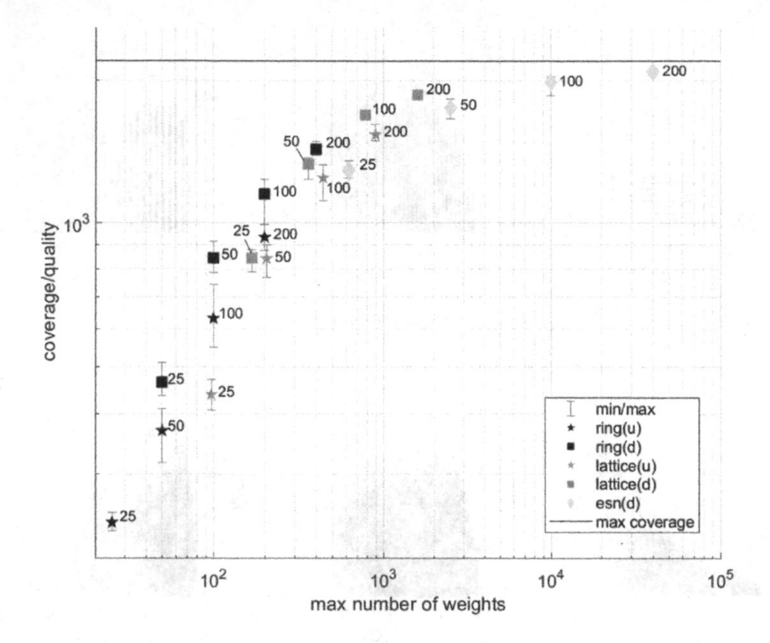

Fig. 6. The quality (coverage) of all network sizes and structures compare to maximum number of weights available by each network.

behaviours than the more complex 25-neuron lattice (undirected). This pattern also continues as both increase in network size.

It is also seen that adding more parameters (weights) does not always lead to more dynamical behaviours, e.g., the 50-neuron ring (directed) and 50-neuron lattice (undirected) produce similar qualities despite the lattice having many more available connections. It then becomes clear that how weights are structured and *directed*, controlling information flow, has a greater affect on quality of the network. This supports similar results using hierarchical networks, where structure and number of parameters also significantly impact performance [6].

7 Conclusion

Assessing and comparing how structure affects dynamical behaviour is often challenging. For unconventional substrates, determining what predefined structure and channels of information exist is even more challenging. Even if structure can be decided at creation, what is a suitable or ideal structure is often limited by physical constraints.

Here we show that the CHARC framework provides a method to assess what effect changes in structure have on computing quality. We use simulated recurrent networks as a model of physical reservoir systems. Our experiments show that networks with low structural complexity can exhibit similar quality and behaviours to more complex structures. However, in almost every case, this is only possible when network size is increased, resulting in a trade-off between size and structure. This result therefore acts as useful guide to the design of new unconventional reservoir computing substrates, where more complex structures tend to be more challenging to implement than simple structures like the ring topology.

Overall, this work showcases one undeveloped area within a much wider challenge: how to better understand the computing properties of substrates. The work demonstrates how the CHARC framework can be used to evaluate changes in design, configuration and representation of unconventional substrates and relate it to a task-independent measure of computing quality. In future work, we intend to further develop the framework to improve the design and fit of computational models, and even substrate design directly. For example, moving beyond the reservoir computing model, leading to a more generic CHARC framework; using the framework as a test-bed to assess unconventional program constructs; and, using quality as an objective to optimise within the substrate design process.

Acknowledgments. This work is part of the SpInspired project, funded by EPSRC grant EP/R032823/1. Jack Dewhirst is funded by an EPSRC DTP PhD studentship.

References

1. Bala, A., Ismail, I., Ibrahim, R., Sait, S.M.: Applications of metaheuristics in reservoir computing techniques: a review. IEEE Access **6**, 58012–58029 (2018)
2. Adamatzky, A.: Game of Life Cellular Automata, vol. 1. Springer, London (2010). https://doi.org/10.1007/978-1-84996-217-9
3. Appeltant, L., et al.: Information processing using a single dynamical node as complex system. Nature Commun. **2**, 468 (2011)
4. Büsing, L., Schrauwen, B., Legenstein, R.: Connectivity, dynamics, and memory in reservoir computing with binary and analog neurons. Neural Comput. **22**(5), 1272–1311 (2010)
5. Crutchfield, J.P.: The calculi of emergence. Physica D **75**(1–3), 11–54 (1994)
6. Dale, M.: Neuroevolution of hierarchical reservoir computers. In: Proceedings of the Genetic and Evolutionary Computation Conference, pp. 410–417. ACM (2018)
7. Dale, M., Miller, J.F., Stepney, S., Trefzer, M.A.: Evolving carbon nanotube reservoir computers. In: Amos, M., Condon, A. (eds.) UCNC 2016. LNCS, vol. 9726, pp. 49–61. Springer, Cham (2016). https://doi.org/10.1007/978-3-319-41312-9_5
8. Dale, M., Miller, J.F., Stepney, S., Trefzer, M.A.: Reservoir computing in materio: an evaluation of configuration through evolution. In: 2016 IEEE Symposium Series on Computational Intelligence (SSCI), pp. 1–8, December 2016
9. Dale, M., Miller, J.F., Stepney, S., Trefzer, M.A.: Reservoir computing in materio: a computational framework for in materio computing. In: 2017 International Joint Conference on Neural Networks (IJCNN), pp. 2178–2185, May 2017
10. Dale, M., Miller, J.F., Stepney, S., Trefzer, M.A.: A substrate-independent framework to characterise reservoir computers. arXiv preprint arXiv:1810.07135 (2018)
11. Gallicchio, C., Micheli, A., Pedrelli, L.: Deep reservoir computing: a critical experimental analysis. Neurocomputing **268**, 87–99 (2017)

12. Goudarzi, A., Lakin, M.R., Stefanovic, D.: DNA reservoir computing: a novel molecular computing approach. In: Soloveichik, D., Yurke, B. (eds.) DNA 2013. LNCS, vol. 8141, pp. 76–89. Springer, Cham (2013). https://doi.org/10.1007/978-3-319-01928-4_6
13. Jaeger, H.: The "echo state" approach to analysing and training recurrent neural networks-with an erratum note. German National Research Center for Information Technology GMD Technical Report 148:34, Bonn, Germany (2001)
14. Jaeger, H.: Short term memory in echo state networks. GMD-Forschungszentrum Informationstechnik (2001)
15. Lavis, D.A.: Equilibrium statistical mechanics of lattice models. Springer, Dordrecht (2015). https://doi.org/10.1007/978-94-017-9430-5
16. Legenstein, R., Maass, W.: Edge of chaos and prediction of computational performance for neural circuit models. Neural Networks **20**(3), 323–334 (2007)
17. Lehman, J., Stanley, K.O.: Exploiting open-endedness to solve problems through the search for novelty. In: ALIFE, pp. 329–336 (2008)
18. Lloyd, S.: Ultimate physical limits to computation. Nature **406**(6799), 1047 (2000)
19. Lukoševičius, M.: A practical guide to applying echo state networks. In: Montavon, G., Orr, G.B., Müller, K.-R. (eds.) Neural Networks: Tricks of the Trade. LNCS, vol. 7700, pp. 659–686. Springer, Heidelberg (2012). https://doi.org/10.1007/978-3-642-35289-8_36
20. Lukoševičius, M., Jaeger, H.: Reservoir computing approaches to recurrent neural network training. Comput. Sci. Rev. **3**(3), 127–149 (2009)
21. Paquot, Y., et al.: Optoelectronic reservoir computing. Scientific Reports, 2 (2012)
22. Pearson, J.E.: Complex patterns in a simple system. Science **261**(5118), 189–192 (1993)
23. Rodan, A., Tiňo, P.: Simple deterministically constructed recurrent neural networks. In: Fyfe, C., Tino, P., Charles, D., Garcia-Osorio, C., Yin, H. (eds.) IDEAL 2010. LNCS, vol. 6283, pp. 267–274. Springer, Heidelberg (2010). https://doi.org/10.1007/978-3-642-15381-5_33
24. Rodan, A., Tino, P.: Minimum complexity echo state network. IEEE Trans. Neural Networks **22**(1), 131–144 (2011)
25. Rodan, A., Tiňo, P.: Simple deterministically constructed cycle reservoirs with regular jumps. Neural Comput. **24**(7), 1822–1852 (2012)
26. Schrauwen, B., Verstraeten, D., Van Campenhout, J.: An overview of reservoir computing: theory, applications and implementations. In: Proceedings of the 15th European Symposium on Artificial Neural Networks. Citeseer (2007)
27. Stepney, S.: The neglected pillar of material computation. Physica D **237**(9), 1157–1164 (2008)
28. Tanaka, G., et al.: Recent advances in physical reservoir computing: a review. arXiv preprint arXiv:1808.04962 (2018)
29. Verstraeten, D., Schrauwen, B., D'Haene, M., Stroobandt, D.: An experimental unification of reservoir computing methods. Neural Networks **20**(3), 391–403 (2007)
30. Xue, Y., Yang, L., Haykin, S.: Decoupled echo state networks with lateral inhibition. Neural Networks **20**(3), 365–376 (2007)

Lindenmayer Systems and Global Transformations

Alexandre Fernandez, Luidnel Maignan, and Antoine Spicher(✉)

Université Paris-Est Créteil, LACL, 61 avenue du Général de Gaulle, 94015 Créteil Cedex, France
{alexandre.fernandez,luidnel.maignan,antoine.spicher}@u-pec.fr

Abstract. *Global transformations*, a category-based formalism for capturing computing models which are simultaneously local, synchronous and deterministic, are introduced through the perspective of *deterministic Lindenmayer systems*, a computing model based on parallel string rewriting. No knowledge of category theory is assumed.

1 Introduction

In [6], the model of *Global Transformation* is introduced with the purpose of capturing the essence of rewriting systems which are simultaneously *local*, *synchronous* and *deterministic*. These properties are meant to express the fact that the application of any given rewriting rule at any given place is decided independently from other applications. All decided applications are thus done in parallel without any restriction. The *global coherence* of these applications is explicitly expressed within the specification of a global transformation at a local level. This emphasis towards *global* application of all the rules is what gave the name to global transformations. This important property draws the line between global transformations and other well-known systems where the application of a rule might prevent another rule to be applied. For instance, in Păun systems [8] the maximally parallel strategy of rule applications is strongly related to the non-determinism which obeys to another paradigm where the transformed elements are thought as a resource and thus can only be used once.

The formalism of global transformations is generic enough to allow any kind of space and any notion of locality, even dynamic structures as often considered in morphogenesis for example or in physics related dynamical systems. To support this claim of genericity, it is useful to see how such systems can be retrieved in the global transformation framework. In this paper, we revisit the computing model of *Deterministic Lindenmayer Systems*. Lindenmayer systems (L systems) provide a mathematical framework to cope with parallel rewriting of strings. Initially, they have been successfully developed for biological modeling purpose where synchronicity captures the fact that the elementary parts of an organism evolve simultaneously [7]. Quickly, L systems have also been studied from the perspective of formal language theory where a parallel rewriting process appears as an interesting alternative to the usual sequential strategy of formal grammar

I. McQuillan and S. Seki (Eds.): UCNC 2019, LNCS 11493, pp. 65–78, 2019.
https://doi.org/10.1007/978-3-030-19311-9_7

derivation [9]. All these studies led to an important taxonomy of L systems. Among them, we focus on *Deterministic* L systems (DL systems) where for any given word, there is only one possible derivation. By expressing DL systems in terms of global transformations, we hope to give a strong insight on each ingredient of global transformations formalism.

In Sect. 2, we gently introduce this formalism in almost full generality. We then proceed to exemplify this formalism in the case of DL systems. To do so, we begin by a formal reminder of these systems in Sect. 3 and then we show how to rearrange their ingredients to present them as global transformations in Sects. 4 and 5. We conclude with a short discussion in Sect. 6.

2 Global Transformations

A global transformation is a very general kind of rewriting system. The basic ingredients of such a system are a collection C of "things" to be transformed[1] and a collection Γ of *rules*, each rule $\gamma \in \Gamma$ coming with a *left hand side* (lhs) $l(\gamma) \in C$ and a *right hand side* (rhs) $r(\gamma) \in C$. Morally, a rule γ can be thought as being of the form $l(\gamma) \Rightarrow r(\gamma)$. The rest is a matter of enriching those sets and functions with the necessary structure to capture locality, synchronicity and determinism. It is done using categorical concepts as explained below, the reader not being assumed to have any prior knowledge of category theory.

2.1 A Structure for Locality

To describe locality, we need to formalize the notion of a thing being a part of another thing. Of course, when a thing a is a part of a thing b which is itself a part of a thing c, then a is a part of c. But this preorder structure is not enough. We also need to track *where* things occur in each other. So when a occurs at a place p in b and b occurs at a place q in c, we need a way to know where a consequently occurs in c. This is exactly formalized by the notion of category. In this formalism, we would talk of three *objects* a, b and c, and two *arrows*[2] $p : a \to b$ and $q : b \to c$ being composed to give $q \circ p : a \to c$.

Definition 1. *A* category $\mathbf{C}$ *consists of:*

- *a set abusively denoted* $\mathbf{C}$ *whose elements are called* objects*;*
- *for any* $a, b \in \mathbf{C}$*, a set*[3] $\mathrm{Arr}_{\mathbf{C}}(a, b)$ *whose elements are called* arrows*, the notations* $f : a \to b$ *and* $a \xrightarrow{f} b$ *meaning* $f \in \mathrm{Arr}_{\mathbf{C}}(a, b)$*;*
- *for every three objects* a, b, c*, a* composition law $\mathrm{Arr}_{\mathbf{C}}(b, c) \times \mathrm{Arr}_{\mathbf{C}}(a, b) \to \mathrm{Arr}_{\mathbf{C}}(a, c)$ *denoted* $\circ$*, i.e., sending any* $f : a \to b$ *and* $g : b \to c$ *to* $g \circ f : a \to c$.

Moreover, the composition law has to be associative *(for any* $f : a \to b$*,* $g : b \to c$ *and* $h : c \to d$*,* $h \circ (g \circ f) = (h \circ g) \circ f$*) with an* identity arrow $\mathrm{id}_x : x \to x$ *for each* $x \in \mathbf{C}$ *(for any* $f : a \to x$ *and* $g : x \to b$*,* $\mathrm{id}_x \circ f = f$ *and* $g \circ \mathrm{id}_x = g$*).*

[1] We consider only one set of "things" because we restrict here to dynamical systems.
[2] In the literature, they are also often called *morphisms* or *homomorphisms*.
[3] These sets are more often denoted $\mathrm{Hom}_{\mathbf{C}}(a, b)$ in the literature.

2.2 Respecting Locality

Given a category $\mathbf{C}$ describing a locality structure, what does it mean for a transformation F on $\mathbf{C}$ to respect locality? Consider a global input b with a local part a appearing at place $p : a \to b$. It is expected from such a transformation to treat a locally, thus producing the output $F(a)$ for this part independently, so that the global output $F(b)$ should contain this local output. Of course, *the way the local output occurs in the global output depends on the way the local part occurs in the global input.* Formally, this means that there should be an associated arrow $F(p) : F(a) \to F(b)$ for each such p in a coherent way. This is exactly captured by homomorphisms of categories, so-called *functors.*

Definition 2. *A* functor $F : \mathbf{C} \to \mathbf{D}$, *with* $\mathbf{C}$ *and* $\mathbf{D}$ *two categories, consists of:*

- *a map abusively denoted* $F : \mathbf{C} \to \mathbf{D}$ *from the set* $\mathbf{C}$ *to the set* $\mathbf{D}$, *and*
- *a map abusively denoted* $F : \mathrm{Arr}_{\mathbf{C}}(a, b) \to \mathrm{Arr}_{\mathbf{D}}(F(a), F(b))$ *for any* $a, b \in \mathbf{C}$

Moreover, these maps have to preserve identity arrows ($\forall a \in \mathbf{C}, F(\mathrm{id}_a) = \mathrm{id}_{F(a)}$) *and compositions* ($\forall a, b, c \in \mathbf{C}, \forall f : a \to b, \forall g : b \to c, F(g \circ f) = F(g) \circ F(f)$).

We certainly want a global transformation to be a functor from $\mathbf{C}$ to $\mathbf{C}$, but since this global transformation is specified through local rules, we must specify things within these rules. Since the signatures of the lhs and rhs functions are $l : \Gamma \to \mathbf{C}$ and $r : \Gamma \to \mathbf{C}$, arrows between lhs must be associated backward[4] through l^{-1} and then forward through r. Thus, for any two rules γ and γ', any arrow $p : l(\gamma) \to l(\gamma')$ in $\mathbf{C}$ is mapped to $r(l^{-1}(p)) : r(\gamma) \to r(\gamma')$ in $\mathbf{C}$. This naturally promotes Γ to a category whose arrows, called *rule inclusions*, are completely determined in terms of those of $\mathbf{C}$ by $l^{-1}(p) : \gamma \to \gamma'$. In other words, the injective function l should be an injective functor *inducing a bijection* between $\mathrm{Arr}_{\Gamma}(\gamma, \gamma')$ and $\mathrm{Arr}_{\mathbf{C}}(l(\gamma), l(\gamma'))$, that is, a *fully faithful* injective functor. The function r is lifted to a functor with no required extra property. This fully specifies the data describing a global transformation.

Definition 3. *A* global transformation T *consists of:*

- *a category* $\mathbf{C}_T$,
- *a category* $\mathbf{\Gamma}_T$ *whose objects and arrows are called* rules *and* rule inclusions,
- *a fully faithful injective functor* $\mathrm{L}_T : \mathbf{\Gamma}_T \to \mathbf{C}_T$ *called the* lhs functor,
- *a functor* $\mathrm{R}_T : \mathbf{\Gamma}_T \to \mathbf{C}_T$ *called the* rhs functor.

A last remark is that this definition is equivalent to requiring $\mathbf{\Gamma}_T$ to be a (full) subcategory of $\mathbf{C}_T$. In this view, R_T specifies the behavior of the transformation on $\mathbf{\Gamma}_T \subseteq \mathbf{C}_T$ only, thus being equivalent to a partial functor from $\mathbf{C}_T$ to $\mathbf{C}_T$. The functor L_T is then the inclusion functor of $\mathbf{\Gamma}_T$ in $\mathbf{C}_T$. This complies with the idea that the behavior of a global transformation on all possible inputs is generated by its behavior on some selected "small" inputs.

[4] Function l has to be injective to respect determinism.

2.3 Designing Locality

Designing a global transformation T consists in choosing the appropriate category $\mathbf{C}_T$ and taking it into account in the choice of the rules. The objects of $\mathbf{C}_T$ are usually clear: at least all possible inputs, outputs, lhs and rhs, often accompanied with their natural sub-parts. The real matter is to design the arrows of $\mathbf{C}_T$. *An arrow specifies that the result of its source has to be found in the result of its target.* This main guideline is illustrated in Sects. 4.1 and 5.

While inclusions between rules are rarely considered in traditional rewriting systems, they are of first importance in the design of rules in a global transformation: they indicate how locality explains the construction of the result. The leitmotiv is indeed that *any local relation between rhs in the result should be justified by a local relation between the lhs in the input.* This relation between two lhs $\mathrm{L}_T(\gamma_1)$ and $\mathrm{L}_T(\gamma_2)$ is not necessarily a direct inclusion and a third lhs $\mathrm{L}_T(\gamma_{12})$ might be needed. If the relation is an overlap, this relation has the form $\gamma_1 \xleftarrow{p} \gamma_{12} \xrightarrow{q} \gamma_2$. If it is an interaction in a common context, the form can be $\gamma_1 \xrightarrow{p} \gamma_{12} \xleftarrow{q} \gamma_2$. The former has been used for illustration in Sects. 4.2 and 5.

Once a global transformation specified by means of its local rules, it is possible to apply it on any given input. This is done by (1) collecting all the occurrences of lhs in the input, then by (2) passing to the corresponding rhs, and (3) by constructing an object containing the required rhs. In this whole process, the arrows need to be conserved, since they specify how the lhs where found in the input, and thus how the rhs must be recomposed. We give here a general description of the operations involved in the specification of the result. The important role of the arrows in this process is only slightly highlighted. We immediately after proceed with DL systems, concrete examples working better to understand their purpose.

2.4 Pattern Matching

The process of finding all occurrences of the lhs in an input object X is often called *pattern matching.* In this process, one considers every place p in X where the lhs of a rule γ occurs, *i.e.*, every pair $\langle \gamma \in \mathbf{\Gamma}_T, p : \mathrm{L}_T(\gamma) \to X\rangle$. We call such a pair a *rule instance* of γ. Note that for any rule inclusion $q : \gamma \to \gamma'$ in $\mathbf{\Gamma}_T$, an instance of γ' in X at some place $p' : \mathrm{L}_T(\gamma') \to X$ necessarily implies an instance of γ at the place $p = p' \circ \mathrm{L}_T(q)$. These data form a category denoted L_T/X called a *comma category*[5].

Definition 4. *The category F/d, with $F : \mathbf{C} \to \mathbf{D}$ a functor and $d \in \mathbf{D}$, has:*

- *for objects, every pair $\langle c \in \mathbf{C}, f : F(c) \to d\rangle$,*
- *for arrows from $\langle c_1, f_1\rangle$ to $\langle c_2, f_2\rangle$, every $g : c_1 \to c_2$ such that $f_1 = f_2 \circ F(g)$.*

This category comes naturally with a projection functor $\mathrm{Proj}_{F/d} : F/d \to \mathbf{C}$ *mapping each object $\langle c \in \mathbf{C}, f : F(c) \to d\rangle$ to c and acting trivially on arrows.*

[5] A comma category is normally defined between two functors to the same category, the case of a single object being represented by a constant functor.

For L_T/X, the functor $\mathrm{Proj}_{L_T/X} : L_T/X \to \mathbf{\Gamma}_T$ maps each rule instance found in the input X to the actual rule. Since R_T gives for each rule its rhs, we compose these functors to obtain $R_T \circ \mathrm{Proj}_{L_T/X} : L_T/a \to \mathbf{C}_T$ that gives the pieces to be patched together, and the way they have to be patched to construct the result.

2.5 Constructing the Result

Taking the simplistic view of a preorder of Sect. 2.1, the result might be described as the smallest way to have all the rule instance rhs patched together; we would talk of least upper bound. In the categorical setting, upper bounds are generalized as *cocones* and least upper bounds as *colimits.*

Upper bounds are usually defined for a *subset of elements.* For categories, arrows and multiplicities must also be taken into account. For this reason, cocones are defined for a *multiset of objects and arrows.* Such a collection is specified as an arbitrary functor into the category of interest, so-called *diagram,* the source category playing the role of an index set. In our case, a cocone can be interpreted as a candidate: it is an object with a set of incoming arrows showing where the rhs of the rule instances appear in the object. Moreover anytime two instance rhs are related by rule inclusion, it is required to also be the case in the candidate.

Definition 5. *A* cocone *u for a diagram $F : \mathbf{I} \to \mathbf{C}$ consists of:*

- *an object of $\mathbf{C}$ abusively denoted u, and*
- *an arrow $\eta_{u,i} : F(i) \to u$ for each object $i \in \mathbf{I}$*

such that $\eta_{u,j} \circ F(f) = \eta_{u,i}$ for any $f : i \to j$ in $\mathbf{I}$.

The *colimit* is the best candidate, that is, the common factor appearing in all candidates. As such, it should appear in any cocone u at the unique exact place devised by the arrows $\{\eta_{u,i}\}_i$ indicating where the parts must be found.

Definition 6. *A* colimit *of a diagram $F : \mathbf{I} \to \mathbf{C}$ is a cocone s such that for any cocone u there is a unique $g_{s,u} : s \to u$ such that $\eta_{u,i} = g_{s,u} \circ \eta_{s,i}$ for any $i \in \mathbf{I}$.*

Note that there might be more than one colimit in some cases. However, any two colimits for the same diagram F are necessarily isomorphic. For this reason, one often talks about "the" colimit $\mathrm{Colim}(F)$ as if it is unique, even when it is not the case. This is either just a loose notation or the choice of one precise colimit in the collection of isomorphic colimits.

Altogether, the result of applying a global transformation T on X is the best object containing all the rhs of the rule instances found in X, arranged as required by rule inclusions.

Definition 7. *Given a global transformation T, the result $T(X)$ of the application of T on an object $X \in \mathbf{C}_T$ is an object $T(X) \cong \mathrm{Colim}(R_T \circ \mathrm{Proj}_{L_T/X})$.*

Notice that the full definition defines the global transformation behavior as a functor. Particularly, this definition is related to a categorical construction called *Kan extension.* We indicate this only as a reference, it is out of the scope of this paper to elaborate on this.

3 Deterministic Lindenmayer Systems

Traditionally, a L system is presented as a triplet giving (1) an alphabet used to build words, (2) a set of rewriting rules called *productions* associating to some subwords a local evolution, and (3) an axiom serving as an initial word[6]. Derivations are then obtained by applying all the productions synchronously wherever they appear. In this context, being deterministic means that all the productions always agree so that there is only one possible result. Well-known classes of such DL systems are *D0L systems* and *DIL systems*, whose definitions given below are strongly inspired by [9].

Notations. For an alphabet Σ and $n \in \mathbb{N}$, Σ^*, Σ^n, $\Sigma^{<n}$ and $\Sigma^{\leqslant n}$ express respectively the set of finite words, of words of length n, lower than n and at most n. Symbol ε denotes the empty word and $|w|$ the length of some word w. Concatenation of a sequence of words $w_1, \ldots, w_n$ is denoted by juxtaposition $w_1 w_2 \ldots w_n$ or by the product $\prod_{i=1}^{n} w_i$. When some set of words are involved in a concatenation, the result is a set in the common way. For any two words u and v, and a non negative integer p, $u \preceq_p v$ denotes that u is included in v at position p, *i.e.*, there exists $\alpha \in \Sigma^p$ and $\beta \in \Sigma^*$ such that $v = \alpha u \beta$.

3.1 D0L Systems

It is the simplest and firstly defined class of L systems. The 0 of D0L reminds that, contrary to DIL systems, the rewriting is done context-free: each letter is replaced independently of the others. This leads to the following definition.

Definition 8. *A* D0L system $\mathcal{L}$ *consists of:*

- *a set* $\Sigma_\mathcal{L}$ *called the* alphabet, *and*
- *a mapping* $\mathrm{P}_\mathcal{L} : \Sigma_\mathcal{L} \to \Sigma_\mathcal{L}^*$.

The fact $\mathrm{P}_\mathcal{L}(a) = w$ *is written* $a \to_\mathcal{L} w$ *and referred to as a* production *of* $\mathcal{L}$. *The* yield *function* $\Rightarrow_\mathcal{L}$ *is given by*

$$w \Rightarrow_\mathcal{L} w' \quad :\Leftrightarrow \quad w = a_1 \ldots a_n \ \wedge \ w' = \mathrm{P}_\mathcal{L}(a_1) \ldots \mathrm{P}_\mathcal{L}(a_n).$$

The yield function $\Rightarrow_\mathcal{L}$ is the monoid endomorphism of $\Sigma_\mathcal{L}^*$ generated by $\to_\mathcal{L}$. Conversely, any endomorphism of the free monoid $\Sigma_\mathcal{L}^*$ defines a D0L system [9].

Example 1. Let us consider the D0L system π defined on the alphabet $\Sigma_\pi = \{a, b\}$ by the productions: $a \to_\pi b$ and $b \to_\pi ab$. For example, the evolution starting from the string *abbab* is given by the iteration of the associated yield function:

$$abbab \Rightarrow_\pi bababbab \Rightarrow_\pi abbabbababbab \Rightarrow_\pi bababbababbabbababbab \Rightarrow_\pi \ldots$$

[6] In the following, we omit the axiom thus allowing any initial word.

This D0L system can be understood as a model of growth of a filamentous organism, where a letter a is interpreted as a young cell, b as an old cell, the rule $a \to_\pi b$ as a young cell getting older, and $b \to_\pi ab$ as an asymmetric cell division. As a formal language, this particular derivation exhibits an interesting recursive property; the word w_i at generation i is the concatenation of the two previous ones: $w_i = w_{i-2}w_{i-1}$.

3.2 DIL Systems

While D0L systems correspond to context-free rewriting, DIL systems provide a context-dependent rewriting where the fate of a letter is influenced by its neighbors. DIL systems are (the deterministic) part of a larger class of L systems called, *L systems with interactions* (IL systems), which were introduced to model biological organisms where cells can interact with each other.

More specifically, a DIL system comes with a signature which consists of two non negative integers, say p and q, so that, the evolution of a letter depends on its p left neighbors and its q right neighbors.

Definition 9. *A* DIL system $\mathcal{L}$ *of* signature $(p, q) \in \mathbb{N}$ *(or shortly a* D(p,q)L system*) consists of:*

- *a set* $\Sigma_\mathcal{L}$ *called the* alphabet, *and*
- *a mapping* $\mathrm{P}_\mathcal{L} : \Sigma_\mathcal{L}^{\leqslant p} \times \Sigma_\mathcal{L} \times \Sigma_\mathcal{L}^{\leqslant q} \to \Sigma_\mathcal{L}^*$.

The fact $\mathrm{P}_\mathcal{L}(\alpha, a, \beta) = w$ *is written* $(\alpha, a, \beta) \to_\mathcal{L} w$ *and referred to as a* production *of* $\mathcal{L}$*. The* yield *function* $\Rightarrow_\mathcal{L}$ *is given by*

$$w \Rightarrow_\mathcal{L} w' \quad :\Leftrightarrow \begin{cases} w = a_1 \dots a_n \\ w' = w_1 \dots w_n \\ (a_{i-p} \dots a_{i-1}, a_i, a_{i+1} \dots a_{i+q}) \to_\mathcal{L} w_i \quad \forall i \in [1, n] \end{cases}$$

where $a_j = \varepsilon$ *for any invalid position* $j \notin [1, n]$.

This definition is very intuitive and extends naturally Definition 8 of D0L systems, identified to D(0,0)L systems. Notice that, in a D(p, q)L system, thanks to the context, a letter is able to evaluate its distance to the left (resp. right) border within the range of its context, that is, by a maximum of p (resp. q).

Example 2. Let us update the D0L system π of Example 1 to forbid any b to divide when it only has a's around, modeling for example an over-consuming of resources by young cells preventing older cells division. The resulting system σ turns out to be of type D(1,1)L for checking neighbor states with productions:

$$(a, b, a) \to_\sigma b \quad (\varepsilon, b, a) \to_\sigma b \quad (a, b, \varepsilon) \to_\sigma b \quad (\_, a, \_) \to_\sigma b \quad (\_, b, \_) \to_\sigma ab$$

where the underscores stand for "any other context". Starting from the same initial string *abbab*, the system σ gives rise to the derivation:

$$abbab \Rightarrow_\sigma bababbb \Rightarrow_\sigma bbbbababab \Rightarrow_\sigma ababababbbbbbb \Rightarrow_\sigma \dots$$

Adding a b at the end of the starting string gives the derivation:

$$abbabb \Rightarrow_\sigma bababbabab \Rightarrow_\sigma bbbbababbbbb \Rightarrow_\sigma ababababbbbabababababab \Rightarrow_\sigma \dots$$

4 D0L Systems as Global Transformations

In this section, our goal is to build a global transformation which reproduces the exact same behavior of a given D0L system $\mathcal{L}$. First, we define in all generality the category $\mathbf{W}_S$ enriching any set of words S to capture the notion of locality for words. Then, we define the global transformation $\overline{\mathcal{L}}$ associated to $\mathcal{L}$. Finally, the soundness of the translation of $\mathcal{L}$ into $\overline{\mathcal{L}}$ is shown.

4.1 The Category of Words $\mathbf{W}_S$

As previously mentioned, the domain category of a global transformation formalizes the notion of locality for the objects to be transformed and their associated results. In the case of D0L systems, these objects are finite words and the arrows (meaning "being a part of") express how a word appears as a subword of another, usually called factors. This is coherent with the guideline indicated in Sect. 2.3. Indeed, for any D0L system on any alphabet Σ, the result of a word always includes the result of any of its subwords. For the sake of brevity, we define the category for an arbitrary subset of words S on an arbitrary alphabet Σ as the discussion here also holds for DIL systems.

Definition 10. *Let $\mathbf{W}_S$, where $S \subseteq \Sigma^*$ for some Σ, be the category having:*

- *for objects all the considered words of S, and*
- *for any two words $u, v \in S$, $\mathrm{Arr}_{\mathbf{W}_S}(u, v) := \{p \in \mathbb{N} \mid u \preceq_p v\}$,*
- *for any two $p : u \to v$ and $q : v \to w$, $q \circ p := (p + q : u \to w)$.*

It is trivial to check that $\mathbf{W}_S$ is a category. Particularly, the composition says that if u is included in v at position p which is itself included in w at position q, then u is included in w at position $p+q$. This operation is definitely associative and the identity inclusion is 0: any word is included in itself at position 0. Notice that if $S \subseteq S'$, $\mathbf{W}_S$ is a full subcategory of $\mathbf{W}_{S'}$.

4.2 Designing the Rules of $\overline{\mathcal{L}}$

In a global transformation, while rules specify, as usual, how a sub-part locally evolves, rule inclusions play the important role of tracking during the rewriting how the input is decomposed and how the output is recomposed. Let us illustrate how it works with the transition $abbab \Rightarrow_\pi bababbab$ of Example 1.

The following diagram presents how $abbab$ is decomposed by pattern matching with the lhs of the productions: each occurrence of each letter is identified by its position through an arrow of $\mathbf{W}_{\Sigma^*}$.

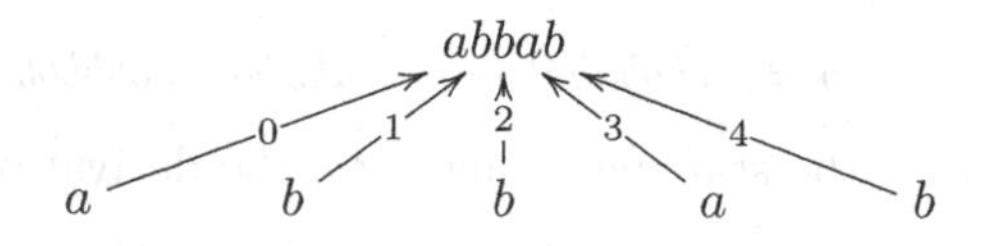

The local application of the productions on each occurrence exhibits all the subwords composing the result.

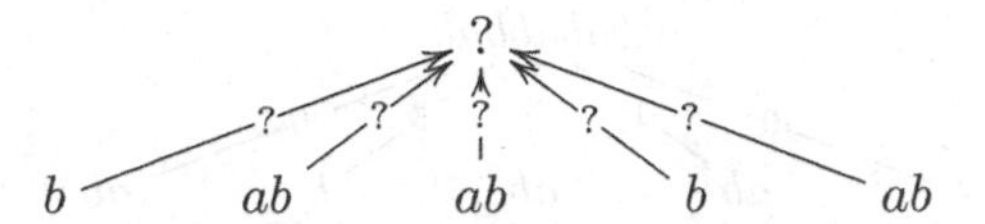

However, without anymore relationship between them, there is no mean to determine in which order they have to be assembled to compose the expected result. As prescribed in Sect. 2.3, the leitmotiv is to track local relations from lhs to rhs. Here, two rhs should be consecutive exactly when their lhs are in the input. This can be captured by realizing that the empty word ε is included between any two adjacent letters so that it can play the role of a common part between lhs that must be mapped to a common part between rhs. We therefore consider an additional rule with ε as lhs. The following diagram is obtained by tracking the occurrences of the lhs (including now ε) in *abbab* together with the relations (*i.e.*, arrows) between these occurrences.

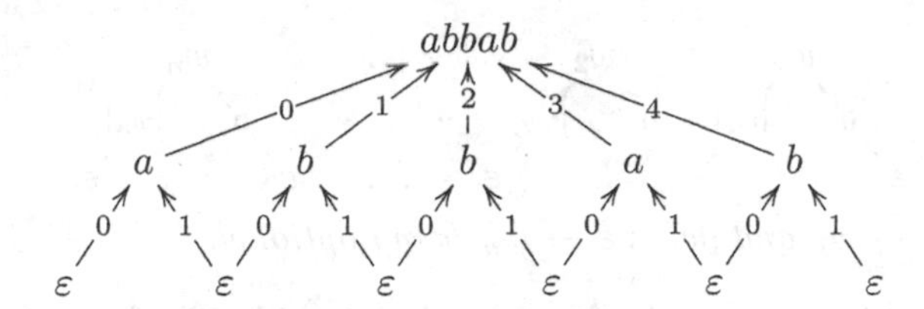

Let us illustrate how to read this diagram by considering the second ε on the left. It represents the occurrence of ε between the first two letters a and b of *abbab*. From this ε, the arrow $1 : \varepsilon \to a$ represents how it is also a sub-occurrence of the first occurrence of a, precisely the empty word occurring on the right of that $a = a\varepsilon$. Similarly, the other arrow $0 : \varepsilon \to b$ shows how it also appears as a sub-occurrence of the occurrence of $b = \varepsilon b$. As such, this diagram provides a representation of the comma category $\mathrm{L}_{\overline{\mathcal{L}}}/abbab$ defined in Sect. 2.4.

Since the yield function of a D0L system is a monoid endomorphism, the additional rule is naturally completed with ε as rhs to get $\varepsilon \Rightarrow_\pi \varepsilon$. The associated inclusion rules are designed as follows: the occurrence $0 : \varepsilon \to a$ of ε in a one-letter word a is sent to the leftmost occurrence $0 : \varepsilon \to \mathrm{P}_\pi(a)$ of ε in the associated rhs $\mathrm{P}_\pi(a)$. Similarly, the occurrence $1 : \varepsilon \to a$ is sent to the rightmost occurrence of ε in $\mathrm{P}_\pi(a)$. In our example, these rules transform the previous pattern matching situation into the following diagram.

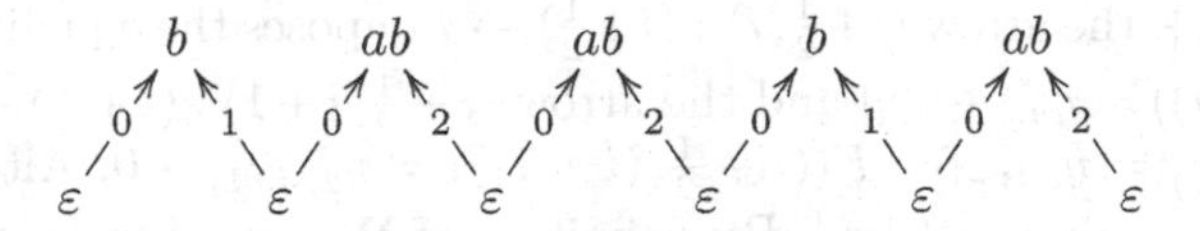

The previous question marks can now be resolved. In categorical terms, the cocone consisting of the word *bababbab*, together with the arrows from the rhs, turns out to be the colimit in $\mathbf{W}_{\Sigma_\pi^*}$ of that diagram, that is, the smallest word containing exactly all the rhs in the right order. Diagrammatically:

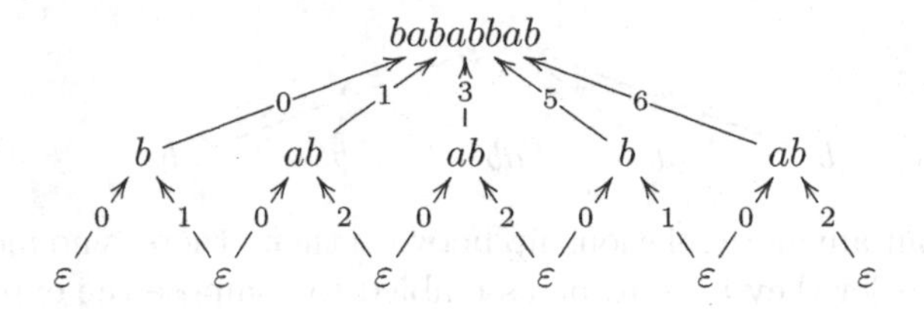

4.3 The Global Transformation $\overline{\mathcal{L}}$

In Definition 8, the concatenation definitely appears as the underlying engine of the yield function and the previous picture illustrates how it is captured as a colimit in $\mathbf{W}_{\Sigma^*}$. This also holds in any $\mathbf{W}_S$ with $S \subseteq \Sigma^*$ as shown in Theorem 1.

Theorem 1. *Let $w_1, \ldots, w_n$ be a non-empty sequence of words of $S \subseteq \Sigma^*$. The word $w_1 \ldots w_n$, if present in S together with ε, is the colimit in $\mathbf{W}_S$ of:*

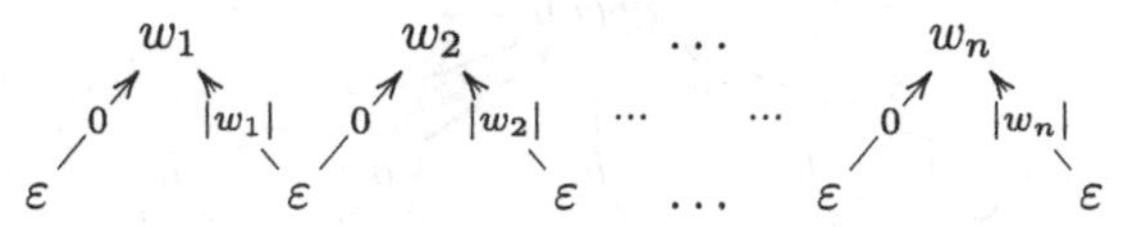

the arrows $0 : \varepsilon \to w_1$ and $|w_n| : \varepsilon \to w_n$ being optional.

Proof. The given diagram can be formally described by the functor $F : \mathbf{I} \to \mathbf{W}_S$ where:

- the set objects of $\mathbf{I}$ is $\{\frac{1}{2}, 1, 1 + \frac{1}{2}, \ldots, n, n + \frac{1}{2}\}$,
- $F(i + \frac{1}{2}) := \varepsilon$ for all $i \in \{0, \ldots, n\}$, and
- $F(i) := w_i$ for all $i \in \{1, \ldots, n\}$,

for the object parts of $\mathbf{I}$ and F and, for their arrow parts:

- $\mathrm{Arr}_{\mathbf{I}}(i, i') := \{(i, i')\}$ if $i' \in \{1, \ldots, n\}$ and $|i' - i| = \frac{1}{2}$,
- $\mathrm{Arr}_{\mathbf{I}}(i, i') := \emptyset$ if $i' \in \{1, \ldots, n\}$ and $|i' - i| > \frac{1}{2}$,
- $F((i - \frac{1}{2}, i)) := 0 \quad : \varepsilon \to w_i$ for all $i \in \{1, \ldots, n\}$,
- $F((i + \frac{1}{2}, i)) := |w_i| : \varepsilon \to w_i$ for all $i \in \{1, \ldots, n\}$.

By Definition 6, a colimit is a particular cocone. For the considered functor, a cocone is given by a word u and a collection of arrows $\{\eta_{u,i} : F(i) \to u\}_{i \in \mathbf{I}}$ respecting the conditions of Definition 5 for each arrows of $\mathbf{I}$. For any $i \in \{0, \ldots, n-1\}$, the arrow $(i+\frac{1}{2}, i) : (i+\frac{1}{2}) \to i$ imposes the equality $\eta_{u,(i+\frac{1}{2})} = \eta_{u,i} \circ F((i + \frac{1}{2}, i)) = \eta_{u,i} + |w_i|$ and the arrow $(i + \frac{1}{2}, i + 1) : (i + \frac{1}{2}) \to (i + 1)$ the equality $\eta_{u,(i+\frac{1}{2})} = \eta_{u,(i+1)} \circ F((i + \frac{1}{2}, (i + 1))) = \eta_{u,(i+1)} + 0$. Altogether, this means that $\eta_{u,(i+1)} = \eta_{u,i} + |w_i|$. By definition of $\mathbf{W}_S$, $\eta_{u,i} : w_i \to u$ is the position of w_i in u so the previous equality means that w_i and w_{i+1} appears consecutively in u. So u contains the word $w_1 \ldots w_n$ at position $\eta_{u,1}$. This immediately shows that $w_1 \dots w_n$ together with the obvious arrows is a cocone. Moreover, it is the excepted colimit since for any cocone u, we have $g_{w,u} = \eta_{u,1} : w_1 \dots w_n \to u$ as required by Definition 6. Finally, the two mentioned arrows are optional in the sense that they do not contribute to the concatenation operation. They simply state that the empty word has to appear at the borders of the colimit, which is in fact trivially true for any word. □

The empty word ε might seem too minimalistic as a common subpart and one might consider this as not natural. This is an extreme case, the case where there is no context. Bigger "common parts" are considered for proper DIL systems. In all cases, the important locality relation transported from lhs to rhs here should seem natural: the relation of subwords being consecutive in a word.

Definition 11. *The global transformation $\overline{\mathcal{L}}$, where $\mathcal{L}$ is a D0L system, is the one where: $\mathbf{C}_{\overline{\mathcal{L}}}$ is $\mathbf{W}_{\Sigma_{\mathcal{L}}^*}$, $\mathbf{\Gamma}_{\overline{\mathcal{L}}} = \mathbf{W}_{\Sigma_{\mathcal{L}} \cup \{\varepsilon\}}$ is the full subcategory of $\mathbf{W}_{\Sigma_{\mathcal{L}}^*}$ with objects $\Sigma_{\mathcal{L}} \cup \{\varepsilon\}$, $\mathrm{L}_{\overline{\mathcal{L}}}$ is the associated inclusion functor, and $\mathrm{R}_{\overline{\mathcal{L}}}$ is given by:*

- $\mathrm{R}_{\overline{\mathcal{L}}}(\varepsilon) := \varepsilon$,
- *for any object $a \in \Sigma_{\mathcal{L}}$, $\mathrm{R}_{\overline{\mathcal{L}}}(a) := \mathrm{P}_{\mathcal{L}}(a)$,*
- *for any $a \in \Sigma_{\mathcal{L}}$, $\mathrm{R}_{\overline{\mathcal{L}}}$ maps the inclusion $0 : \varepsilon \to a$ to $0 : \varepsilon \to \mathrm{P}_{\mathcal{L}}(a)$,*
- *for any $a \in \Sigma_{\mathcal{L}}$, $\mathrm{R}_{\overline{\mathcal{L}}}$ maps the inclusion $1 : \varepsilon \to a$ to $|\mathrm{P}_{\mathcal{L}}(a)| : \varepsilon \to \mathrm{P}_{\mathcal{L}}(a)$.*

This definition is sound in the sense that $\overline{\mathcal{L}}$ mimics $\mathcal{L}$ step by step.

Theorem 2. *For any $w \in \Sigma_{\mathcal{L}}^*$, $w \Rightarrow_{\mathcal{L}} \overline{\mathcal{L}}(w)$.*

Proof. Suppose that $w = a_1 \dots a_n$ with $n = |w|$. We want to show that $\overline{\mathcal{L}}(w) = \mathrm{P}_{\mathcal{L}}(a_1) \dots \mathrm{P}_{\mathcal{L}}(a_n)$. The pattern matching $\mathrm{L}_{\overline{\mathcal{L}}}/w$ of Sect. 2.4 takes the following form.

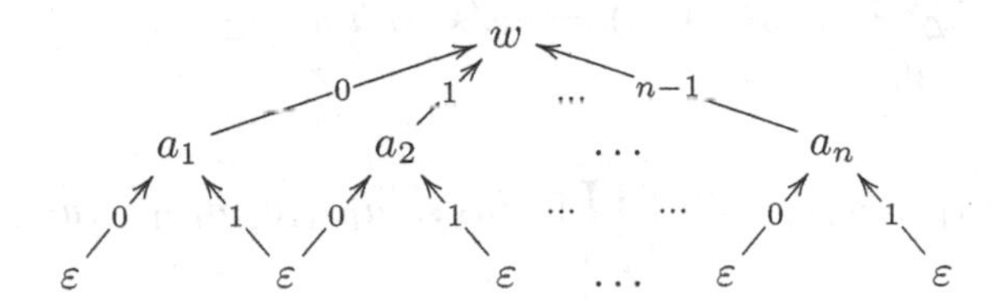

$\overline{\mathcal{L}}(w)$ is defined as the colimit of the rhs diagram $\mathrm{R}_{\overline{\mathcal{L}}} \circ \mathrm{Proj}_{\mathrm{L}_{\overline{\mathcal{L}}}/w}$:

$\mathrm{P}_{\mathcal{L}}(a_1)$ $\mathrm{P}_{\mathcal{L}}(a_2)$ ... $\mathrm{P}_{\mathcal{L}}(a_n)$

0 l_1 0 l_2 ... ... 0 l_n

ε ε ε ... ε ε

where l_i denotes the length of $\mathrm{P}_{\mathcal{L}}(a_i)$. This diagram is similar to the one of Theorem 1 which states that the colimit, a.k.a. $\overline{\mathcal{L}}(w)$, is the concatenation $\mathrm{P}_{\mathcal{L}}(a_1) \dots \mathrm{P}_{\mathcal{L}}(a_n)$ as expected. Notice finally that the result still holds when $n = 0$. In this case, the diagram reduces to a unique object ε. The colimit of such a one-object diagram is the object itself, that is ε so that $\overline{\mathcal{L}}(\varepsilon) = \varepsilon$. □

5 DIL Systems as Global Transformations

The translation of a DIL system into a global transformation is an exercise very similar to the previous one. Definition 11 only needs to be updated to deal with contexts, particularly at the borders. For this, we introduce an extra symbol $\star \notin \Sigma$ to represent the border, so as to be able to identify if a subword lays near the border of a word. For any $\alpha \in \Sigma^*$, $\star\alpha\star$ denotes the *word* α with its two borders, $\star\alpha$ (resp. $\alpha\star$) is the *subword* α stuck to the left (resp. right) border, and α is a proper *subword* without any border constraints. These situations are exactly captured by the category $\mathbf{W}_{\{\star,\varepsilon\}\Sigma^*\{\star,\varepsilon\}}$. Notice how words and subwords are distinguished in this setting. Particularly, a word $\star\alpha\star$ has no arrow to any other word $\star\beta\star$ because the symbol $\star$ can not be found in the middle of these words. This lack of arrows is coherent with DIL systems since the result of a word $\star\beta\star$ is not required to include the result of $\star\alpha\star$ anytime α is a subword of β. This is because the letters near the borders can evolve in different ways, as it is illustrated in the two derivations of Example 2. When considering a D(p, q)L system, the most that can be said about a word $\star u\star$ containing a subword $vwx \in \Sigma^p\Sigma^*\Sigma^q$ is that the result for u includes the result of w. The arrows of the category $\mathbf{W}_{\{\star,\varepsilon\}\Sigma^*\{\star,\varepsilon\}}$ are thus completely coherent with this situation.

Definition 12. *The global transformation* $\overline{\mathcal{L}}$, *where* $\mathcal{L}$ *is a* $D(p,q)L$ *system for some* $p, q \in \mathbb{N}$, *is the one where* $\mathbf{C}_{\overline{\mathcal{L}}}$ *is* $\mathbf{W}_{\{\star,\varepsilon\}\Sigma_{\mathcal{L}}^*\{\star,\varepsilon\}}$, $\mathbf{\Gamma}_{\mathcal{L}} = \mathbf{W}_S$ *is the full subcategory of* $\mathbf{W}_{\{\star,\varepsilon\}\Sigma_{\mathcal{L}}^*\{\star,\varepsilon\}}$ *with:*

$$S := \star{\Sigma_{\mathcal{L}}}^{<(p+q)}\star \ \cup\ \star{\Sigma_{\mathcal{L}}}^{p+q} \ \cup\ {\Sigma_{\mathcal{L}}}^{p+q}\star \ \cup\ {\Sigma_{\mathcal{L}}}^{p+q} \ \cup\ {\Sigma_{\mathcal{L}}}^{p+q+1},$$

$L_{\mathcal{L}}$ *is the associated inclusion functor, and* $R_{\mathcal{L}}$ *is given by:*

1. *For any* $w \in {\Sigma_{\mathcal{L}}}^{<(p+q)}$, $R_{\mathcal{L}}(\star w\star) := \star w'\star$ *with* $w \Rightarrow_{\mathcal{L}} w'$.
2. *For any* $a_1, \ldots, a_{p+q} \in \Sigma_{\mathcal{L}}$,

$$R_{\mathcal{L}}(\star a_1 \ldots a_{p+q}) := \star\left(\prod_{i=1}^{p} \mathrm{P}_{\mathcal{L}}(a_1 \ldots a_{i-1}, a_i, a_{i+1} \ldots a_{i+q})\right),$$

$$R_{\mathcal{L}}(a_{p+q} \ldots a_1\star) := \left(\prod_{i=q}^{1} \mathrm{P}_{\mathcal{L}}(a_{i+p} \ldots a_{i+1}, a_i, a_{i-1} \ldots a_1)\right)\star.$$

3. *For any* $(\alpha, a, \beta) \in {\Sigma_{\mathcal{L}}}^p \times \Sigma_{\mathcal{L}} \times {\Sigma_{\mathcal{L}}}^q$, $R_{\mathcal{L}}(\alpha a\beta) := \mathrm{P}_{\mathcal{L}}(\alpha, a, \beta)$.
4. *For any* $w \in {\Sigma_{\mathcal{L}}}^{p+q}$, $R_{\mathcal{L}}(w) := \varepsilon$.
5. *For any* $\alpha \in {\Sigma_{\mathcal{L}}}^{p+q}$ *and* $a \in \Sigma_{\mathcal{L}} \cup \{\star\}$, $R_{\mathcal{L}}$ *maps the inclusion* $0 : \alpha \to \alpha a$ *to* $0 : \varepsilon \to R_{\mathcal{L}}(\alpha a)$ *and the inclusion* $1 : \alpha \to a\alpha$ *to* $|R_{\mathcal{L}}(\alpha a)| : \varepsilon \to R_{\mathcal{L}}(\alpha a)$.

In this definition, case (1) pays a particular attention to the finite set of words with length lower than $p + q$ which are explicitly transformed. Longer words are managed by cases (2–5). Let us illustrate these cases with the following picture describing the derivation of $abbab \Rightarrow_\sigma bababbb$ of Example 2 as computed by its categorical counterpart $\overline{\sigma}$.

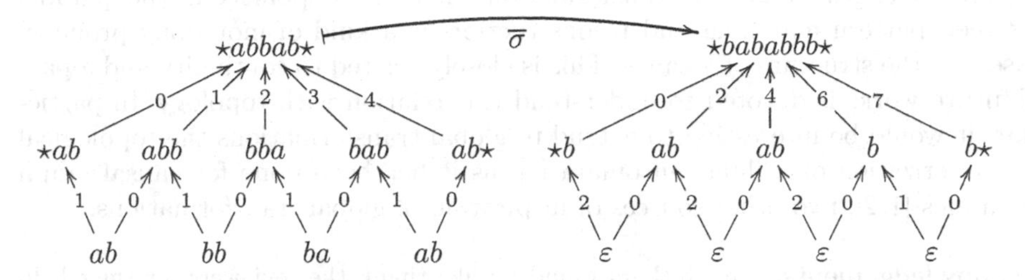

Case (2) deals with left and right borders respectively: a unique rule transforms all letters at a distance lower than p (resp. q) of the left (resp. right) border. Here, the two such border matchings are $0 : \star ab \to \star abbab\star$ and $4 : ab\star \to \star abbab\star$. These rules are particularly responsible of the presence of the left and right symbols $\star$ in the result. For example, $\mathrm{R}_{\overline{\sigma}}(\star ab) = \star \mathrm{P}_\sigma(\varepsilon, a, b) = \star b$. Case (3) is in charge of proper internal subwords using $\mathrm{P}_\mathcal{L}$ directly. The matching $2 : bba \to \star abbab\star$ gives raise to the rhs $\mathrm{R}_{\overline{\sigma}}(bba) = \mathrm{P}_\sigma(b, b, a) = ab$. Finally, case (4) deals with the common part of the previous rules to prepare the concatenation. Each such common part, *e.g.*, $bba \xleftarrow{1} ba \xrightarrow{0} bab$, is mapped to ε with respect to the inclusion rules of case (5). In our example, it is mapped to $ab \xleftarrow{2} \varepsilon \xrightarrow{0} b$ which specifies that ab and b have to be consecutive in the result: abb at position 4. As easily observed, the whole rhs situation coincides with the expectations of Theorem 1 (without the optional arrows) so that the output is the concatenation of the rhs in the right order. The soundness of Definition 12 holds in the exact same way as for Theorem 2; the proof is not reproduced here.

6 Conclusion

We have introduced global transformations, a categorical formalism capturing deterministic synchronous rewriting systems. DL systems have been presented as a particular case illustrating the whys and wherefores of the formalism.

The elegance of L systems made them popular as a kind of revolution in the domain of modeling. A lot of works were then devoted to extend L systems in order to escape from the linearity of strings [5,10]. In a way, global transformations, strongly inspired by L systems in their genesis, are part of the same effort.

The genericity gained with global transformations comes from the use of category theory and the present paper (string rewriting) together with the seminal one [6] (mesh rewriting) show how customizable they are. Pattern matching has already been seen as a generic operation [11]. On the contrary, the reconstruction part is often an *ad hoc* operation coming with the underlying data structure (*e.g.*, the concatenation for strings). In global transformations, the genericity of reconstruction is captured by colimits, a very general notion of quotient/amalgamation. Such a use of colimits has already been investigated in

SPO and DPO techniques for sequential and parallel graph rewriting [3]. Global transformations can be compared to these techniques and their extensions [6].

The strength of global transformations is then to point out the parallel between pattern matching and reconstruction as a kind of monotony property based on the structure of locality. This is closely related to continuity and a part of future works is devoted to understand this relation with topology. In particular, it would be interesting to extend to global transformations the topological characterization of cellular automata [4], as it has been done for causal graph dynamics [1,2], two other sources of inspiration of global transformations.

Acknowledgements. The authors would be like thank the reviewers for their help in improving the quality of this paper. This work was partly supported by the DIM RFSI project *Theory and Pratice of Global Transformations*, Région Île-de-France.

References

1. Arrighi, P., Dowek, G.: Causal graph dynamics. In: Czumaj, A., Mehlhorn, K., Pitts, A., Wattenhofer, R. (eds.) ICALP 2012. LNCS, vol. 7392, pp. 54–66. Springer, Heidelberg (2012). https://doi.org/10.1007/978-3-642-31585-5_9
2. Arrighi, P., Martiel, S., Nesme, V.: Cellular automata over generalized Cayley graphs. Math. Struct. Comput. Sci. **28**(3), 340–383 (2018)
3. Grzegorz, R., Hartmut, E., Hans-jorg, K.: Handbook of Graph Grammars and Computing by Graph Transformations, vol. 3: Concurrency, Parallelism, and Distribution. World Scientific (1999)
4. Hedlund, G.A.: Endomorphisms and automorphisms of the shift dynamical system. Math. Syst. Theor. **3**(4), 320–375 (1969)
5. Kurth, W., Kniemeyer, O., Buck-Sorlin, G.: Relational growth grammars – a graph rewriting approach to dynamical systems with a dynamical structure. In: Banâtre, J.-P., Fradet, P., Giavitto, J.-L., Michel, O. (eds.) UPP 2004. LNCS, vol. 3566, pp. 56–72. Springer, Heidelberg (2005). https://doi.org/10.1007/11527800_5
6. Maignan, L., Spicher, A.: Global graph transformations. In: GCM@ ICGT, pp. 34–49 (2015)
7. Prusinkiewicz, P., Lindenmayer, A.: The Algorithmic Beauty of Plants. Springer, New York (2012)
8. Păun, G.: From cells to computers: computing with membranes (P systems). Biosystems **59**(3), 139–158 (2001)
9. Rozenberg, G., Salomaa, A.: The Mathematical Theory of L Systems, vol. 90. Academic press, New York (1980)
10. Smith, C., Prusinkiewicz, P., Samavati, F.: Local specification of surface subdivision algorithms. In: Pfaltz, J.L., Nagl, M., Böhlen, B. (eds.) AGTIVE 2003. LNCS, vol. 3062, pp. 313–327. Springer, Heidelberg (2004). https://doi.org/10.1007/978-3-540-25959-6_23
11. Spicher, A., Giavitto, J.-L.: Interaction-based programming in MGS. In: Adamatzky, A. (ed.) Advances in Unconventional Computing. ECC, vol. 22, pp. 305–342. Springer, Cham (2017). https://doi.org/10.1007/978-3-319-33924-5_13

Swarm-Based Multiset Rewriting Computing Models

Kaoru Fujioka(✉)

International College of Arts and Sciences, Fukuoka Women's University,
1-1-1 Kasumigaoka, Higashi-ku, Fukuoka 813-8529, Japan
kaoru@fwu.ac.jp

Abstract. Swarm-based computing models and multi-agent-based models have been investigated using parallel processing computation. Based on the preceding models and multiset computings, we propose new computing systems, called swarm systems, that are multiset rewriting systems to formalize swarms' behaviors. A configuration in a swarm system is expressed by a multiset of agents to simulate swarm movements. Swarm automata are also introduced based on swarm systems, which accept strings by considering the configuration sequences. Transition rules in a swarm automaton are labeled by elements of an alphabet and when a configuration consists of final agents then the corresponding sequences of rule symbols are accepted. Position information for agents is added to those swarm models. We show that swarm automata with position information are universal if transition rules are applied in parallel. On the other hand, swarm automata without position information are computationally equivalent to finite state automata.

Keywords: Swarm grammar · Swarm behavior · Rewriting system · Multi-agent system · Cellular automaton

1 Introduction

The Boids model is a well-known multi-agent-based model to simulate a flocking behavior of birds introduced by Reynolds [8]. Each bird independently decides its next position from the neighboring birds through three simple rules. We can often see sophisticated and complicated swarm movement in nature by interacting with simple agent reactions like the Boids model. The cellular automaton is a discrete parallel model using multiple cells which are in one of a finite number of states [4]. Each cell decides its next state according to the states of the cells in its neighborhood and updates its state simultaneously with other cells. In spite of its simple definition, the computational power of cellular automata is known to be universal. To unify some parallel processing ideas, swarm grammars and swarm graph grammars were introduced consisting of a rewriting system (probabilistic L-system) and a set of agent specifications [2,3].

I. McQuillan and S. Seki (Eds.): UCNC 2019, LNCS 11493, pp. 79–93, 2019.
https://doi.org/10.1007/978-3-030-19311-9_8

On the other hand, using the concept of multiset, multiset automaton was introduced in [1] based on multiset addition and subtraction, then a Chomsky-like hierarchy was provided by multiset automaton. P automata in [6] accepts a string using a membrane system which describes a sequence of multisets of objects using membrane structure to simulate living cells.

Inspired by swarm-based and multi-agent-based computing models, we introduce swarm systems to formalize and specify the process of agent movements by using multisets to express multi-agents. When we focus on interactions of agents in a swarm, we see that they have an effect on the neighboring agents and environment, which leads to the next interactions of agents. Therefore, first we introduce swarm systems to formalize the process of agent movement which is defined from the information of the agent itself and surrounding agents autonomously. Then, we introduce position information into agents to add a concept of distance among agents.

Based on swarm systems, swarm automata are introduced which accept strings corresponding to the labels of transition rules to analyze the effects of swarm movement. In a similar way to swarm systems, position information of agents is introduced in a swarm automaton and computing power of a swarm automaton with and without position information is analyzed.

This paper is organized as follows. We recall basic facts from formal language and automata theory and provide terminologies for multisets in Sect. 2. Swarm systems and swarm automata are introduced and some characterizations of swarm automata are also presented in Sect. 3. In Sect. 4, we introduce position information into a swarm system and a swarm automaton. The computational powers of swarm automaton both with and without position information are considered. Finally, Sect. 5 concludes the paper.

2 Preliminaries

We assume that the reader is familiar with the basic notions in formal language theory (see, e.g., [7,9]). A Turing machine is a tuple $M = (Q, \Sigma, \Gamma, \delta, q_I, F)$, where Q is a finite set of states, Σ is a finite set of input symbols, $\Gamma \supseteq \Sigma$ is a finite set of tape symbols, q_I is the initial state in Q, and $F \subseteq Q$ is a finite set of final states. $\delta : Q \times \Gamma \to 2^{Q \times \Gamma \times \{-1,0,1\}}$ is a transition function. We assume that each rule has a unique label and all labels of rules in δ are denoted by $Lab(\delta)$. For $\delta(q, a) \ni (p, b, j)$, M reads the symbol a in the state q, changes to the state p, replaces a with b, and moves the head to the direction indicated by j. A language accepted by M consists of strings $w \in \Sigma^*$, where starting from the configuration $q_I w$, M reaches a configuration with a state in F.

A pushdown automaton is a tuple $M = (Q, \Sigma, \Gamma, \delta, q_I, Z_0, F)$, where Q, Σ, q_I, and F are the same as before. Γ is a finite set of stack symbols and Z_0 is the initial stack symbol in Γ. $\delta : Q \times (\Sigma \cup \{\epsilon\}) \times \Gamma \to 2^{Q \times \Gamma^*}$ is a transition function. A transition $(q, ax, Z\beta) \vdash_{pd} (p, x, \gamma\beta)$ is defined if $\delta(q, a, Z) \ni (p, \gamma)$ for $p, q \in Q$, $a \in \Sigma \cup \{\epsilon\}$, $x \in \Sigma^*$, $Z \in \Gamma$, $\beta, \gamma \in \Gamma^*$. The reflexive and transitive closure of $\vdash_{pd}$ is denoted by $\vdash_{pd}^*$. A language accepted by M is $L(M) = \{w \in$

$\Sigma^* \mid (q_0, w, Z_0) \vdash_{pd}^* (f, \epsilon, \beta),\ f \in F,\ \beta \in \Gamma^*\}$. A language is said to be *linear* if it is accepted by a *one-turn pushdown automaton*, where there is at most one *turn* that switches from non-decreasing the stack height to non-increasing for any $w \in \Sigma^*$.

A finite state automaton M is a tuple $M = (Q, \Sigma, \delta, q_I, F)$, where Q, Σ, q_I, and F are the same as before. $\delta : Q \times (\Sigma \cup \{\epsilon\}) \to 2^Q$ is a transition function. A transition $(q, ax) \vdash (p, x)$ is defined if $\delta(q, a) \ni p$ for $p, q \in Q$, $a \in \Sigma \cup \{\epsilon\}$, $x \in \Sigma^*$. The reflexive and transitive closure of $\vdash$ is denoted by $\vdash^*$. A language accepted by M is $L(M) = \{w \in \Sigma^* \mid (q_0, w) \vdash^* (f, \epsilon),\ f \in F\}$. A language L is said to be *regular* if there is a finite state automaton M such that $L = L(M)$.

A *multiset* over A is a mapping $\sigma : A \to \mathbb{N}$ and $\sigma(a)$ denotes the number of copies of the symbol a in the multiset σ, where $\mathbb{N}$ is the set of non-negative natural numbers. For a finite set A, $A^{\#}$ is the set of all multisets over A including the empty multiset $\emptyset$ which satisfies that $\emptyset(a) = 0$ for all $a \in A$. The *weight* of a multiset σ over A is $||\sigma|| = \sum_{a \in A} \sigma(a)$. We also use $||\sigma||_a$ instead of $\sigma(a)$.

The union of two multisets σ_1 and σ_2 over A denoted by $\sigma_1 + \sigma_2$ is a mapping $A \to \mathbb{N}$ defined by $(\sigma_1 + \sigma_2)(a) = \sigma_1(a) + \sigma_2(a)$ for any $a \in A$. For $t \in \mathbb{N}$ and a multiset σ over A, the multiplication of σ by t denoted by $\sigma \times t$ is a mapping $A \to \mathbb{N}$ defined by $\sigma \times t(a) = \sigma(a) \times t$ for any $a \in A$.

For multisets σ_1 and σ_2 over A, we define inclusion relation $\sigma_1 \subseteq \sigma_2$ if and only if $\sigma_1(a) \leq \sigma_2(a)$ for any $a \in A$. The difference of two multisets σ_1 and σ_2 denoted by $\sigma_1 - \sigma_2$ is a mapping $A \to \mathbb{N}$ defined by $(\sigma_1 - \sigma_2)(a) = \sigma_1(a) - \sigma_2(a)$ for any $a \in A$, provided that $\sigma_2 \subseteq \sigma_1$.

3 Swarm-Based Models

A swarm consists of a number of agents which interact with agents in the swarm and changes itself. We introduce swarm systems to formalize the interaction between agents under multiset operations. Roughly speaking, an agent in a multiset changes itself by applying transition rules based on the condition of some of the other agents in the multiset starting from the initial configuration. Then, we introduce swarm automata which is a computing model to accept strings based on swarm systems.

3.1 Swarm System

A swarm system is a tuple $\Gamma = (A, R, \tau_0)$, where A is a finite set of *agents*, τ_0 is an *initial configuration* in $A^{\#}$, and R is a finite set of *transition rules* over A. A transition rule in R is of the form $(a, \sigma) \to \sigma'$, where $a \in A$ and $\sigma, \sigma' \in A^{\#}$.

We define two transition relations $\Longrightarrow_s$ and $\Longrightarrow_p$ as follows (see Fig. 1).

- A *sequential transition* $\Longrightarrow_s$ is defined such that $\tau \Longrightarrow_s \tau'$ for τ, $\tau' \in A^{\#}$ if and only if $\tau' = (\tau - \{a\}) + \sigma'$, $(a, \sigma) \to \sigma'$ in R, $\sigma \subseteq \tau - \{a\}$.
- A *parallel transition* $\Longrightarrow_p$ is defined such that $\tau \Longrightarrow_p \tau'$ for τ, $\tau' \in A^{\#}$ if and only if $\tau = \{a_1\} + \ldots + \{a_m\}$, $\tau' = \sigma_1' + \ldots + \sigma_m'$, $(a_i, \sigma_i) \to \sigma_i'$ in R, $\sigma_i \subseteq \tau - \{a_i\}$, for $1 \leq i \leq m$.

Intuitively, the sequential transition implies that if the multiset τ includes the multiset σ besides the agent a, then the agent a is transformed into σ' by the transition rule $(a, \sigma) \rightarrow \sigma'$. Similarly, the parallel transition implies that if the multiset τ includes both the multiset σ_i and the agent a_i, then the agent a_i is transformed into σ_i' by the transition rule $(a_i, \sigma_i) \rightarrow \sigma_i'$, for any $1 \leq i \leq m$.

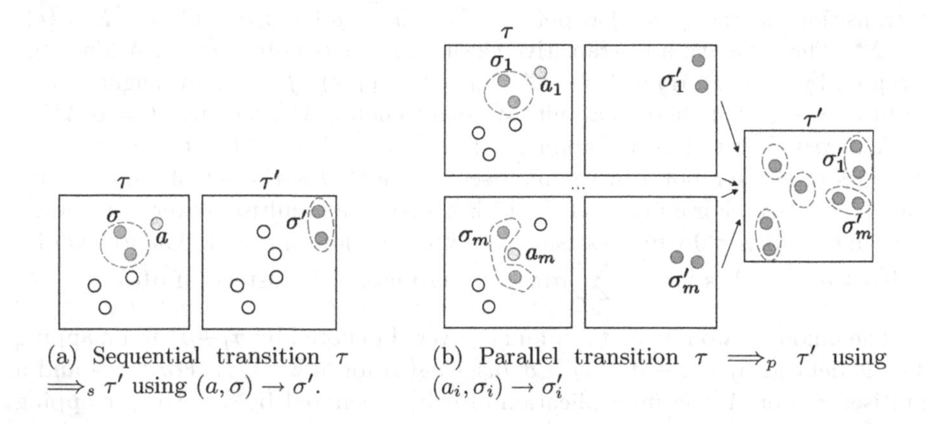

(a) Sequential transition $\tau \Longrightarrow_s \tau'$ using $(a, \sigma) \rightarrow \sigma'$.

(b) Parallel transition $\tau \Longrightarrow_p \tau'$ using $(a_i, \sigma_i) \rightarrow \sigma_i'$

Fig. 1. Description of transition relations in a swarm system.

A *configuration sequence* of Γ starts from the initial configuration τ_0, that is $\tau_0 \Longrightarrow_m \tau_1 \Longrightarrow_m \cdots \Longrightarrow_m \tau_{n-1} \Longrightarrow_m \tau_n$ with $m \in MODE$, where $MODE = \{s, p\}$ and τ_i $(0 \leq i \leq n)$ in $A^{\#}$ is a *configuration*. The reflexive and transitive closure of $\Longrightarrow_s$ and $\Longrightarrow_p$ is denoted by $\Longrightarrow_s^*$ and $\Longrightarrow_p^*$, respectively.

3.2 Swarm Automaton

We introduce a swarm automaton which is a kind of an extension of swarm systems and accepts strings corresponding to the labels of transition rules. A swarm automaton $\Pi = (A, \Sigma, R, \tau_0, F)$ is a tuple, where A, and τ_0 are the same as before. Σ is a finite set of symbols called an *alphabet*. R is a finite set of transition rules of the form $(a, \sigma) \xrightarrow{w} \sigma'$, where $a \in A$, $\sigma, \sigma' \in A^{\#}$, and $w \in \Sigma \cup \{\epsilon\}$. $F \subseteq A$ is a finite set of *final agents*.

A swarm automaton $\Pi = (A, \Sigma, R, \tau_0, F)$ is said to be *strictly deterministic* if for any a in A and w in Σ, there exists at most one transition rule $(a, \sigma) \xrightarrow{w} \sigma'$ in R and there is no transition rule $(a, \sigma) \xrightarrow{\epsilon} \sigma'$ in R for any $\sigma, \sigma' \in A^{\#}$. A swarm automaton $\Pi = (A, \Sigma, R, \tau_0, F)$ is said to be *1-context* if any transition rule $(a, \sigma) \xrightarrow{w} \sigma'$ in R satisfies $||\sigma|| = 1$.

In a similar way to the case of swarm systems, we define two transition relations $\stackrel{w}{\Longrightarrow}_s$ and $\stackrel{w}{\Longrightarrow}_p$ as follows.

- A *sequential transition* $\stackrel{w}{\Longrightarrow}_s$ is defined such that $\tau \stackrel{w}{\Longrightarrow}_s \tau'$ for τ, $\tau' \in A^{\#}$ if and only if $\tau' = (\tau - \{a\}) + \sigma'$, $(a, \sigma) \xrightarrow{w} \sigma'$ in R, $\sigma \subseteq \tau - \{a\}$.

– A *parallel transition* $\overset{w}{\Longrightarrow}_p$ is defined such that $\tau \overset{w}{\Longrightarrow}_p \tau'$ for τ, $\tau' \in A^{\#}$ if and only if $\tau = \{a_1\} + \ldots + \{a_m\}$, $\tau' = \sigma'_1 + \ldots + \sigma'_m$, $(a_i, \sigma_i) \overset{w}{\rightarrow} \sigma'_i$ in R, $\sigma_i \subseteq \tau - \{a_i\}$, for $1 \leq i \leq m$.

For a *configuration sequence* $\tau_0 \overset{w_1}{\Longrightarrow}_m \tau_1 \overset{w_2}{\Longrightarrow}_m \cdots \overset{w_{n-1}}{\Longrightarrow}_m \tau_{n-1} \overset{w_n}{\Longrightarrow}_m \tau_n$ with $m \in MODE$, we simply write $\tau_0 \overset{w_1 \ldots w_n}{\Longrightarrow}_m \tau_n$. The two configuration sequences are illustrated in Fig. 2.

A *language accepted by a swarm automaton Π in the m-transition mode* with $m \in MODE$ is

$$L_m(\Pi) = \{w_1 \ldots w_n \in \Sigma^* \mid \tau_0 \overset{w_1 \ldots w_n}{\Longrightarrow}_m \tau_n,\ \tau_n \in F^{\#}\}.$$

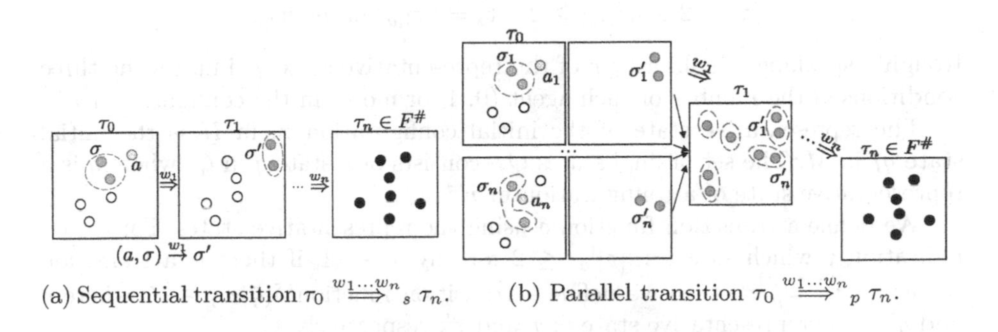

(a) Sequential transition $\tau_0 \overset{w_1 \ldots w_n}{\Longrightarrow}_s \tau_n$. (b) Parallel transition $\tau_0 \overset{w_1 \ldots w_n}{\Longrightarrow}_p \tau_n$.

Fig. 2. Description of configuration sequences in a swarm automaton.

A swarm automaton Π *with single agent configuration* in the m-transition mode with $m \in MODE$ satisfies that any configuration in Π consists of a single agent. That is, for any configuration sequence $\tau_0 \overset{w_1}{\Longrightarrow}_m \tau_1 \overset{w_2}{\Longrightarrow}_m \cdots \overset{w_{n-1}}{\longrightarrow}_m \tau_{n-1} \overset{w_n}{\Longrightarrow}_m \tau_n$ with $m \in MODE$ in Π, $||\tau_i|| = 1$ holds for any $0 \leq i \leq n$.

Lemma 1. *A language accepted by a swarm automaton with single agent configuration in the m-transition mode with $m \in MODE$ is accepted by a finite state automaton.*

For a swarm automaton Π with single agent configuration in the m-transition mode, we can directly construct a finite state automaton M such that $L_m(\Pi) = L(M)$ by simulating the configuration sequences in Π. We omit the details here.

Corollary 1. *A language L is accepted by a swarm automaton with single agent configuration in the m-transition mode with $m \in MODE$ if and only if L is regular.*

For a regular language $L = L(M)$ with a finite state automaton M, we can directly construct a swarm automaton Π with single agent configuration in the m-transition mode such that $L(M) = L_m(\Pi)$. We omit the details here.

From Corollary 1, a swarm automaton with single agent configuration is equivalent to finite state automaton regardless of transition mode.

We now focus on a language accepted by a swarm automaton in the p-transition mode.

Lemma 2. *Consider a swarm automaton Π which is strictly deterministic and 1-context. There exists a finite state automaton M such that $L(M) = L_p(\Pi)$.*

Proof. Consider a swarm automaton $\Pi = (A, \Sigma, R, \tau_0, F)$ which is strictly deterministic, 1-context and $A = \{a_0, \dots, a_{k-1}\}$. We construct a finite state automaton $M = (Q, \Sigma, \delta, q_I, Q_F)$, where $Q = \{q_0, \dots, q_{3^k-1}\}$. For a configuration τ, a *representative state* of τ is q_r, where $r = \mathtt{t}_0 3^0 + \mathtt{t}_1 3^1 + \dots + \mathtt{t}_{k-1} 3^{k-1}$, for $0 \le i \le k-1$, $\mathtt{t}_i$ satisfies

$$\mathtt{t}_i = 2 \text{ if } ||\tau||_{a_i} \ge 2, \quad \mathtt{t}_i = ||\tau||_{a_i} \text{ otherwise.}$$

Roughly speaking, a subscript r of the representative state q_r implies the three conditions of the number of each agent (0, 1, or more) in the configuration τ.

The representative state of the initial configuration τ_0 in Π is the initial state q_I in M. The set of final states Q_F consists of a state q in Q, where q is a representative state of a configuration in $F^{\#}$.

We define a transition function δ using the representative states. For a configuration τ which satisfies $||\tau||_{a_i} \le 2$ for any $a_i \in A$, if there is a transition sequence $\tau \stackrel{w}{\Longrightarrow}_p \tau'$, then we define a transition function $\delta(\bar{q}, w) \ni \bar{q}'$, where $\bar{q}$ and $\bar{q}'$ is the representative state of τ and τ', respectively.

We show that if there is a configuration sequence $\tau_0 \stackrel{w_1 \dots w_n}{\Longrightarrow}_p \tau_n$, then there exists $\delta(q_I, w_1 \dots w_n) \ni \bar{q}_n$, where $\bar{q}_n$ is the representative state of τ_n by induction. Assume that the claim holds for n. Consider a configuration sequence $\tau_0 \stackrel{w_1 \dots w_n}{\Longrightarrow}_p \tau_n \stackrel{w}{\Longrightarrow}_p \tau_{n+1}$. Then, $\delta(q_I, w_1 \dots w_n) \ni \bar{q}_n$ holds for the representative state $\bar{q}_n$ of τ_n by induction hypothesis. For the configuration τ_n, there is a configuration $\bar{\tau}_n$ which satisfies $||\bar{\tau}_n||_{a_i} \le 2$ for any $a_i \in A$ and the representative state of $\bar{\tau}_n$ is $\bar{q}_n$.

From the definition of strictly deterministic, for w in Σ and an agent a_i in A, there is at most one transition rule $(a_i, \sigma_i) \stackrel{w}{\rightarrow} \sigma_i'$ in R $(0 \le i \le k-1)$. Then, $\tau_{n+1} = \sigma_0' \times ||\tau_n||_{a_0} + \dots + \sigma_{k-1}' \times ||\tau_n||_{a_{k-1}}$ holds. Also, there is a transition $\bar{\tau}_n \stackrel{w}{\Longrightarrow}_p \bar{\tau}_{n+1}$, where $\bar{\tau}_{n+1} = \sigma_0' \times \mathtt{t}_0 + \dots + \sigma_{k-1}' \times \mathtt{t}_{k-1}$ for $\mathtt{t}_i = ||\tau_n||_{a_i}$. Then the representative state of τ_{n+1} is the representative state of $\bar{\tau}_{n+1}$. From the construction of δ, for a transition $\bar{\tau}_n \stackrel{w}{\Longrightarrow}_p \bar{\tau}_{n+1}$, there is a transition function $\delta(\bar{q}_n, w) \ni \bar{q}_{n+1}$, where $\bar{q}_{n+1}$ is the representative state of $\bar{\tau}_{n+1}$ and τ_{n+1}. Therefore, $\delta(q_I, w_1 \dots w_n w) \ni \bar{q}_{n+1}$ holds.

We note that from the definition of 1-context, in the parallel transition $\tau_n \stackrel{w}{\Longrightarrow}_p \tau_{n+1}$, an agent a_i in τ_n can be applied a transition rule $(a_i, \sigma_i) \stackrel{w}{\rightarrow} \sigma_i'$ with $\sigma_i = \{a'\}$ so long as $||\tau_n - \{a_i\}||_{a'} \ge 1$ holds. Especially for a transition rule $(a_i, \{a_i\}) \stackrel{w}{\rightarrow} \sigma_i'$, a configuration τ_n needs to satisfy $||\tau_n||_{a_i} \ge 2$. This means that the representative states are enough to check the presence of agents in τ_n.

On the other hand, from the definition of δ, for $\delta(q_I, w_1 \dots w_n) \ni \bar{q}_n$, it is straightforward that there is a transition sequence $\tau_0 \stackrel{w_1 \dots w_n}{\Longrightarrow} \tau_n$ such that the representative state of τ_n is $\bar{q}_n$. From the definition of Q_F, there is a configuration $\tau_0 \stackrel{w}{\Longrightarrow} \tau_n$ with $\tau_n \in F^{\#}$ if and only if $\delta(q_0, w) \ni \bar{q}_n$, where $\bar{q}_n$ is a representative state of τ_n and $\bar{q}_n \in Q_F$. □

Based on Lemma 2, we consider the following lemmas for a language accepted by a swarm automaton in the p-transition mode.

Lemma 3. *Consider a strictly deterministic swarm automaton* $\Pi = (A, \Sigma, R, \tau_0, F)$. *There exists a finite state automaton* M *such that* $L(M) = L_p(\Pi)$.

Proof. Consider a strictly deterministic swarm automaton $\Pi = (A, \Sigma, R, \tau_0, F)$, where $A = \{a_0, \ldots, a_{k-1}\}$. Let l be a natural number such that $l = \max\{||\sigma|| \mid (a, \sigma) \stackrel{w}{\to} \sigma' \in R\}$. Then, we construct a finite state automaton $M = (Q, \Sigma, \delta, q_I, Q_F)$, where $Q = \{q_0, \ldots, q_h\}$ with $h = (l+2)^k - 1$. For a configuration τ, a *representative state* in Q of a configuration τ is q_r, where $r = \mathtt{t}_0(l+2)^0 + \mathtt{t}_1(l+2)^1 + \ldots + \mathtt{t}_{k-1}(l+2)^{k-1}$, for $0 \le i \le k-1$, $\mathtt{t}_i$ satisfies

$$\mathtt{t}_i = l+1 \text{ if } ||\tau||_{a_i} \ge l+1, \quad \mathtt{t}_i = ||\tau||_{a_i} \text{ otherwise.}$$

In a similar way that is used to prove Lemma 2, for a configuration τ which satisfies $||\tau||_{a_i} \le l+1$ for any $a_i \in A$, if there is a transition sequence $\tau \stackrel{w}{\Longrightarrow}_p \tau'$, then we define a transition function $\delta(\bar{q}, w) \ni \bar{q}'$, where $\bar{q}$ and $\bar{q}'$ is the representative state of τ and τ', respectively.

Let q_I be the initial state which is a representative state of the initial configuration τ_0. The set of final states Q_F is a set of state q in Q, where q is a representative state of a configuration in $F^{\#}$.

In a similar way that is used to prove Lemma 2, we can prove that there is a configuration $\tau_0 \stackrel{w_1 \ldots w_n}{\Longrightarrow}_p \tau_n$ if and only if $\delta(q_0, w_1 \ldots w_n) \ni \bar{q}_n$, where $\bar{q}_n$ is a representative state of τ_n. From the definition of Q_F, $L_p(\Pi) = L(M)$ holds. □

Lemma 4. *Consider a swarm automaton* $\Pi = (A, \Sigma, R, \tau_0, F)$. *There exists a finite state automaton* M *such that* $L(M) = L_p(\Pi)$.

Proof. Consider a swarm automaton $\Pi = (A, \Sigma, R, \tau_0, F)$. In a similar way to Lemma 3, we can construct a finite state automaton $M = (Q, \Sigma, \delta, q_I, Q_F)$. Let l be a natural number such that $l = \max\{||\sigma|| \mid (a, \sigma) \stackrel{w}{\to} \sigma' \in R\}$. For a configuration τ, where for any a in A, $||\tau||_a \le l+1$ holds, consider all the possible configurations $\tau \stackrel{w}{\Longrightarrow}_p \tau'$ and construct the nondeterministic transition function $\delta(\bar{q}, w) \ni \bar{q}'$, where $\bar{q}$ and $\bar{q}'$ is the representative state of τ and τ', respectively. In a similar way that is used to prove Lemmas 2 and 3, we can prove that $L_p(\Pi) = L(M)$. □

From Lemma 4 and Corollary 1, we have the following corollary.

Corollary 2. *A class of languages accepted by a swarm automaton in the* p*-transition mode is equivalent to the class of regular languages.*

On the other hand, there is the following result for the s-transition mode.

Lemma 5. *There exists a linear language which is generated by a swarm automaton in the s-transition mode.*

Proof. Consider a swarm automaton $\Pi = (A, \Sigma, R, \tau_0, F)$, where $A = \{s, a_0, a_1, x, y, \$\}$, $\Sigma = \{x, y\}$, $\tau_0 = \{a_0, s\}$, and $F = \{a_1\}$. R consists of the following transition rules:

$$
\begin{array}{ll}
(a_0, \{s\}) \xrightarrow{x} \{\$\}, & (s, \{a_0\}) \xrightarrow{x} \{x\}, \\
(a_0, \{x\}) \xrightarrow{x} \{a_0, x\}, & (x, \sigma) \xrightarrow{x} \{\$\} \ (\sigma \in \{\{a_0\}, \{x\}\}), \\
(a_0, \{x\}) \xrightarrow{y} \{a_1\}, & (x, \sigma) \xrightarrow{y} \{\$\} \ (\sigma \in \{\{a_0\}, \{x\}\}), \\
(a_1, \{x\}) \xrightarrow{y} \{\$\}, & (x, \sigma) \xrightarrow{y} \emptyset \ (\sigma \in \{\{a_1\}, \{x\}\}), \\
(\$, \sigma) \xrightarrow{w} \{\$\} \ (\sigma \in A^{\#}, \ ||\sigma|| = 1, \ w \in \Sigma). &
\end{array}
$$

From the transition rules for $\$ \in A - F$, no configuration including the agent $\$$ leads to a configuration in $F^{\#}$. Thus, we ignore any configuration sequence starting from the configuration including $.

Consider a configuration sequence for the input x^3y^4 in $L(\Pi)$ such that

$$
\begin{aligned}
&\{a_0, \underline{s}\} \xRightarrow{x}_s \{\underline{a_0}, x\} \xRightarrow{x}_s \{\underline{a_0}, x, x\} \xRightarrow{x}_s \{\underline{a_0}, x, x, x\} \\
&\xRightarrow{y}_s \{a_1, x, x, \underline{x}\} \xRightarrow{y}_s \{a_1, x, \underline{x}\} \xRightarrow{y}_s \{a_1, \underline{x}\} \xRightarrow{y}_s \{a_1\}.
\end{aligned}
$$

In the configuration sequence, a transition rule is applied to the underlined agents. Figure 3 illustrates the configuration sequences with inputs xy^2, x^2y^3, and x^3y^4.

$$
\begin{array}{ccccccc}
 & {}^{x}\!\nearrow & \$, s & {}^{x,y}\!\nearrow & a_0, \$ & {}^{x,y}\!\nearrow & a_0, x, \$ \\
a_0, s & \xRightarrow{x} & a_0, x & \xRightarrow{x} & a_0, x, x & \xRightarrow{x} & a_0, x, x, x \\
 & & \Downarrow y & & \Downarrow y & & \Downarrow y \\
a_1 & \xLeftarrow{y} & a_1, x & \xLeftarrow{y} & a_1, x, x & \xLeftarrow{y} & a_1, x, x, x \\
 & & \Downarrow y & & \Downarrow y & & \Downarrow y \\
 & & \$, x & & \$, x, x & & \$, x, x, x
\end{array}
$$

Fig. 3. Configuration sequences for accepting xy^2, x^2y^3, and x^3y^4.

For an input x, an agent a_0 generates an agent x after checking the presence of the agent x by the rule $(a_0, \{x\}) \xrightarrow{x} \{a_0, x\}$. Further, an agent x will be replaced by the empty set by the input y after checking the presence of a_1 or x by the rule $(x, \sigma) \xrightarrow{y} \emptyset$ $(\sigma \in \{\{a_1\}, \{x\}\})$. Once an agent a_0 in τ_0 is transformed into a_1 by the rule $(a_0, \{x\}) \xrightarrow{y} \{a_1\}$, there is no possibility that the agent a_1 is transformed into a_0.

From the fact that $F = \{a_1\}$, we see that $L_s(\Pi) = \{x^n y^{n+1} \mid n \geq 1\}$, which is linear. □

From Lemma 5, Corollary 2, and Corollary 1, the class of languages accepted by swarm automaton in the s-transition mode properly contains the class of languages accepted by swarm automaton in the p-transition mode.

4 Introducing Position Information

As you see in the Boids model and so forth, an agent in a swarm is quite often affected by nearby agents. In a cellular automaton, each cell changes its own state based on its neighboring cells. To introduce those concept of distance among agents, we add position information for each agent to the swarm-based models introduced in the previous section.

An *agent* a *in* A *with position* d in a *domain* D is denoted by $a[d]$. It is possible that there are multiple agents with the same position. $A[D]$ denotes the set of $a[d]$ with $a \in A$ and $d \in D$. For example, if we consider one-dimensional (resp., two-dimensional) space, an agent position can be expressed by a natural number in $D = \mathbb{N}$ or a real number in $D = \mathbb{R}$ (resp., a pair in $D = \mathbb{N} \times \mathbb{N}$ or $D = \mathbb{R} \times \mathbb{R}$). Similarly, we can express an agent position in x-dimensional space ($x \geq 1$).

We define the following operation for agents with position based on the multiset operations in Sect. 2. A multiset over $A[D]$ is a mapping $\tau : A[D] \rightarrow \mathbb{N}$ and $\tau(a[d])$ denotes the number of $a[d]$ in the multiset τ. $A[D]^{\#}$ is the set of all multisets over $A[D]$ including the empty multiset $\emptyset$ which satisfies that $\emptyset(a[d]) = 0$ for all $a \in A$, $d \in D$. The union of two multisets of agents with position τ_1 and τ_2 in $A[D]^{\#}$ denoted by $\tau_1 + \tau_2$ is a mapping $A[D] \rightarrow \mathbb{N}$ defined by $(\tau_1 + \tau_2)(a[d]) = \tau_1(a[d]) + \tau_2(a[d])$ for any $a \in A$ and $d \in D$. The difference of two multisets of agents with position τ_1 and τ_2 in $A[D]^{\#}$ denoted by $\tau_1 - \tau_2$ is a mapping $A[D] \rightarrow \mathbb{N}$ defined by $(\tau_1 - \tau_2)(a[d]) = \tau_1(a[d]) - \tau_2(a[d])$, provided that $\tau_1(a[d]) \geq \tau_2(a[d])$, for any $a \in A$ and $d \in D$.

For a multiset τ in $A[D]^{\#}$, let $agt(\tau) \in A^{\#}$ be a multiset of agents in τ. For example, for $\tau = \{a_1[d_1], a_1[d_2], a_2[d_1], a_2[d_2]\}$, $agt(\tau) = \{a_1, a_1, a_2, a_2\}$. Let the weight of τ be $||\tau|| = \sum_{a \in A,\ d \in D} \tau(a[d])$.

For a multiset τ in $A[D]^{\#}$ with $||\tau|| \geq r$, the set of neighboring agents from the position d with r agents denoted by $\tau_d^r \subseteq A[D]^{\#}$ consists of r nearest agents with position in τ from the position d determined by agent position. For example, for $\tau = \{a_1[d_1], a_1[d_2], a_2[d_1], a_2[d_2]\}$ with $d_1, d_2 \in \mathbb{N}$ and $0 < d_1 < d_2$, $\tau_0^3 = \{\{a_1[d_1], a_2[d_1], a_1[d_2]\}, \{a_1[d_1], a_2[d_1], a_2[d_2]\}\}$.

A *swarm system with position* is a tuple $\Gamma = (A, R, \tau_0, D)$, where A is the same as before and D is a domain of agents. An initial configuration τ_0 is the multiset of agents with position in $A[D]^{\#}$. R is a finite set of rules such that $(a, \sigma) \rightarrow \sigma'\langle v \rangle$, where $a \in A$, $\sigma, \sigma' \in A^{\#}$, v is a vector over D. Roughly speaking, an agent a with position d is transformed into σ' where agents in σ' are with position $d + v$ by the rule $(a, \sigma) \rightarrow \sigma'\langle v \rangle$.

We define transition relations with position $\Longrightarrow_{sp}$ and $\Longrightarrow_{pp}$ as follows.

- A *sequential transition with position* $\Longrightarrow_{sp}$ is defined such that $\tau \Longrightarrow_{sp} \tau'$ for τ, $\tau' \in A[D]^{\#}$ if and only if
  $(a, \sigma) \rightarrow \sigma'\langle v \rangle$ in R,
  $\tau' = (\tau - a[d]) + \tau''$, $\tau'' = \{b[d + v] \mid b \in \sigma'\}$ with $agt(\tau'') = \sigma'$,
  and there is τ_* such that $\tau_* \in (\tau - a[d])_p^{||\sigma||}$ and $\sigma \subseteq agt(\tau_*)$.

– A *parallel transition with position* $\Longrightarrow_{pp}$ is defined such that $\tau \Longrightarrow_{pp} \tau'$ for $\tau,\ \tau' \in A[D]^{\#}$ if and only if, for $1 \leq i \leq m$,
$\tau = \{a_1[d_1]\} + \ldots + \{a_m[d_m]\}$, $(a_i, \sigma_i) \rightarrow \sigma_i'\langle v_i \rangle$ in R,
$\tau' = \tau_1'' + \ldots + \tau_m''$, $\tau_i'' = \{b[d_i + v_i] \mid b \in \sigma_i'\}$ with $agt(\tau_i'') = \sigma'$,
and there is τ_{*i} such that $\tau_{*i} \in (\tau - a_i[d_i])_{d_i}^{||\sigma_i||}$ and $\sigma_i \subseteq agt(\tau_{*i})$.

The two transition relations are illustrated in Fig. 4.

A configuration sequence of Γ starts from the initial configuration τ_0, that is $\tau_0 \Longrightarrow_m \tau_1 \Longrightarrow_m \cdots \Longrightarrow_m \tau_{n-1} \Longrightarrow_m \tau_n$ with $m \in PMODE$, where $PMODE = \{sp, pp\}$ and τ_i $(0 \leq i \leq n)$ in $A[D]^{\#}$ is a configuration. The reflexive and transitive closure of $\Longrightarrow_{sp}$ and $\Longrightarrow_{pp}$ is denoted by $\Longrightarrow_{sp}^*$ and $\Longrightarrow_{pp}^*$, respectively.

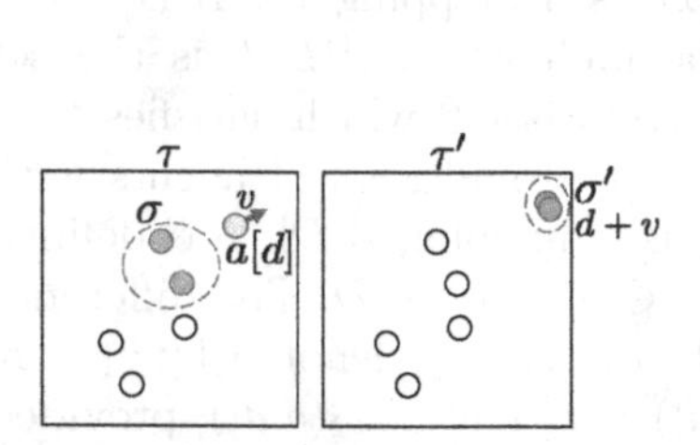

(a) Sequential transition with position $\tau \Longrightarrow_{sp} \tau'$ using $(a, \sigma) \rightarrow \sigma'\langle v \rangle$.

(b) Parallel transition with position $\tau \Longrightarrow_{pp} \tau'$ using $(a_i, \sigma_i) \rightarrow \sigma_i'\langle v_i \rangle$ for $1 \leq i \leq m$.

Fig. 4. Description of transition relations in a swarm system with position.

A *swarm automaton with position* is a tuple $\Pi = (A, \Sigma, R, \tau_0, F, D)$, where A, Σ, F and D are the same as before. R is a finite set of transition rules of the form $(a, \sigma) \xrightarrow{w} \sigma'\langle v \rangle$, where $a \in A$, $\sigma, \sigma' \in A^{\#}$, $w \in \Sigma \cup \{\epsilon\}$, and v is a vector over P. Especially for $\sigma' = \emptyset$, a transition rule is of the form $(a, \sigma) \xrightarrow{w} \emptyset$. τ_0 is an *initial configuration* in $A[D]^{\#}$.

We define two transition relations with position $\stackrel{w}{\Longrightarrow}_{sp}$ and $\stackrel{w}{\Longrightarrow}_{pp}$ as follows.

1. A *sequential transition with position* $\stackrel{w}{\Longrightarrow}_{sp}$ is defined such that $\tau \stackrel{w}{\Longrightarrow}_{sp} \tau'$ for $\tau,\ \tau' \in A[D]^{\#}$ if and only if
$(a, \sigma) \xrightarrow{w} \sigma'\langle v \rangle$ in R,
$\tau' = (\tau - a[d]) + \tau''$, $\tau'' = \{b[d + v] \mid b \in \sigma'\}$ with $agt(\tau'') = \sigma'$,
and there is τ_* such that $\tau_* \in (\tau - a[d])_d^{||\sigma||}$ and $\sigma \subseteq agt(\tau_*)$.
2. A *parallel transition with position* $\stackrel{w}{\Longrightarrow}_{pp}$ is defined such that $\tau \stackrel{w}{\Longrightarrow}_{pp} \tau'$ for $\tau,\ \tau' \in A[D]^{\#}$ if and only if, for $1 \leq i \leq m$,
$\tau = \{a_1[d_1]\} + \ldots + \{a_m[d_m]\}$, $(a_i, \sigma_i) \xrightarrow{w} \sigma_i'\langle v_i \rangle$ in R,
$\tau' = \tau_1'' + \ldots + \tau_m''$, $\tau_i'' = \{b[d_i + v_i] \mid b \in \sigma_i'\}$ with $agt(\tau_i'') = \sigma_i'$,
and there is τ_{*i} such that $\tau_{*i} \in (\tau - a_i[d_i])_{d_i}^{||\sigma_i||}$ and $\sigma_i \subseteq agt(\tau_{*i})$.

We define a *language accepted by a swarm automaton with position Π in the m-transition mode* with $m \in PMODE$ as

$$L_m(\Pi) = \{w_1 \dots w_n \in \Sigma^* \mid \tau_0 \overset{w_1 \dots w_n}{\Longrightarrow}_m \tau_n,\ agt(\tau_n) \in F[D]^\#\}.$$

We focus on a language accepted by a swarm automaton with position in the *pp*-transition mode.

Lemma 6. *There exists a linear language which is accepted by a swarm automaton with position in the pp-transition mode.*

Proof. Consider a swarm automaton $\Pi = (A, \Sigma, R, \tau_0, F, D)$, where $A = \{s, a_0, a_1, x, y\}$, $\Sigma = \{x, y\}$, $\tau_0 = \{a_0[0], s[0]\}$, $F = \{a_1\}$, and $D = \mathbb{N}$. R consists of the following transition rules:

$$\begin{array}{ll}
(a_0, \{s\}) \overset{x}{\rightarrow} \{a_0\}\langle 0\rangle, & (s, \{a_0\}) \overset{x}{\rightarrow} \{x\}\langle 0\rangle, \\
(a_0, \{x\}) \overset{x}{\rightarrow} \{a_0, x\}\langle 0\rangle, & (x, \sigma) \overset{x}{\rightarrow} \{x\}\langle 1\rangle\ (\sigma \in \{\{a_0\}, \{x\}\}), \\
(a_0, \{x\}) \overset{y}{\rightarrow} \{a_1\}\langle 0\rangle, & (x, \sigma) \overset{y}{\rightarrow} \{x\}\langle 0\rangle\ (\sigma \in \{\{a_0\}, \{x\}\}), \\
(a_1, \{x\}) \overset{y}{\rightarrow} \{a_1\}\langle 1\rangle, & (x, \{a_1\}) \overset{y}{\rightarrow} \emptyset.
\end{array}$$

Consider a configuration sequence for the input string x^3y^4 in $L_{pp}(\Pi)$ such that

$$\begin{aligned}
&\{a_0[0], s[0]\} \\
\overset{x}{\Longrightarrow}_{pp}\ &\{a_0[0], x[0]\} \overset{x}{\Longrightarrow}_{pp} \{a_0[0], x[0], x[1]\} \overset{x}{\Longrightarrow}_{pp} \{a_0[0], x[0], x[1], x[2]\} \\
\overset{y}{\Longrightarrow}_{pp}\ &\{a_1[0], x[0], x[1], x[2]\} \overset{y}{\Longrightarrow}_{pp} \{a_1[1], x[1], x[2]\} \overset{y}{\Longrightarrow}_{pp} \{a_1[2], x[2]\} \\
\overset{y}{\Longrightarrow}_{pp}\ &\{a_1[3]\}.
\end{aligned}$$

Figure 5 illustrates the configuration sequences with inputs xy^2, x^2y^3, and x^3y^4.

For the input x, the agent a_0 generates a new agent x after checking the present agent x by the rule $(a_0, \{x\}) \overset{x}{\rightarrow} \{a_0, x\}\langle 0\rangle$. Further, the agent x will be the empty set by the input y after checking the present agent a_1 by the rule $(x, \{a_1\}) \overset{y}{\rightarrow} \emptyset$. Once an agent a_0 in τ_0 is transformed into a_1 by the rule $(a_0, \{x\}) \overset{y}{\rightarrow} \{a_1\}\langle 0\rangle$, there is no possibility that a_1 is transformed into a_0.

From the fact that $F = \{a_1\}$, we see that $L_{pp}(\Pi) = \{x^n y^{n+1} \mid n \geq 1\}$, which is linear. □

We show the equivalence of the accepting powers between swarm automaton with position in the *pp*-transition mode and Turing machines as follows.

Lemma 7. *A language accepted by a Turing machine is accepted by a swarm automaton with position in the pp-transition mode.*

Proof. Consider a Turing machine $M = (Q, \Sigma, \Gamma, \delta, q_0, F)$. We construct a swarm automaton with position $\Pi = (A, \Sigma, R, \tau_0, F', \mathbb{N})$ as follows. For the new symbols $s, s_1, ¥, \$, \# \notin \Gamma \cup Q \cup Lab(\delta)$, we set $A = \Gamma \cup Q \cup Lab(\delta) \cup \{s, s_1, ¥, \$, \#\}$ and $F' = F \cup \Gamma \cup \{\$, \#\}$. The initial configuration is $\tau_0 = \{\$[0], s[0], \#[1]\}$.

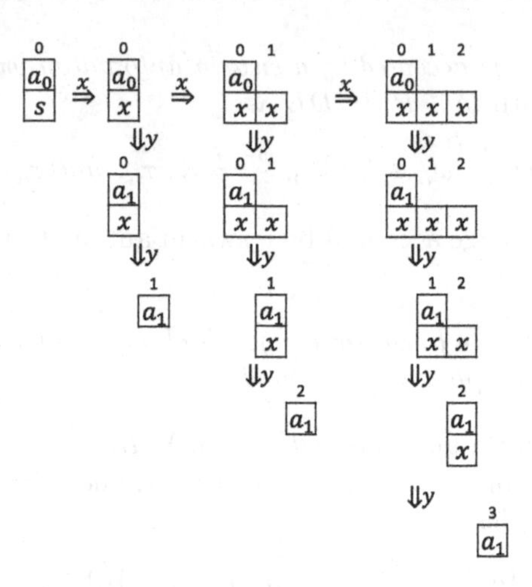

Fig. 5. Configuration sequences for accepting xy^2, x^2y^3, and x^3y^4. To avoid overlapping, agents in the same column have the same position in $\mathbb{N}$ denoted above the array.

R consists of the following three categories:

1. $(s,\sigma) \xrightarrow{x} \{s,x,¥\}\langle 1\rangle$ for $\sigma \in \Sigma^{\#} \cup \{\{\$\}\}$, $||\sigma|| = 1$, $x \in \Sigma$,
   $(\#,\sigma) \xrightarrow{x} \{\#\}\langle 1\rangle$ for $\sigma \in \Sigma^{\#} \cup \{\{\$\},\{s\}\}$, $||\sigma|| = 1$, $x \in \Sigma$,
   $(\$,\sigma) \xrightarrow{x} \{\$\}\langle 0\rangle$ for $\sigma \in \Sigma^{\#} \cup \{\{s\}\}$, $||\sigma|| = 1$, $x \in \Sigma$,
   $(a,\sigma) \xrightarrow{x} \{a\}\langle 0\rangle$ for $\sigma \in \Sigma^{\#} \cup \{\{s\}\}$, $||\sigma|| = 1$, $x \in \Sigma$, $a \in \Sigma \cup \{¥\}$,
2. $(s,\sigma) \xrightarrow{\epsilon} \{s_1\}\langle -1\rangle$ for $\sigma \in A^{\#}$,
   $(s_1,\sigma) \xrightarrow{\epsilon} \{s_1\}\langle -1\rangle$ for $\sigma \in \Sigma^{\#}$, $||\sigma|| = 1$,
   $(s_1,\{\$\}) \xrightarrow{\epsilon} \{q_0\}\langle 1\rangle$,
   $(a,\sigma) \xrightarrow{\epsilon} \{a\}\langle 0\rangle$ for $\sigma \in \Sigma^{\#} \cup \{\{s\}\}$, $||\sigma|| = 1$, $a \in \Sigma \cup \{¥,\#,\$\}$,
3. $(q,\{a\}) \xrightarrow{\epsilon} \{r,b,¥\}\langle 0\rangle$ for $r : \delta(q,a) \ni (p,b,j)$, $r \in Lab(\delta)$,
   $(a,\{q\}) \xrightarrow{\epsilon} \emptyset$ for $r : \delta(q,a) \ni (p,b,j)$, $r \in Lab(\delta)$,
   $(r,\sigma) \xrightarrow{\epsilon} \{p\}\langle j\rangle$ for $r : \delta(q,a) \ni (p,b,j)$, $r \in Lab(\delta)$, $\sigma \in A^{\#}$,
   $(¥,\sigma) \xrightarrow{\epsilon} \emptyset$ for $\sigma \in Q^{\#}$, $||\sigma|| = 1$,
   $(q,\{¥\}) \xrightarrow{\epsilon} \{q\}\langle 0\rangle$ for $q \in Q \cup \Gamma$.

For the simulation of a Turing machine M with an input $w = w_1 \dots w_n$, the swarm automaton Π behaves as follows in the above three categories, which is illustrated in Fig. 6:

1. Π generates an agent w_i in Σ on the position i together with the dummy agent ¥ by the agent s.
2. After the agent s is transformed into s_1, the agent s_1 moves left to the agent \$ on the position 0. Then, s_1 is transformed into q_0 on the position 1.
3. An agent q in Q simulates a transition rule δ using an agent in $Lab(\delta)$. For a transition rule $r : \delta(q,a) \ni (p,b,j)$, the agent q with neighboring agent a is transformed into agents r and b, then the agent r is transformed into the

agent p assigned its position according to j. Through this process, the dummy agent ¥ plays a role that only an agent in Γ located in the same position with q is transformed into $\emptyset$.

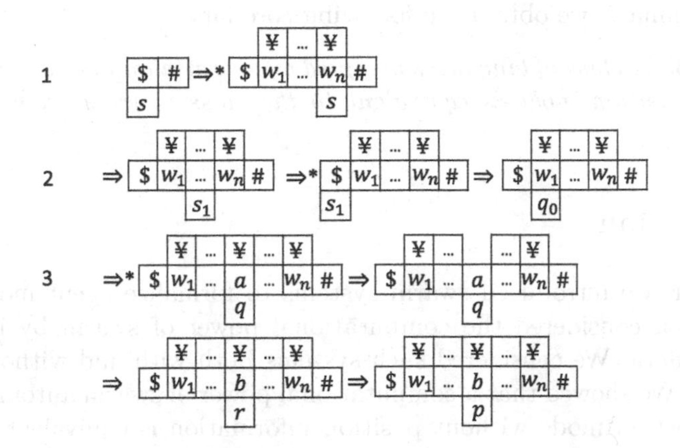

Fig. 6. Configuration diagram of Π to simulate M with an input $w_1 \ldots w_n$. Π generates agents $w_1, \ldots, w_n$, and ¥, then simulates a transition rule $r : \delta(q, a) \ni (p, b, j)$ with $j = 0$. To avoid overlapping, agents in the same column have the same position in $\mathbb{N}$.

We note that there is no possibility that undesired transition rule is applied. As for a configuration derived by using transition rules in Category 1, as long as there exists the agent s in the configuration, there is no $q \in Q$ nor $r \in Lab(\delta)$ in the configuration, then no transition rule in Category 3 is applied.

After applying the rules in Category 1, the agent s is transformed into s_1 which changes to the agent q_0 by the transition rules in Category 2. In this process of Category 2, as long as there exists the agent s_1, no transition rule in Category 1 or 3 is applied. Once a configuration includes an agent in Q, we can apply transition rules related to the transition function δ. As long as there exists an agent in $Q \cup Lab(\delta)$ in a configuration in Category 2 or 3, no transition rule in Category 1 is applied.

In the process of Category 3, there exists at most one agent in $Q \cup Lab(\delta)$ in a configuration, which is implied from the definition of transition rules in Category 3. Now we suppose that there is an agent q in Q and a in Γ at the position ρ. For the transition function $r : \delta(q, a) \ni (p, b, j)$, there is $(q, \{a\}) \xrightarrow{\epsilon} \{r, b, c\}\langle 0 \rangle$ in R. Even if there is an agent ¥ on ρ, after applying the transition rule $(q, \{¥\}) \xrightarrow{\epsilon} \{q\}\langle 0 \rangle$, the agent q can be transformed into $\{r, b, ¥\}$ by checking the existence of the agent a. Then the agent r is uniquely transformed into p by the transition rule $(r, \sigma) \xrightarrow{\epsilon} \{p\}\langle j \rangle$. In this process, all agents located other than ρ are unchanged by the rule $(q, \{¥\}) \xrightarrow{\epsilon} \{q\}\langle 0 \rangle$. This is because the nearest agent of the agent in Γ is not an agent in Q but ¥.

Thus, only transition rule which simulates the transition function δ is applied in a configuration sequence in Π. From the definition of F', a configuration τ_n such that $\tau_0 \overset{w_1 \dots w_n}{\Longrightarrow}_{pp} \tau_n$ satisfies $agt(\tau_n) \in F^{\#}$ if and only if M enters a final state for the input $w_1 \dots w_n$. Therefore, $L(M) = L_{pp}(\Pi)$ holds. □

From Lemma 7, we obtain the following corollary.

Corollary 3. *A class of languages accepted by a swarm automaton with position in the pp-transition mode is equivalent to the class of recursively enumerable languages.*

5 Conclusion

In this paper, we introduced swarm systems to formalize agent movements in a swarm, then considered the computational power of swarm by introducing swarm automata. We considered such systems both with and without position information. We showed that a computational power of swarm automaton in the parallel transition mode without position information is equivalent to the one of finite state automaton. In contrast to without position information, a swarm automaton in the parallel transition mode with position information is shown to be equivalent to Turing machine.

From the definition, a swarm system with position information is a kind of generalized cellular automaton. On the other hand, a variety of computing models using multisets has been investigated such as P system, P automaton, reaction system, and reaction automaton [5,6]. Our future work is to explore the relationship between swarm computing models and those models.

References

1. Csuhaj-Varjú, E., Martín-Vide, C., Mitrana, V.: Multiset automata. In: Calude, C.S., Păun, G., Rozenberg, G., Salomaa, A. (eds.) WMC 2000. LNCS, vol. 2235, pp. 69–83. Springer, Heidelberg (2001). https://doi.org/10.1007/3-540-45523-X_4
2. von Mammen, S., Jacob, C.: Evolutionary swarm design of architectural idea models. In: Proceedings of the 10th Annual Conference on Genetic and Evolutionary Computation, GECCO 2008, pp. 143–150. ACM, New York (2008)
3. von Mammen, S., Phillips, D., Davison, T., Jacob, C.: A graph-based developmental swarm representation and algorithm. In: Dorigo, M., et al. (eds.) ANTS 2010. LNCS, vol. 6234, pp. 1–12. Springer, Heidelberg (2010). https://doi.org/10.1007/978-3-642-15461-4_1
4. von Neumann, J.: Theory of Self-reproducing Automata. University of Illionois Press, Champain (1966)
5. Okubo, F., Kobayashi, S., Yokomori, T.: Reaction automata. Theor. Comput. Sci. **429**, 247–257 (2012)
6. Paun, G., Rozenberg, G., Salomaa, A.: The Oxford Handbook of Membrane Computing. Oxford University Press, Inc., New York (2010)

7. Puaun, G., Rozenberg, G., Salomaa, A.: DNA Computing - New Computing Paradigms. Springer, Heidelberg (1998). https://doi.org/10.1007/978-3-662-03563-4
8. Reynolds, C.W.: Flocks, herds and schools: a distributed behavioral model. SIGGRAPH Comput. Graph. **21**(4), 25–34 (1987)
9. Rozenberg, G., Salomaa, A. (eds.): Handbook of Formal Languages. Springer, Heidelberg (1997)

DNA Origami Words and Rewriting Systems

James Garrett, Nataša Jonoska, Hwee Kim(✉), and Masahico Saito

Department of Mathematics and Statistics, University of South Florida,
4202 E. Fowler Ave., Tampa, FL 33620, USA
{jgarrett1,jonoska,hweekim}@mail.usf.edu, saito@usf.edu

Abstract. We classify rectangular DNA origami structures according to their scaffold and staples organization by associating a graphical representation to each scaffold folding. Inspired by well studied Temperley-Lieb algebra, we identify basic modules that form the structures. The graphical description is obtained by 'gluing' basic modules one on top of the other. To each module we associate a symbol such that gluing of molecules corresponds to concatenating the associated symbols. Every word corresponds to a graphical representation of a DNA origami structure. A set of rewriting rules defines equivalent words that correspond to the same graphical structure. We propose two different types of basic module structures and corresponding rewriting rules. For each type, we provide the number of all possible structures through the number of equivalence classes of words. We also give a polynomial time algorithm that computes the shortest word for each equivalence class.

1 Introduction

Self-assembly is a process where smaller components (usually molecules) autonomously assemble to form a larger structure. This process is essential in building biomolecular structures and high order polymers [16]. Applications of self-assembly range from electric circuits at nano level [1,5] to smart drug delivery systems [10,15]. A well-known self-assembly variant is the DNA origami system introduced by Rothemund [12]. In DNA origami, a single-stranded DNA plasmid, called the *scaffold*, outlines a shape, while short DNA strands, called *staples*, connect different parts of the scaffold, fixing the terminal rigid structure. The left side of Fig. 1 shows a segment of schematic DNA origami where the scaffold is depicted by a black line while staples are represented by colored lines with arrows. Experimental results of several DNA origami shapes from Rothemund's original paper [12] are shown to the right of Fig. 1.

Theoretical approaches to analyze DNA origami have been mainly focused on efficient sequence design of staples as well as synthetic scaffolds that fold into the target shape [11,14]. However, the same outlined shape can be obtained by different scaffold and staple organizations. In this paper, we use graphical description to describe different scaffold/staple organization within the same

I. McQuillan and S. Seki (Eds.): UCNC 2019, LNCS 11493, pp. 94–107, 2019.
https://doi.org/10.1007/978-3-030-19311-9_9

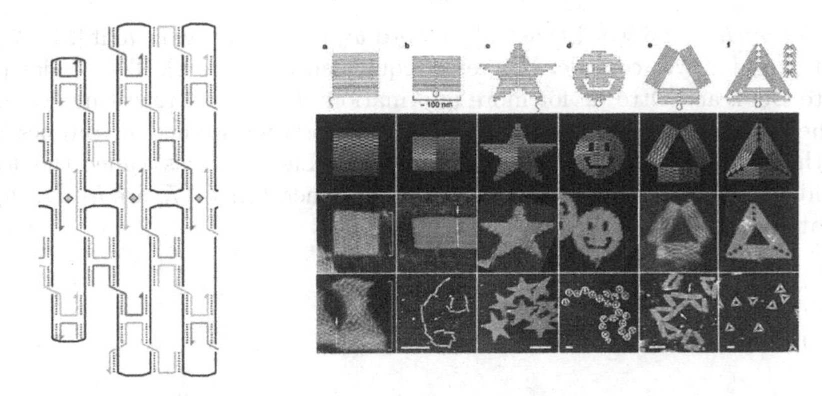

Fig. 1. (Left) A schematic representation of a DNA origami structure. The scaffold is a black line and staples are colored lines with arrows. (Right) Various shapes made by DNA origami. Both figures are from Rothemund [12]. (Color figure online)

origami shape. We identify unit building blocks (modules) for the graphical representations whose composition (one on top of another) through connecting the corresponding staple/scaffold strands builds up a larger structure. The unit blocks correspond to symbols in an alphabet, and concatenation of symbols corresponds to composition of the modules. We observe that the unit structures within DNA origami resemble the diagram representation of the generators of the Jones monoid, a monoid variant of the well studied Temperley-Lieb algebras [3,7,8]. We assign symbols to the unit structures, and define rewriting rules that provide equivalence of words corresponding to their graphical representation equivalence. We propose two types of basic module structures and their corresponding rewriting rules. For each type, we provide the number of distinct equivalence classes of words, which corresponds to the possible DNA origami structures. We also compute the size of the shortest word within each class.

2 Preliminaries

An alphabet Σ is a non-empty finite set of symbols. A word $w = w_1 w_2 \cdots w_n \in \Sigma^n$ is a finite sequence of n symbols over Σ, and $|w| = n$ denotes the size of the word. We use ϵ to denote the empty word. A *subword*, or a *factor*, of a word $w = w_1 w_2 \cdots w_n$ is $w' = w_i \cdots w_j$ where $1 \leq i \leq j \leq n$. We use Σ^* to denote the set of all words over Σ. Concatenation of two words x and y is denoted by $x \cdot y$, or simply xy.

A word rewriting system (Σ, R) consists of an alphabet Σ and a set $R \subseteq \Sigma^* \times \Sigma^*$ of rewriting rules. In this paper, both Σ and R are finite. An element (x, y) of R is called a rewriting rule, and is written as $x \to y$. In general, we rewrite uxv as uyv for $u, v \in \Sigma^*$ if $(x, y) \in R$, and denote such rewriting by $uxv \to uyv$. For a sequence of words $u = x_1 \to x_2 \to \cdots \to x_n = v$ in a rewriting system (Σ, R), we write $u \to_* v$. We consider the equivalence closure of $\to_*$, and denote an *equivalence class* of a word w as $[w]$. A word $w_0 \in [w]$ is *irreducible* if $|w_0| \leq |w'|$ for all $w' \in [w]$. We consider the set of equivalence classes $\mathcal{O}$. The reader may refer to Book and Otto [2] for more information about word rewriting systems.

The Temperley-Lieb algebra has been extensively studied in physics and knot theory [8]. A monoid version of Temperley-Lieb algebras, called the Jones monoid $\mathcal{J}_n$, has also been studied [3,7,9]. The generators of $\mathcal{J}_n$ are $h_1, \ldots, h_{n-1}$ and satisfy three classes of relations:

1. $h_i h_j h_i = h_i$ for $|i - j| = 1$
2. $h_i h_i = h_i$
3. $h_i h_j = h_j h_i$ for $|i - j| \geq 2$

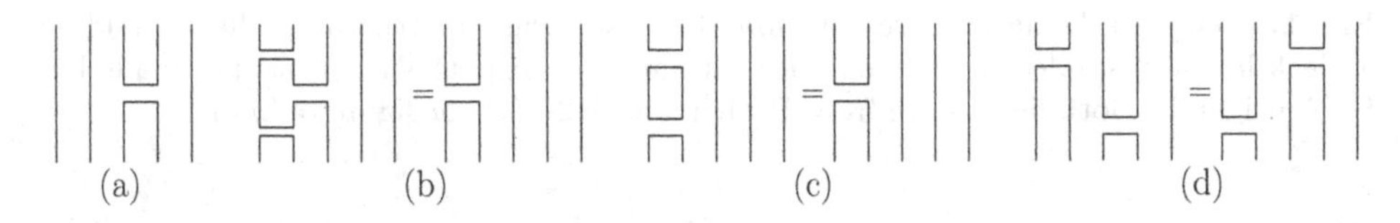

Fig. 2. Graphical representation of the Jones monoid $\mathcal{J}_5$. (a) The generator h_3. (b) The relation $h_1 h_2 h_1 = h_1$. (c) The relation $h_1 h_1 = h_1$. (d) The relation $h_1 h_3 = h_3 h_1$.

The generators and relations can be represented graphically as in Fig. 2 [9]. There are n endpoints at the top and the bottom of graphical representations of elements of $\mathcal{J}_n$. A generator h_i connects the ith and $i{+}1$st top endpoints and the ith and $i{+}1$st bottom endpoints, while other endpoints are connected by vertical lines. The generator h_3 in $\mathcal{J}_5$ is presented in Fig. 2 (a), connecting the top 3rd and 4th and the bottom 3rd and 4th points, respectively. Multiplication of two elements corresponds to concatenation of diagrams, placing the diagram of the first element on top of the second, and removing closed loops. The relations 1, 2 and 3 can also be expressed graphically as in Fig. 2 (b)–(d), respectively. Two elements in the Jones monoid are equal if their graphical representations are equivalent, that is, they have the same set of top-bottom connecting segments after deleting internal loops. For any two words that have equivalent diagrams, one word can be rewritten to the other using the sequence of relations 1 to 3. In simplification of the DNA origami structure, we take a similar approach where we only take into account the endpoints of scaffolds and staples that are visible at the top and the bottom borderline of the whole structure. Thus, we use the Jones monoid as a base to construct DNA origami words and corresponding rewriting systems.

3 DNA Origami Words and Rewriting Systems

3.1 DNA Origami Words

We focus on rectangular DNA origami structures. They can be formed by a variety of scaffold-strand folds and connecting staples. We introduce an algebraic way to distinguish these different folds yielding the same overall shape. We use basic unit structures (modules) that build the shape and associate symbols (generators) to these basic modules. Based on graphical diagrams, and inspired by the Jones monoid diagrams, we define equivalence of two origami structures, and define corresponding rewriting rules that realize the equivalence in the graphical diagrams.

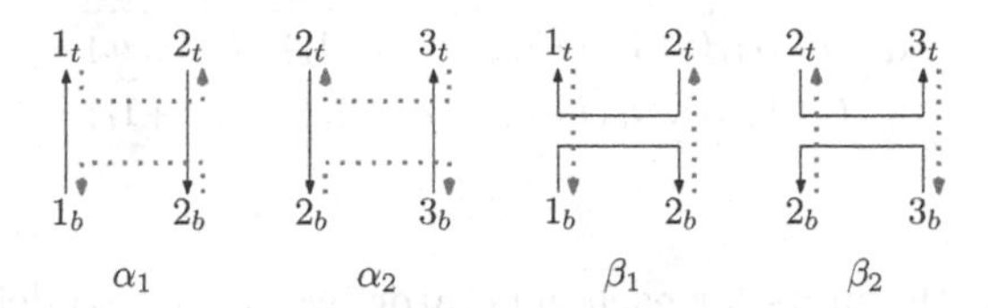

Fig. 3. Graphical representation of units of α_i and β_i ($i = 1, 2$). Scaffolds are represented by black lines and staples are represented by red dotted lines. For better visibility, staples are shifted right. (Color figure online)

In this schematics of the DNA origami structure, we consider columns made of scaffolds, and staples that go along the scaffolds as follows: There are places where two adjacent scaffolds connect the two columns, and also places where two adjacent staples connect the two columns. In addition, because DNA is oriented, the scaffolds and staples have directions: adjacent scaffolds are anti-parallel, and a staple is anti-parallel to a scaffold it connects to. A graphical structure corresponding to DNA origami is thus presented with types of directed segments and the corresponding end-point connections. In addition, in order to define composition of structures when some parts of the structures are missing, we consider 'virtual' staples and scaffold. We use $p = i_t$ (i_b) to represent a point at the top (bottom) of the ith column. We assume that scaffolds at the ith column go downward if i is odd, and upward if i is even. A *graphical structure* is a tuple $(\mathcal{R}_{sca}, \mathcal{V}_{sca}, \mathcal{R}_{sta}, \mathcal{V}_{sta})$ of sets of ordered pairs (p, q) of points $p, q \in \{i_x \mid 1 \leq i \leq n, x = \text{t or b}\}$. The set $\mathcal{R}_{sca}$ ($\mathcal{R}_{sta}$) contains ordered pairs (p, q) of points, each pair representing a *real* scaffold (staple) starting from p and ending at q, respectively. The set $\mathcal{V}_{sca}$ ($\mathcal{V}_{sta}$) contains ordered pairs that represent *virtual* scaffolds (staples). Namely, in order to define concatenation of structures, for columns without scaffolds (staples), we assume that there exist straight scaffolds (staples) in the columns of lacking scaffolds (staples) which are not visible, for convenience of definition of concatenation. For an ordered pair (p, q) of endpoints, we define the *reversal* pair as (q, p).

Given n as the width of the structure, we define basic modules and corresponding generators $\Sigma_n = \{\alpha_i, \beta_i \mid 1 \leq i \leq n-1\}$ as an alphabet for DNA origami words with the order $\alpha_1 < \cdots < \alpha_{n-1} < \beta_1 < \cdots < \beta_{n-1}$. We say that α_i is *complementary* to β_i, and vice versa. For each generator α_i, β_i, Table 1 shows the set of pairs of scaffolds and staples that describe their structures between the ith and the i+1st columns. The four pairs that describe α_i (resp. β_i) are called *units* for α_i (resp. β_i). The units of the generators α_i and β_i ($i = 1, 2$) are shown in Fig. 3.

Table 1. Units for generators of odd i's (pairs are reversed for even i's).

	$\mathcal{R}_{sca}$	$\mathcal{R}_{sta}$
α_i	$(i_b, i_t), (i{+}1_t, i{+}1_b)$	$(i_t, i{+}1_t), (i{+}1_b, i_b)$
β_i	$(i{+}1_t, i_t), (i_b, i{+}1_b)$	$(i_t, i_b), (i{+}1_b, i{+}1_t)$

In addition to the units for each generator, we must also define pairs corresponding to the columns left of the ith and right of the i+1st column. Unlike the Jones monoid, the choice of structure for these surrounding columns is not trivial. We define three possible systems for this surrounding structure through choices of real or virtual scaffolds and staples. The structure of each generator $\gamma_i \in \Sigma_n$ has a *context* $\mathcal{C}(\gamma_i)$ which consists of pairs (k_t, k_b) and their reverses where $k \notin \{i, i+1\}$. The context $\mathcal{C}(\gamma_i)$ can have real or virtual pairs. Depending on the choice of virtual versus real context, there can be different structural descriptions. Table 2 defines three situations that can be used for three different descriptions of graphical structures $\mathcal{G}_{max(n)}$, $\mathcal{G}_{sta(n)}$, $\mathcal{G}_{min(n)}$, each representing a possible choice for the generator γ_i. We note that in $\mathcal{G}_{max(n)}$ the context $\mathcal{C}(\gamma_i)$ has both $\mathcal{V}_{sca}$ and $\mathcal{V}_{sta}$ empty, i.e. the whole context is real. In $\mathcal{G}_{mid(n)}$, the context $\mathcal{C}(\gamma_i)$ has $\mathcal{V}_{sca} = \emptyset$, that is, the scaffold context is real but the staple context is virtual. In the case of $\mathcal{G}_{min(n)}$, the whole context $\mathcal{C}(\gamma_i)$ is virtual. The corresponding graphical structures of α_2's in different $\mathcal{G}$'s are shown in Fig. 4.

Table 2. Definition of three real and virtual scaffold and staple contexts for γ_i when i is odd. The pairs are reversed for even i's.

	$k \notin \{i, i+1\}$	$\mathcal{R}_{sca}$	$\mathcal{V}_{sca}$	$\mathcal{R}_{sta}$	$\mathcal{V}_{sta}$
$\mathcal{G}_{max(n)}$	odd k	(k_b, k_t)		(k_t, k_b)	
	even k	(k_t, k_b)		(k_b, k_t)	
$\mathcal{G}_{mid(n)}$	odd k	(k_b, k_t)			(k_t, k_b)
	even k	(k_t, k_b)			(k_b, k_t)
$\mathcal{G}_{min(n)}$	odd k		(k_b, k_t)		(k_t, k_b)
	even k		(k_t, k_b)		(k_b, k_t)

Concatenation of words and the corresponding graphical structure is defined similarly as in the Jones monoid diagrams. Graphical structures that correspond to words in Σ_n^* are obtained by joining graphical structures of generators as explained below. The graphical structure corresponding to concatenation of two words is obtained by placing the graphical structure of the first word on top of the graphical structure of the second and connect the vertical lines that meet. In the case of virtual staples or scaffolds the connection follows the rule: If a real scaffold (staple) meets a virtual scaffold (staple), then the virtual scaffold (staple) becomes real. This process simulates the real structure extending through the empty space represented by the virtual structure.

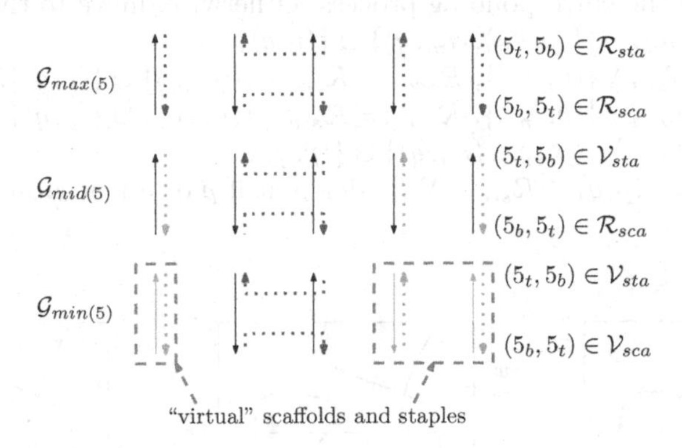

Fig. 4. Different graphical structures of α_2's in $\mathcal{G}_{max(5)}$, $\mathcal{G}_{mid(5)}$ and $\mathcal{G}_{min(5)}$. Virtual scaffolds and staples are colored in gray.

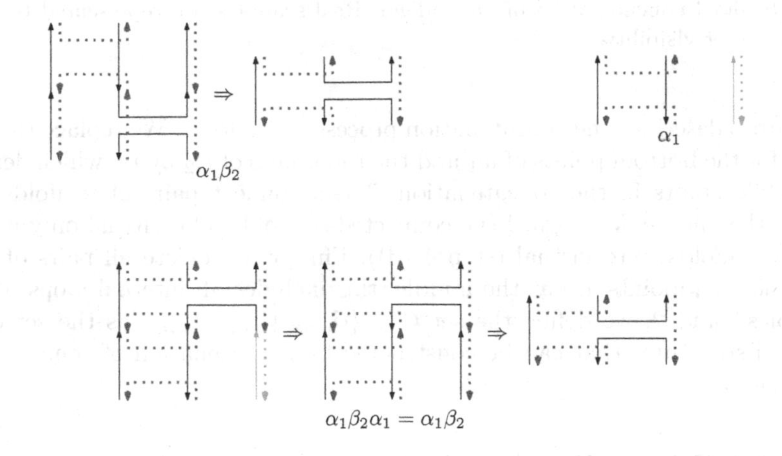

Fig. 5. Concatenation of $\alpha_1\beta_2$ and α_1 under $\mathcal{G}_{min(3)}$

Figure 5 shows concatenation of $\alpha_1\beta_2$ and α_1 under $\mathcal{G}_{min(3)}$. Formally, the graphical structure of a word $w = w_1w_2$ is defined as follows: Suppose $G(w_1) = (\mathcal{R}_{sca1}, \mathcal{V}_{sca1}, \mathcal{R}_{sta1}, \mathcal{V}_{sta1})$ and $G(w_2) = (\mathcal{R}_{sca2}, \mathcal{V}_{sca2}, \mathcal{R}_{sta2}, \mathcal{V}_{sta2})$ are graphical structures of w_1 and w_2, respectively. The graphical structure $G(w) = G(w_1w_2) = (\mathcal{R}_{sca}, \mathcal{V}_{sca}, \mathcal{R}_{sta}, \mathcal{V}_{sta})$ is obtained with the following: The scaffold sets ($\mathcal{R}_{sca}$ and $\mathcal{V}_{sca}$) are obtained with the procedure (the staples follow an equivalent procedure):

1. For all ordered pairs in $\mathcal{R}_{sca1}$ and $\mathcal{V}_{sca1}$, replace the subscript b by m.
2. For all ordered pairs in $\mathcal{R}_{sca2}$ and $\mathcal{V}_{sca2}$, replace the subscript t by m.
3. Let $\mathcal{R}_{sca} = \mathcal{R}_{sca1} \cup \mathcal{R}_{sca2}$ and $\mathcal{V}_{sca} = \mathcal{V}_{sca1} \cup \mathcal{V}_{sca2}$.
4. (Connecting scaffolds) While possible, find one of the following pairs of scaffolds and do the corresponding process. Otherwise, move to the next step.
   (a) $\mathcal{R}_{sca} := \mathcal{R}_{sca} \setminus \{(p, i_m), (i_m, q)\} \cup \{(p, q)\}$.
   (b) $\mathcal{V}_{sca} := \mathcal{V}_{sca} \setminus \{(i_m, q)\}$, $\mathcal{R}_{sca} := \mathcal{R}_{sca} \setminus \{(p, i_m)\} \cup \{(p, q)\}$.
   (c) $\mathcal{V}_{sca} := \mathcal{V}_{sca} \setminus \{(p, i_m)\}$, $\mathcal{R}_{sca} := \mathcal{R}_{sca} \setminus \{(i_m, q)\} \cup \{(p, q)\}$.
   (d) $\mathcal{V}_{sca} := \mathcal{V}_{sca} \setminus \{(p, i_m), (i_m, q)\} \cup \{(p, q)\}$.
5. For every pair $(p, q) \in \mathcal{R}_{sca} \cup \mathcal{V}_{sca}$, delete it if p or q has subscript m.

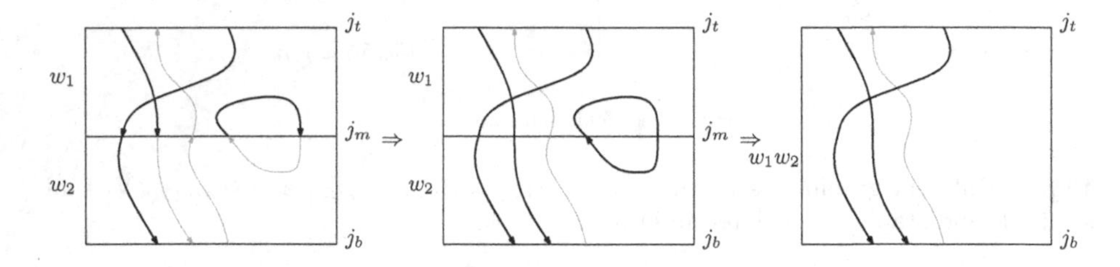

Fig. 6. Scaffold concatenation of w_1 and w_2. Real scaffolds are represented by thick lines for better visibility.

Figure 6 describes the concatenation process of scaffolds. We replace the subscripts for the bottom points of w_1 and the top points of w_2 by m, which denotes the middle points in the concatenation. Then, connect pairs of scaffolds that meet in the middle. We regard the connected scaffold to be virtual only if both original scaffolds were virtual (step 4 (d)). Finally, we delete all pairs of scaffolds whose endpoints are at the middle, this includes all internal loops. Based on Tables 1 and 2, we define the set $\mathcal{G} \in \{\mathcal{G}_{max}, \mathcal{G}_{mid}, \mathcal{G}_{min}\}$ as the set of all graphical structures that can be constructed by concatenation of generators in the model $\mathcal{G}$.

3.2 DNA Origami Rewriting Systems

It is straightforward that for any alphabet, given two words, the graphical structure of their concatenation is unique. Thus, if $G(w_1) = G(w_2)$ under a model $\mathcal{G}$, we say $w_1 \sim w_2$ in $\mathcal{G}$ and define a rewriting rule $w_1 \leftrightarrow w_2$, in both directions, between two equivalent words. Due to difference of context structures, rewriting rules for $\mathcal{G}_{max(n)}$, $\mathcal{G}_{mid(n)}$ and $\mathcal{G}_{min(n)}$ differ from each other. For each structure, we find the set of basic rewriting rules that generate the equivalence and analyze the set of distinct equivalence classes.

$\mathcal{G}_{max(n)}$ Case. We first observe that all staples and scaffolds in $\mathcal{G}_{max(n)}$ are real. We observe that except for added directions, and their concatenation results in a bijection between scaffolds and staples without lasting conflict of the directions. Moreover, scaffolds in α_i (and staples in β_i) are straight and do not affect the structure of scaffolds (staples) when concatenated. For convenience, we use γ and δ to represent an arbitrary generator, and $\overline{\gamma}$ to denote the complementary generator of γ. By concatenating generators we obtain the following rules that form the set $R_{max(n)}$ (Fig. 7 shows the inter-commutation rule):

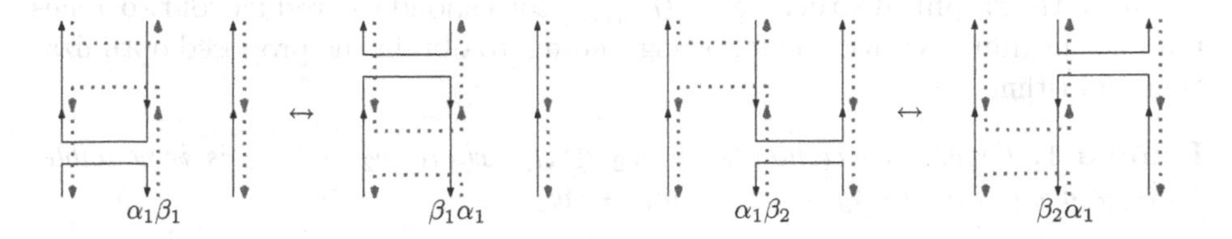

Fig. 7. Inter-commutation rewriting rule for $\mathcal{G}_{max(3)}$

1. (inter-commutation rule) $\gamma_i\overline{\gamma_j} \leftrightarrow \overline{\gamma_j}\gamma_i$
2. (idempotency rule) $\gamma_i\gamma_i \leftrightarrow \gamma_i$
3. (intra-commutation rule) $\gamma_i\gamma_j \leftrightarrow \gamma_j\gamma_i$ for $|i-j| \geq 2$
4. (TL relation rule) $\gamma_i\gamma_j\gamma_i \leftrightarrow \gamma_i$ for $|i-j| = 1$

Based on the set $R_{max(n)}$ we define the set $\mathcal{O}_{max(n)}$ of equivalence classes, i.e. $\mathcal{O}_{max(n)} = \Sigma^* \setminus R_{max(n)}$. We say that a rule is *non-increasing* if the resulting word is not longer than the original word. In the above rules, rules 1, 3 and right directions of rules 2 and 4 form the set of non-increasing rules. A sequence of rewriting rules with only non-increasing rules is said to be a non-increasing rewriting.

Let $\Sigma_{(\alpha)n} = \{\alpha_1, \ldots, \alpha_{n-1}\}$ and $\Sigma_{(\beta)n} = \{\beta_1, \ldots, \beta_{n-1}\}$. Using the inter-commutation rule, we may rewrite any word w to $w_a w_b$, where $w_a \in \Sigma^*_{(\alpha)n}$ and $w_b \in \Sigma^*_{(\beta)n}$. We say that such a word $w_a w_b$ is in an inter-commutation-free form. Also, using intra-commutation rules, we may set additional conditions for $w_a = (\alpha_{i_1}\alpha_{i_1-1}\cdots\alpha_{k_1})(\alpha_{i_2}\alpha_{i_2-1}\cdots\alpha_{k_2})\cdots(\alpha_{i_p}\alpha_{i_p-1}\cdots\alpha_{k_p})$ where i_p is the maximum subscript in w_a, $i_{j+1} > i_j$ and $k_{j+1} > k_j$ for $1 \leq k < p$, and a similar condition for w_b. Such w_a and w_b are unique [7], and we call such $w_a w_b$ a commutation-free form of w.

We regard the graphical structure of a word as pairs of scaffolds and staples, which can be seen as two independent Jones monoid diagrams. Knowing that the relations 1 to 3 of the Jones monoid can sufficiently describe equivalence of diagrams, we have the following theorem:

Theorem 1. *For all $w_1, w_2 \in \Sigma^*_n$, $G(w_1) = G(w_2)$ under $\mathcal{G}_{max(n)}$ if and only if $w_1 \rightarrow_* w_2$ using $R_{max(n)}$. In other words, there exists a bijection between $\mathcal{G}_{max(n)}$ and $\mathcal{O}_{max(n)}$.*

Given n, the number of elements of $\mathcal{J}_n$ is equal to the Catalan number $C_n = \frac{1}{n+1}\binom{2n}{n}$ [9], and the maximum size of an element in $\mathcal{J}_n$ is $\left\lfloor \frac{n^2}{4} \right\rfloor$ [4,7]. Thus, the following remark holds.

Remark 1. Given n, $|\mathcal{O}_{max(n)}| = \left(\frac{1}{n+1}\binom{2n}{n}\right)^2$, and the maximum size of a minimum irreducible word in $\mathcal{O}_{max(n)}$ is $2\left\lfloor \frac{n^2}{4} \right\rfloor$.

Since the graphical structures in $\mathcal{G}_{max(n)}$ correspond to products of two Jones monoid diagrams, we use the following Lemma to obtain the proposed optimization algorithm.

Lemma 1. *Given two elements $w_1, w_2 \in \mathcal{J}_n$ where $w_2 \in [w_1]$ is irreducible, there is a non-increasing rewriting $w_1 \to_* w_2$.*

Theorem 2. *Given a word $w_0 \in \Sigma_n^*$ of size m, an irreducible word in $[w_0]$ can be obtained within $O(nm^2)$ time.*

$\mathcal{G}_{mid(n)}$ Case. Similar to the $\mathcal{G}_{max(n)}$ case, we have the following rewriting rules:

1. (Inter-commutation) $\gamma_i\overline{\gamma_j} \leftrightarrow \overline{\gamma_j}\gamma_i$
2. (Idempotency) $\gamma_i\gamma_i \leftrightarrow \gamma_i$
3. (Intra-commutation) $\gamma_i\gamma_j \leftrightarrow \gamma_j\gamma_i$ for $|i-j| \geq 2$

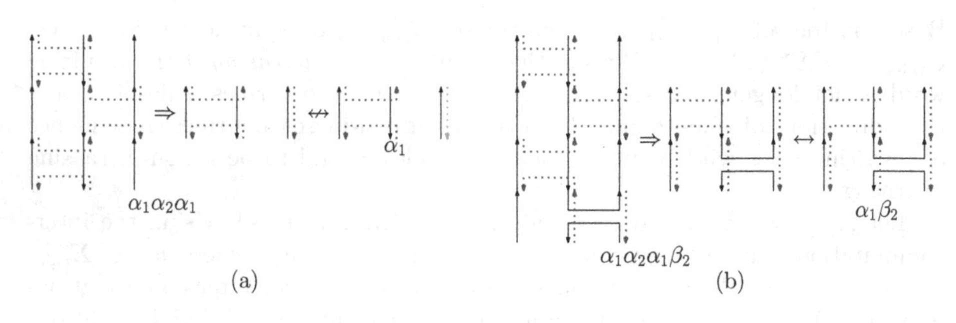

Fig. 8. Examples of equivalence in $\mathcal{G}_{mid(3)}$. (a) $\alpha_1\alpha_2\alpha_1 \nsim \alpha_1$ (b) $\alpha_1\alpha_2\alpha_1\beta_2 \sim \alpha_1\beta_2$

Due to the lack of default real staples in generators, we cannot directly introduce the rewriting rule $\gamma_i\gamma_j\gamma_i \leftrightarrow \gamma_i$ for $|i-j| = 1$. For example, we cannot rewrite $\alpha_1\alpha_2\alpha_1$ as α_1, since $\alpha_1\alpha_2\alpha_1$ has a straight real staple at the third column while α_1 does not (see Fig. 8 (a)). We introduce the *span* of a word w as the set of columns $span(w) = \bigcup_{\gamma_j \text{ in } w} \{j, j+1\}$.

Lemma 2. *Under $\mathcal{G}_{mid(n)}$, the span equals to the set of columns where real staples exist.*

Directly from Lemma 2, it follows that two equivalent words have the same span, and rewriting rules in the Jones monoid can be applied when both sides have the same span. For example, $\alpha_1\alpha_2\alpha_1\beta_2 \leftrightarrow \alpha_1\beta_2$ holds as in Fig. 8 (b) because β_2 adds 3 to $span(\alpha_1\beta_2)$ and $span(\alpha_1\alpha_2\alpha_1\beta_2) = span(\alpha_1\beta_2)$. In general, we have the following additional rewriting rules, where $\delta \in \{\alpha, \beta\}$ and $v \in \Sigma_n^*$:

4*. $\delta_j v \gamma_i \gamma_{i-1} \gamma_i \leftrightarrow \delta_j v \gamma_i$ if $j = i-1$ or $i-2$
5*. $\delta_j v \gamma_i \gamma_{i+1} \gamma_i \leftrightarrow \delta_j v \gamma_i$ if $j = i+1$ or $i+2$
6*. $\gamma_i \gamma_{i-1} \gamma_i v \delta_j \leftrightarrow \gamma_i v \delta_j$ if $j = i-1$ or $i-2$
7*. $\gamma_i \gamma_{i+1} \gamma_i v \delta_j \leftrightarrow \gamma_i v \delta_j$ if $j = i+1$ or $i+2$

The symbol δ_j in the right hand side of rules 4* to 7* ensures that the span of the two words is the same. There is a finite set B of words, such that rules 4* to 7* for any $v \in \Sigma^*$ can be derived using rules 1 to 3 and the subset of rules 4* to 7* using only words $v \in B$. In order to determine the subset B, we define a *zig-zag word* $w \in \Sigma_n^*$ to be a word where each pair of adjacent generators in w have adjacent indices. We call a maximal subword of increasing (decreasing) indices as zig (zag). For example, $w = \alpha_3\alpha_4\alpha_3\alpha_2\alpha_1\alpha_2$ is a zig-zag word with a zig-zag-zig sequence in the word: $\alpha_3\alpha_4$ being the zig, $\alpha_4\alpha_3\alpha_2\alpha_1$ being the zag, and $\alpha_1\alpha_2$ being the zig. Using rules 2 to 7*, we can rewrite any zig-zag word that consists of single generators type either in $\Sigma_{(\alpha)n}$, or $\Sigma_{(\beta)n}$, as a zig-zag word with at most three zigs or zags, which we call the *zig-zag normal form.*

Theorem 3. *For $w_1 = \delta_j v \gamma_i \gamma_{i-1} \gamma_i$ and $w_2 = \delta_j v \gamma_i$ where $j = i-1$ or $i-2$, we can rewrite w_1 as w_2 using rules 1 to 3 and the following rule: $\gamma_j v \gamma_i \gamma_{i-1} \gamma_i \leftrightarrow \gamma_j v \gamma_i$ where $v \in \Sigma_{(\gamma)n}^*$ and $v = \epsilon$ or $\gamma_j v \gamma_i$ is in the zig-zag normal form.*

Similar simplification works for rules 4* to 7*. From Theorem 3, $B \subseteq \Sigma_n^*$ is the set of all words in zig-zag normal form. Thus the rules 4* to 7* reduce to the following rewriting rules for $v \in \Sigma_{(\gamma)n}^*$ such that either $v = \epsilon$ or $\gamma_j v \gamma_i \in B$ in rule 4* and 5* (or $\gamma_i v \gamma_j \in B$ in rule 6* and 7*):

4. $\gamma_j v \gamma_i \gamma_{i-1} \gamma_i \leftrightarrow \gamma_j v \gamma_i$ if $j = i-1$ or $i-2$
5. $\gamma_j v \gamma_i \gamma_{i+1} \gamma_i \leftrightarrow \gamma_j v \gamma_i$ if $j = i+1$ or $i+2$
6. $\gamma_i \gamma_{i-1} \gamma_i v \gamma_j \leftrightarrow \gamma_i v \gamma_j$ if $j = i-1$ or $i-2$
7. $\gamma_i \gamma_{i+1} \gamma_i v \gamma_j \leftrightarrow \gamma_i v \gamma_j$ if $j = i+1$ or $i+2$

For given n, let the set $R_{mid(n)}$ of rewriting rules consist of the above seven kinds of rules for $1 \leq i, j \leq n$. As observed, the set $R_{mid(n)}$ is finite because Σ_n and B are finite. Then $\mathcal{O}_{mid(n)} = \Sigma_n^* \setminus R_{mid(n)}$.

Theorem 4. *For all $w_1, w_2 \in \Sigma_n^*$, $G(w_1) = G(w_2)$ under $\mathcal{G}_{mid(n)}$ if and only if $w_1 \rightarrow_* w_2$ using $R_{mid(n)}$. In other words, there exists a bijection between $\mathcal{G}_{mid(n)}$ and $\mathcal{O}_{mid(n)}$.*

We compute the number of equivalence classes of words in $\mathcal{O}_{mid(n)}$. We use a binary string of length n to represent the graphical structures in $\mathcal{G}_{mid(n)}$ such that the ith bit equals 1 if and only if the ith staple and the ith scaffold is a straight line. Each binary string is uniquely determined with a tuple $(a_1, b_1, \ldots, a_k, b_k)$ where a_i (b_i) represents the number of ith consecutive 0's (1's). For example, the 8-bit binary string 00111000 corresponds to a tuple $(2, 3, 3, 0)$. In particular, the bit 0 corresponds to $(1, 0)$ and the bit 1 to $(0, 1)$. Let $\mathbb{N}^0 = \mathbb{N} \cup \{0\}$. The set of tuples corresponding to binary strings of length n is denoted $T_n = \{p = (a_1, b_1, \ldots, a_k, b_k) \mid k \geq 1, \text{ for all i}, a_i, b_i \in \mathbb{N}^0 \text{ and } \sum_{i=1}^{k}(a_i + b_i) = n\}$. Note that T_n is the set of partitions of n in $2k$ summands.

Theorem 5. *Given* $n \in \mathbb{N}^0$, *for each tuple* $p \in \bigcup_{1 \leq i \leq n} T_i$, *let* $D(p) \in \mathbb{N}^0$ *be recursively defined as follows:*

- *for* $p \in T_0$, $D(0, 0) = 1$.
- *for* $p \in T_1$, $D(p) = 1$ *if* $p = (0, 1)$ *and* $D(p) = 0$ *if* $p = (1, 0)$.
- *for* $p = (a_1, b_1, \ldots, a_k, b_k) \in T_n$, $(n > 0)$ *we have* $D(p) = \prod_{i=1}^{k} D(a_i, 0)$.
- *for* $n > 1$, *we have* $D(0, n) = 1$ *and*

$$D(n, 0) = \left(\frac{1}{n+1}\binom{2n}{n}\right)^2 - \sum_{p \in T_n \setminus \{(n,0)\}} D(p).$$

Then, $|\mathcal{O}_{mid(n)}|$ *is given as*

$$|\mathcal{O}_{mid(n)}| = d(n) = \sum_{p \in T_n} [D(p) \times x(p)] - n$$

where

$$x(a_1, b_1, \ldots, a_k, b_k) = \begin{cases} (b_1 + 1) & \text{if } k = 1, \\ (b_1 + 1) \cdot \prod_{i=2}^{k-1} \left(\frac{b_i(b_i + 1)}{2} + 1\right) \cdot (b_k + 1) & \text{if } k \neq 1, a_1 = 0, \\ \prod_{i=1}^{k-1} \left(\frac{b_i(b_i + 1)}{2} + 1\right) \cdot (b_k + 1) & \text{if } k \neq 1, a_1 > 0. \end{cases}$$

For each graphical structure in $\mathcal{G}_{max(n)}$, we may assign a unique binary number of size n, where the ith digit is 1 if the ith column has both straight vertical scaffolds and staples, and 0 otherwise. Then, $D(p)$ is the number of graphical structures in $\mathcal{G}_{max(n)}$ whose assigned binary number corresponds to p. In particular, the sum of all $D(p)$ for $p \in T_n$ is equal to the square of the nth Catalan number by Remark 1. For each p, the term $x(p)$ gives the number of possible combinations of virtual straight staples within the segment of the graphical representation that consist of only vertical straight lines. The sequence $d(n)$ for $1 \leq n \leq 10$ is

$$1, 4, 31, 253, 2247, 21817, 227326, 2499598, 28660639, 339816259.$$

It is not listed in the OEIS [13] list of sequences, and the non-recursive formula of $d(n)$ is still open.

Theorem 6. *An upper bound of the size $u(n)$ of a minimum irreducible word in $\mathcal{O}_{mid(n)}$ is given by $2^n - 2$.*

The bound in Theorem 6 is not tight, and the exact size of a maximum irreducible word is open.

Theorem 7. *Given a word $w_0 \in [w_0] \in \mathcal{O}_{mid(n)}$ of size m, we can find an irreducible word of $[w_0]$ within $O(nm^2)$ time.*

The proofs of Lemma 1 and Theorem 2 work similarly for Theorem 7. We first rewrite w_0 as w in the inter-commutation-free form. Then we repeatedly find one of the following conditions in w if possible and rewrite w accordingly:

1. If $w = v_1\gamma_i v_2\gamma_i v_3$ where $v_1, v_2, v_3 \in \Sigma_n^*$ and v_2 does not have γ_{i+1}, γ_i and γ_{i-1}, then rewrite w as $v_1v_2\gamma_i v_3$.
2. If $w = v_1\gamma_i v_2\gamma_{i+1}v_3\gamma_i v_4$ where $v_1, v_2, v_3, v_4 \in \Sigma_n^*$, v_2, v_3 do not have γ_{i+1}, γ_i and γ_{i-1}, there exists δ_{i+1} in v_1 or v_4, or δ_{i+2} in v_1, v_2, v_3 or v_4, then rewrite w as $v_1v_2\gamma_i v_3 v_4$.
3. If $w = v_1\gamma_i v_2\gamma_{i-1}v_3\gamma_i v_4$ where $v_1, v_2, v_3, v_4 \in \Sigma_n^*$, v_2, v_3 do not have γ_{i+1}, γ_i and γ_{i-1}, there exists δ_{i-1} in v_1 or v_4, or δ_{i-2} in v_1, v_2, v_3 or v_4, then rewrite w as $v_1v_2\gamma_i v_3 v_4$.

3.3 Concluding Remarks

We have proposed modules and corresponding generators for DNA origami structures, defined concatenation of words and rewriting rules, and analyzed equivalence classes based on graphical equivalence. One model that we have not discussed is $\mathcal{G}_{min(n)}$. For $\mathcal{G}_{min(n)}$, seven types of rewriting rules for $\mathcal{G}_{mid(n)}$ hold. Moreover, we may prove that Theorem 4 holds for $\mathcal{G}_{min(n)}$ using the similar proof. It turns out that there is bijection between $\mathcal{G}_{mid(n)}$ and $\mathcal{G}_{min(n)}$, and $\mathcal{O}_{mid(n)} = \mathcal{O}_{min(n)}$.

Graphical structures corresponding to generators α_i's and β_i's in Fig. 3 describe crossing of scaffolds and staples in DNA origami well, while using only two types of generators. Here we explore possible further development of generators that are more plausible to DNA origami.

The first observation on the current generators is that they are vertically and horizontally symmetric (without directions), which causes the graphical structure to always have a cup-shaped fragment of a real scaffold at the top as in Fig. 9 (a). DNA origami does not have such fragments at the border of the structure, which leads us to revise generators to define such borders. Figure 9 (b) proposes four different generators that are used to substitute α_1. In these generators, we introduce asymmetric structures that can be used to construct borders of the structure. We may define generators for β similarly. Under the assumption that we use the same concatenation procedure, for a graphical structure that corresponds to α_1, we can make an arbitrary number of scaffolds and staples virtual by concatenation of four new generators as in Fig. 9 (c). Now, suppose we define the rewriting system based on equivalence under such generators. For each pair of diagrams of $\mathcal{J}_n$, we have $2n$ staples and scaffolds which can become virtual. From analysis similar to the proof of Theorem 5, the size of the set of equivalence classes becomes $\left(\left(\frac{1}{n+1}\binom{2n}{n}\right)^2 - 1\right) \cdot 2^{2n} + 1$. The set of rewriting rules that are sufficient to describe equivalence under such generators is open.

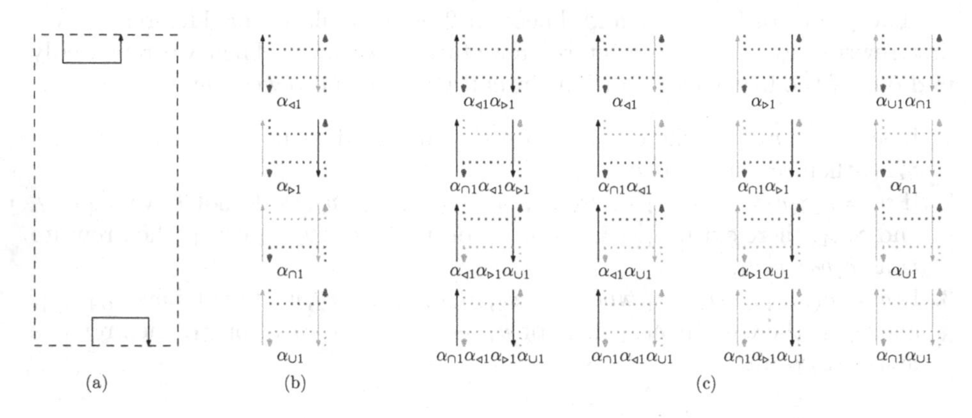

Fig. 9. (a) In a graphical structure generated from α_i's and β_i's, there always exists a cup-shaped fragment of a real scaffold at the top (and cap-shaped fragment at the bottom). (b) Four revised generators that substitutes α_1. Virtual scaffolds and staples are colored in gray. (c) We can make arbitrary staples and scaffolds in α_1 virtual.

The second observation on the current generators is that we do not consider which side of the scaffold the staple is on. In the DNA origami structure, staples can be on the left or the right of the scaffold, and these two cases are distinguished. Moreover, for two adjacent staples at the opposite side of the same scaffold, they either disconnect or connect by crossing the scaffold. To model this observation, we may introduce revised graphical structures for α and β as in Fig. 10 (a). Staples are either at the left or the right of the scaffold, and some staple ends are extending which can be connected to other staples regardless of the side. We assume that two adjacent staples can be connected except when two are non-extending ends and at the opposite side. This additional condition for staple connection changes some of the commutation rewriting rules—for example, $\alpha_1\beta_1 \not\leftrightarrow \beta_1\alpha_1$ as in Fig. 10 (b). Algebraic analysis on relations based on such generators is done by Garrett et al. [6]. The set of rewriting rules that are sufficient to describe equivalence under such generators is still open.

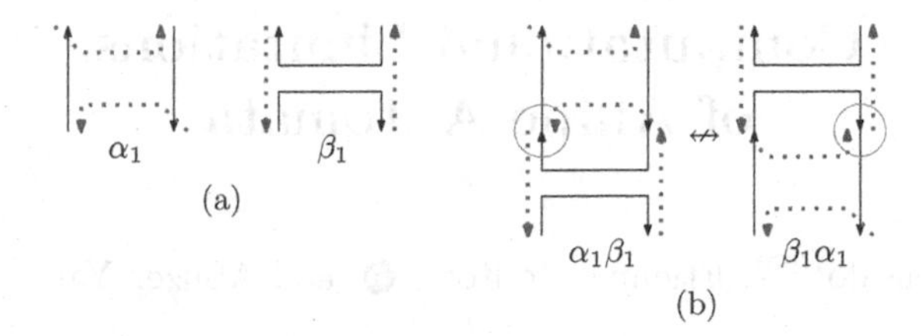

Fig. 10. (a) Revised generators α_1 and β_1. Two diagonal ends of staples in α_1 represent extending staple-ends. (b) Adjacent staples can be connected except when two are non-extending ends and at the opposite side.

Acknowledgment. This work is partially supported by NIH R01GM109459, and by NSF's CCF-1526485, DMS-1800443 and DMS-1764366.

References

1. Bhuvana, T., Smith, K.C., Fisher, T.S., Kulkarni, G.U.: Self-assembled CNT circuits with ohmic contacts using Pd hexadecanethiolate as in situ solder. Nanoscale **1**(2), 271–275 (2009)
2. Book, R.V., Otto, F.: String-Rewriting Systems. Springer, New York (1993). https://doi.org/10.1007/978-1-4613-9771-7
3. Borisavljević, M., Došen, K., Petric, Z.: Kauffman monoids. J. Knot Theor. Ramifications **11**(2), 127–143 (2002)
4. Dolinka, I., East, J.: The idempotent-generated subsemigroup of the Kauffman monoid. Glasgow Math. J. **59**(3), 673–683 (2017)
5. Eichen, Y., Braun, E., Sivan, U., Ben-Yoseph, G.: Self-assembly of nanoelectronic components and circuits using biological templates. Acta Polym. **49**(10–11), 663–670 (1998)
6. Garrett, J., Jonoska, N., Kim, H., Saito, M.: Algebraic systems for DNA origami motivated from Temperley-Lieb algebras. CoRR, abs/1901.09120 (2019)
7. Jones, V.F.R.: Index for subfactors. Inventiones Math. **72**, 1–25 (1983)
8. Kauffman, L.H.: Knots and Physics. World Scientific, New York (2001)
9. Lau, K.W., FitzGerald, D.G.: Ideal structure of the Kauffman and related monoids. Commun. Algebra **34**(7), 2617–2629 (2006)
10. Li, J., Fan, C., Pei, H., Shi, J., Huang, Q.: Smart drug delivery nanocarriers with self-assembled DNA nanostructures. Adv. Mater. **25**(32), 4386–4396 (2013)
11. Rothemund, P.W.K.: Design of DNA origami. In: Proceedings of 2005 International Conference on Computer-Aided Design, pp. 471–478 (2005)
12. Rothemund, P.W.K.: Folding DNA to create nanoscale shapes and patterns. Nature **440**(7082), 297–302 (2006)
13. The on-line encyclopedia of integer sequences. https://oeis.org/
14. Veneziano, R., et al.: Designer nanoscale DNA assemblies programmed from the top down. Science **352**(6293), 1534 (2016)
15. Verma, G., Hassan, P.A.: Self assembled materials: design strategies and drug delivery perspectives. Phys. Chem. Chem. Phys. **15**(40), 17016–17028 (2013)
16. Whitesides, G.M., Boncheva, M.: Beyond molecules: self-assembly of mesoscopic and macroscopic components. Proc. Nat. Acad. Sci. U.S.A. **99**(8), 4769–4774 (2002)

Computational Limitations of Affine Automata

Mika Hirvensalo[1](✉), Etienne Moutot[1,2], and Abuzer Yakaryılmaz[3]

[1] Department of Mathematics and Statistics, University of Turku, 20014 Turku, Finland
mikhirve@utu.fi

[2] LIP, ENS de Lyon – CNRS – UCBL – Université de Lyon, École Normale Supérieure de Lyon, Lyon, France
etienne.moutot@ens-lyon.org

[3] Center for Quantum Computer Science, Faculty of Computing, University of Latvia, Rīga, Latvia
abuzer@lu.lv

Abstract. We present two new results on the computational limitations of affine automata. First, we show that the computation of bounded-error rational-valued affine automata is simulated in logarithmic space. Second, we give an impossibility result for algebraic-valued affine automata. As a result, we identify some unary languages (in logarithmic space) that are not recognized by algebraic-valued affine automata with cutpoints.

1 Introduction

Finite automata are an interesting model to study since they express the very natural limitation of finite memory. They are also good computational models, since they are simpler than many others machines like pushdown automata or Turing machines. Due to this simplicity, there exists many different models of finite automata, all trying to express different computational settings. Deterministic [14], probabilistic [12] and quantum [2] finite automata (DFAs, PFAs, and QFAs, respectively) have been studied to try to understand better the computational limitations inherent to all these cases.

Recently, Díaz-Caro and Yakaryılmaz introduced a new model, called *affine computation* [3]. As a non-physical model, the goal of affine computation is to investigate the power of interference caused by negative amplitudes in the computation, like in the quantum case. But unlike QFAs, affine finite automata (AfAs) have unbounded state set and the final operation corresponding to quantum measurement cannot be interpreted as linear. The final operation in AfAs is analogous to renormalization in Kondacs-Watrous [9] or Latvian [1] quantum automata models.

AfAs and their certain generalizations have been investigated in a series of works [3,6,7,18]. In most of the cases, affine models (e.g., bounded-error and unbounded-error AfAs, zero-error affine OBDDs, zero-error affine counter

I. McQuillan and S. Seki (Eds.): UCNC 2019, LNCS 11493, pp. 108–121, 2019.
https://doi.org/10.1007/978-3-030-19311-9_10

automata, etc.) have been shown more powerful than their classical or quantum counterparts. On the other hand, we still do not know too much regarding the computational limitations of AfAs. Towards this direction, we present two new results. First, we show that the computation of bounded-error rational-valued affine automata is simulated in logarithmic space, and so we answer positively one of the open problems in [3]. Second, we give an impossibility result for algebraic-valued AfAs, and, as a result, we identify some unary languages (in logarithmic space) that are not recognized by algebraic-valued AfAs with cutpoints.

2 Preliminaries

For a given word w, w_i represents its i-th letter. For any given class C, $\mathsf{C}_{\mathbb{Q}}$ and $\mathsf{C}_{\mathbb{A}}$ denotes the classes defined by the machines restricted to have rational-valued and algebraic-valued components, respectively. The logarithmic and polynomial space classes are denoted as L and PSPACE, respectively. We assume that the reader is familiar with the basics of automata theory.

2.1 Models

As a *probability distribution* (also known as a *stochastic vector*) we understand a (column) vector with nonnegative entries summing up to one, and a *stochastic matrix* (also known as a *Markov matrix*) here stands for a square matrix whose all columns are probability distributions.

Definition 1 (PFA). *A k-state probabilistic finite automaton (PFA) P over alphabet Σ is a triplet*

$$P = (\boldsymbol{x}, \{M_i \mid i \in \Sigma\}, \boldsymbol{y})$$

where $\boldsymbol{x} \in \mathbb{R}^k$ is a stochastic vector called initial distribution, *each $M_i \in \mathbb{R}^{k\times k}$ is a stochastic matrix, and $\boldsymbol{y} \in \{0,1\}^k$ is the final vector (each 1 in $\boldsymbol{y}$ represents an accepting state).*

For any input word $w \in \Sigma^*$ with length n, P has a probability distribution of states as follows: $M_w\boldsymbol{x} = M_{w_n} \cdots M_{w_1}\boldsymbol{x}$. The *accepting probability* corresponds to the probability of P being in an accepting state after reading w, which is given by

$$f_P(w) = \boldsymbol{y}^T M_{w^R}\boldsymbol{x}. \tag{1}$$

Affine finite automaton (AfA) is a generalization of PFA allowing negative transition values. Only allowing negative values in the transition matrices does not add any power (generalized PFAs are equivalent to PFAs, see [16]), but affine automata introduce also a non-linear behaviour. The automaton acts like a generalized probabilistic automaton until the last operation, which is a non-linear operation called a *weighting operation.*

Definition 2. *A vector $\boldsymbol{v} \in \mathbb{R}^k$ is an affine vector if and only if its coordinates sums up to* 1. *A matrix M is an affine matrix if and only if all its columns are affine vectors.*

The following property is straightforward to verify, and it will ensure that affine automata are well defined.

Property 1. *If M and N are affine matrices, then MN is also an affine matrix. In particular, if $\boldsymbol{v}$ is an affine vector, then $M\boldsymbol{v}$ is also an affine vector.*

Definition 3 (AfA). *A k-state* AfA *A over alphabet Σ is a triplet*

$$A = (\boldsymbol{x}, \{M_i \mid i \in \Sigma\}, F)$$

where $\boldsymbol{x}$ is an initial affine vector, each M_i is an affine transition matrix, and $F = \mathrm{diag}(\delta_1, \ldots, \delta_n)$ is the final projection matrix, where each $\delta_i \in \{0,1\}$ for $1 \le i \le n$.

The value computed by an affine automaton can be most conveniently defined via the following notion:

Definition 4. *Notation $|\boldsymbol{v}| = \sum_i |v_i|$ stands for the usual L^1 norm.*

Now, the final value of the affine automaton A of Definition 3 is

$$f_A(w) = \frac{|FM_w\boldsymbol{v}_0|}{|M_w\boldsymbol{v}_0|}. \tag{2}$$

Clearly $f_A(w) \in [0,1]$ for any input word $w \in \Sigma^*$.

Remark 1. Notice that the final value for PFAs (1) is defined as matrix product $\boldsymbol{v}_f \mapsto \boldsymbol{y}^T\boldsymbol{v}_f$, which is a linear operation on $\boldsymbol{v}_f$. On the other hand, computing final value from $\boldsymbol{v}_f$ as in (2) involves nonlinear operations $\boldsymbol{v}_f \mapsto \dfrac{|F\boldsymbol{v}_f|}{|\boldsymbol{v}_f|}$ such as L^1-norm and normalization (division).

2.2 Cutpoint Languages

Given a function $f : \Sigma^* \to [0,1]$ computed by an automaton (stochastic or affine), there are different ways of defining the language of recognized by this automaton.

Definition 5 (Cutpoint languages). *A language $L \subseteq \Sigma^*$ is recognized by an automaton A with cutpoint $\lambda \in [0,1)$ if and only if*

$$L = \{w \in \Sigma^* \mid f_A(w) > \lambda\}.$$

These languages are called cutpoint languages. In the case of probabilistic (resp., affine automata), the set of cut-point languages are called stochastic languages *(resp.,* affine languages*) and denoted by* SL *(resp.,* AfL*).*

We remark that fixing the cutpoint in the interval $(0, 1)$ does not change the classes SL and AfL [3,12].

Definition 6 (Exclusive cutpoint languages). *A language $L \subseteq \Sigma^*$ is recognized by an automaton A with exclusive cutpoint $\lambda \in [0, 1]$ if and only if*

$$L = \{w \in \Sigma^* \mid f_A(w) \neq \lambda\}.$$

These languages are called exclusive cutpoint languages. In the case of probabilistic (resp., affine automata), the set of exclusive cut-point languages are called exclusive stochastic languages *(resp.,* exclusive affine languages*) and denoted by* $\mathsf{SL}^{\neq}$ *(resp.,* $\mathsf{AfL}^{\neq}$*). The complement of* $\mathsf{SL}^{\neq}$ *(resp.,* $\mathsf{AfL}^{\neq}$*) is* $\mathsf{SL}^{=}$ *(resp.,* $\mathsf{AfL}^{=}$*).*

Again, we remark that fixing the cutpoint in the interval $(0, 1)$ does not change the classes $\mathsf{SL}^{\neq}$, $\mathsf{SL}^{=}$, $\mathsf{AfL}^{\neq}$, and $\mathsf{AfL}^{=}$ [3,11,12].

A stronger condition is to impose that accepted and rejected words are separated by a gap: the cutpoint is said to be isolated.

Definition 7 (Isolated cutpoint or bounded error). *A language L is recognized by an automaton A with* isolated cutpoint *λ if and only if there exist $\delta > 0$ such that $\forall w \in L, f_A(w) \geq \lambda + \delta$, and $\forall w \notin L, f_A(w) \leq \lambda - \delta$. The set of languages recognized with* bounded error *(or isolated cutpoint) affine automata is denoted by* BAfL.

A classical result by Rabin [13] shows that isolated cutpoint stochastic languages are regular. Rabin's proof essentially relies on two facts: (1) the function mapping the final vector into $[0, 1]$ is a contraction, and (2) the state vector set is bounded. By modifying Rabin's proof, it is possible to show that also many quantum variants of stochastic automata obey the same principle [2]: bounded-error property implies the regularity of the accepted languages. In fact, Jeandel generalized Rabin's proof by demonstrating that the compactness of the state vector set together with the continuity of the final function are sufficient to guarantee the regularity of the accepted language if the cutpoint is isolated [8].

3 Logarithmic Simulation

Macarie [10] proved that $\mathsf{SL}^{=}_{\mathbb{Q}} \subseteq \mathsf{L}$ and $\mathsf{SL}_{\mathbb{Q}} \subseteq \mathsf{L}$. That is, the computation of any rational-valued probabilistic automaton can be simulated by an algorithm using only logarithmic space. However, this logarithmic simulation cannot be directly generalized for rational-valued affine automata due to the non-linearity of their last operation. In order to understand why, we will first reproduce the proof.

Before that, let us introduce the most important space-saving technique:

Definition 8. *Notation $(b \mod c)$ stands for the least nonnegative integer a satisfying $a \equiv b \pmod c$. If $\boldsymbol{x} = (x_1, \ldots, x_r)$ and $\boldsymbol{n} = (n_1, \ldots, n_r) \in \mathbb{Z}^r$, we define $\boldsymbol{x} \pmod{\boldsymbol{n}} = ((x_1 \bmod n_1), \ldots, (x_r \bmod n_r))$. Analogously, for any matrix $A \in \mathbb{Z}^{k \times k}$, we define $(A \pmod n)_{ij} = (A_{ij} \mod n)$.*

The problem of recovering x from the residue representation $((x \bmod n_1), \ldots, (x \bmod n_r))$ is practically resolved by the following well-known theorem.

Theorem 2 (The Chinese Remainder Theorem). *Let $n_1, \ldots, n_r$ be pairwise coprime integers, $a_1, \ldots, a_r$ be arbitrary integers, and $N = n_1 \cdots n_r$. Then there exists an integer x such that*

$$x \equiv a_1 \pmod{n_1}, \ldots, x \equiv a_r \pmod{n_r}, \tag{3}$$

and any two integers x_1 and x_2 satisfying (3) satisfy also $x_1 \equiv x_2 \pmod{N}$.

Remark 2. The above remarks and the Chinese Remainder Theorem imply that the integer ring operations $(+, \cdot)$ can be implemented using the residue representation, and that the integers can be uncovered from the residue representations provided that (1) $\boldsymbol{n} = (n_1, \ldots, n_r)$ consists of pairwise coprime integers and (2) the integers stay in interval of length $N - 1$, where $N = n_1 \cdots n_r$.

Remark 3. In order to ensure that $\boldsymbol{n} = (n_1, \ldots, n_r)$ consists of pairwise coprime integers, we select numbers n_i from the set of prime numbers. For the reasons that will become obvious later, we will however omit the first prime 2.

Definition 9. $\boldsymbol{p}_r$ *is an r-tuple $\boldsymbol{p}_r = (3, 5, 7, \ldots, p_r)$ consisting of r first primes by excluding 2. For this selection, a consequence of the prime number theorem is that, asymptotically, $P_r = 3 \cdot 5 \cdot 7 \cdot \cdots \cdot p_r = \frac{1}{2}e^{(1+o(1))r \ln r}$.*

Theorem 3 (Macarie [10]). $\mathsf{SL}_{\mathbb{Q}}^{=} \subseteq \mathsf{L}$.

Proof. For a given alphabet Σ, let $L \in \Sigma^*$ be a language in $\mathsf{SL}_{\mathbb{Q}}^{=}$ and $P = (\boldsymbol{x}, \{M_i \mid i \in \Sigma\}, \boldsymbol{y})$ be a k-state rational-valued PFA over Σ such that

$$L = \left\{ w \in \Sigma^* \mid f_P(w) = \frac{1}{2} \right\}.$$

We remind that, for any input word $w = w_1 \cdots w_n \in \Sigma^*$, we have

$$f_P(w) = \boldsymbol{y}^T M_{w_n} \cdots M_{w_1} \boldsymbol{x}. \tag{4}$$

Since each $M_i \in \mathbb{Q}^{k \times k}$, there exists a number $D \in \mathbb{N}$ providing that each matrix $M_i' = DM_i \in \mathbb{Z}^{k \times k}$, and (4) can be rewritten as

$$f_P(w) = \frac{1}{D^n} \underbrace{\boldsymbol{y}^T M_{w_n}' \ldots M_{w_1}' \boldsymbol{x}}_{f_{P'}(w)},$$

and the language L can be characterized as

$$L = \{ w \in \Sigma^* \mid 2 f_{P'}(w) = D^n \}. \tag{5}$$

Since the original matrices M_i are stochastic, meaning that their entries are in $[0, 1]$, it follows that each matrix $M_i' = DM_i$ has integer entries in $[0, D]$.

Moreover, $f_P(w) \in [0,1]$ implies that $f_{P'}(w) \in [0, D^n]$ for every input word $w \in \Sigma^n$. As now $f_{P'}(w)$ can be computed by multiplying $k \times k$ integer matrices, the residue representation will serve as a space-saving technique.

We will fix r later, but the description of the algorithm is as follows: For each entry p of $\boldsymbol{p}_r = (3, 5, 7, \ldots, p_r)$, we let $M_i^{(p)} = M_i' \mod p$, and compute

$$(2f_{P'}(w) \mod p) = \boldsymbol{y}^T M_{w_n}^{(p)} \cdots M_{w_1}^{(p)} \boldsymbol{x} \tag{6}$$

as all the products are computed modulo p, $k^2 \log p$ bits are needed to compute (6). Likewise, $(D^n \bmod p)$ can be computed in space $O(\log p)$ for each coordinate p of $\boldsymbol{p}_r$. The comparison $2f_{P'}(w) \equiv D^n \pmod p$ can hence done in $O(\log p)$ space.

Reusing the space, the comparison can be made sequentially for each coordinate of $\boldsymbol{p}_r$, and if any comparison gives a negative outcome, we can conclude that $2P'(w) \neq D^n$.

To conclude the proof, it remains to fix r so that both $2f_{P'}(w)$ and D^n are smaller than $P_r = 3 \cdot 5 \cdot 7 \cdot \cdots \cdot p_r$. If no congruence test is negative, then the Chinese Remainder Theorem ensures that $2f_{P'}(w) = D^n$. Since $2f_{P'}(w) \leq D^n$, we need to select r so that $\frac{1}{2}e^{(1+o(1))r \ln r} > 2D^n$, which is equivalent to $\log \frac{1}{2} + (1 + o(1))r \ln r > \log 2 + n \log D$. This inequality is clearly satisfied with $r = n$ for large enough n, and for each $n \geq 1$ by choosing $r = c \cdot n$, where c is a positive constant (depending on D).

As a final remark let us note that $p_{\lfloor cn \rfloor}$, the $\lfloor cn \rfloor$-th prime, can be generated in logarithmic space and the prime number theorem implies that $O(\log n)$ bits are enough to present $p_{\lfloor cn \rfloor}$, since c is a constant. □

To extend the above theorem to cover $\mathsf{SL}_{\mathbb{Q}}$ as well, auxiliary results are used.

Lemma 1 (Macarie [10]). *If N is an odd integer and x, $y \in [0, N-1]$ are also integers, then $x \geq y$ iff $x - y$ has the same parity as $((x - y) \bmod N)$.*

Proof. As x, $y \in [0, N-1]$, it follows that

$$(x - y \mod N) = \begin{cases} x - y \text{ if } x \geq y \\ N + x - y \text{ if } x < y, \end{cases}$$

which shows that the parity changes in the latter case since N is odd. □

The problem of using the above lemma is that, in modular computing, numbers x and y are usually known only by their residue representations $\mathrm{Res}_{\boldsymbol{p}_r}(x)$ and $\mathrm{Res}_{\boldsymbol{p}_r}(y)$, and it is not straightforward to compute the parity from the modular representation in logarithmic space. Macarie solved this problem not only for parity but also for a more general modulus (not necessarily equal to 2).

Lemma 2 (Claim modified from [10]). *For any integer x and modulus $\boldsymbol{p}_r = (3, 5, 7, \ldots, p_r)$, there is a deterministic algorithm that given $\mathrm{Res}_{\boldsymbol{p}_r}(x)$ and $M \in \mathbb{Z}$ as input, produces the output $x \pmod M$ in space $O(\log p_r + \log M)$.*

As a corollary of the previous lemma, Macarie presented a conclusion which implies the logarithmic space simulation of rational stochastic automata.

Lemma 3 (Claim modified from [10]). *Let* $\boldsymbol{p}_r = (3, 5, 7, \ldots, p_r)$ *and* $P_r = 3 \cdot 5 \cdot 7 \cdot \cdots \cdot p_r$. *Given the residue representations of integers* x, $y \in [0, P_r - 1]$, *the decisions* $x > y$, $x = y$ *or* $x < y$ *can be made in* $O(\log p_r)$ *space.*

Proof. The equality test can be done as in the proof Theorem 3, testing the congruence sequentially for each prime. Testing $x \geq y$ is possible by Lemmata 1 and 2: First compute $\mathrm{Res}_{\boldsymbol{p}_r}(z) = \mathrm{Res}_{\boldsymbol{p}_r}(x) - \mathrm{Res}_{\boldsymbol{p}_r}(y) \pmod{\boldsymbol{p}_r}$, then compute the parities of x, y, z using Lemma 2 with $M = 2$. □

The following theorem is a straightforward corollary from the above:

Theorem 4. $\mathsf{SL}_{\mathbb{Q}} \subseteq \mathsf{L}$.

When attempting to prove an analogous result to affine automata, there is at least one obstacle: computing the final value includes the absolute values, but the absolute value is not even a well-defined operation in the modular arithmetic. For example, $2 \equiv -3 \pmod 5$, but $|2| \not\equiv |-3| \pmod 5$. This is actually another way to point out that, in the finite fields, there is no order relation compatible with the algebraic structure.

Hence for affine automata with matrix entries of both signs, another approach must be adopted. One obvious approach is to present an integer n as a pair $(|n|, \mathrm{sgn}(n))$, and apply modular arithmetic to $|n|$. The signum function and the absolute value indeed behave smoothly with respect to the product, but not with the sum, which is a major problem with this approach, since to decide the sign of the sum requires a comparison of the absolute values, which seems impossible without having the whole residue representation. The latter, in its turn seems to cost too much space resources to fit the simulation in logarithmic space.

Hence the logspace simulation for automata with matrices having both positive and negative entries seems to need another approach. It turns out that we can use the procedure introduced by Turakainen already in 1969 [15,16].

Theorem 5. $\mathsf{AfL}_{\mathbb{Q}} \subseteq \mathsf{L}$.

Proof. For a given alphabet Σ, let $L \in \Sigma^*$ be a language in $\mathsf{AfL}_{\mathbb{Q}}$ and $A = (\boldsymbol{x}, \{M_i \mid i \in \Sigma\}, F)$ be a k-state rational-valued AfA over Σ such that

$$L = \left\{ w \in \Sigma^* \mid f_A(w) > \frac{1}{2} \right\}.$$

For each $M_i \in \mathbb{Q}^{k \times k}$, we define a new matrix as $B_i = \begin{pmatrix} 0 & \boldsymbol{0}^T & 0 \\ \boldsymbol{c}_i & M_i & \boldsymbol{0} \\ e_i & \boldsymbol{d}_i^T & 0 \end{pmatrix}$, where $\boldsymbol{c}_i$, $\boldsymbol{d}_i$, and e_i are chosen so that the column and row sums of B_i are zero. We define $\boldsymbol{x}' = \begin{pmatrix} 0 \\ \boldsymbol{x} \\ 0 \end{pmatrix}$ as the new initial state. For the projection matrix F, we define an extension $F' = \begin{pmatrix} 0 & 0 & 0 \\ 0 & F & 0 \\ 0 & 0 & 0 \end{pmatrix}$. It is straightforward to see that $|B_w \boldsymbol{v}_0'| = |M_w v_0|$ as well as $|F' B_w \boldsymbol{v}_0'| = |F M_w v_0|$.

For the next step, we introduce a $(k+2) \times (k+2)$ matrix $\mathbb{E}$, whose each element is 1. It is then clear that $\mathbb{E}^n = (k+2)^{n-1}\mathbb{E}$ and $B_i\mathbb{E} = \mathbb{E}B_i = \mathbf{0}$. Now we define

$$C_i = B_i + m\mathbb{E},$$

where $m \in \mathbb{Z}$ is selected large enough to ensure the nonnegativity of the matrix entries of each C_i. It follows that

$$C_w = B_w + m^{|w|}(k+2)^{|w|-1}\mathbb{E},$$

and

$$C_w \boldsymbol{x}' = B_w \boldsymbol{x}' + m^{|w|}(k+2)^{|w|-1}\mathbb{E}\boldsymbol{x}'.$$

Similarly,

$$F' C_w \boldsymbol{x}' = F' B_w \boldsymbol{x}' + m^{|w|}(k+2)^{|w|-1}F'\mathbb{E}\boldsymbol{x}'.$$

Now

$$\frac{|F M_w \boldsymbol{v}_0|}{|M_w \boldsymbol{v}_0|} = \frac{|F' B_w \boldsymbol{v}_0|}{|B_w \boldsymbol{v}_0|} = \frac{\left|F' C_w \boldsymbol{v}_0' - m^{|w|}(k+2)^{|w|-1}F'\mathbb{E}\boldsymbol{x}'\right|}{\left|C_w \boldsymbol{x}' - m^{|w|}(k+2)^{|w|-1}\mathbb{E}\boldsymbol{x}'\right|}$$

which can further be modified by expanding the denominators away: For an integer g large enough all matrices $D_i = gC_i$ will be integer matrices and the former equation becomes

$$\frac{|F M_w \boldsymbol{x}|}{|M_w \boldsymbol{x}|} = \frac{|F' B_w \boldsymbol{x}|}{|B_w \boldsymbol{x}|} = \frac{\left|F' D_w \boldsymbol{x}' - m^{|w|}(k+2)^{|w|-1}g^{|w|+1}F'\mathbb{E}\boldsymbol{x}'\right|}{\left|D_w \boldsymbol{x}' - m^{|w|}(k+2)^{|w|-1}g^{|w|+1}\mathbb{E}\boldsymbol{x}'\right|}. \quad (7)$$

Hence the inequality

$$\frac{|F M_w \boldsymbol{x}|}{|M_w \boldsymbol{x}|} \geq \frac{1}{2}$$

is equivalent to

$$2\left|F' D_w \boldsymbol{x}' - m^{|w|}(k+2)^{|w|-1}g^{|w|+1}F'\mathbb{E}\boldsymbol{x}'\right| \geq \left|D_w \boldsymbol{x}' - m^{|w|}(k+2)^{|w|-1}g^{|w|+1}\mathbb{E}\boldsymbol{x}'\right|. \quad (8)$$

In order to verify inequality (8) in logarithmic space, it sufficient to demonstrate that the residue representations of both sides can be obtained in logarithmic space.

For that end, the residue representation of vector $\boldsymbol{a} = F' D_w \boldsymbol{x}' \in \mathbb{R}^{k+2}$ can be obtained in logarithmic space as in the proof of Theorem 3.

Trivially, the residue representation of $\boldsymbol{b} = m^{|w|}(k+2)^{|w|-1}g^{|w|+1}F'\mathbb{E}\boldsymbol{x}' \in \mathbb{R}^{k+2}$ can be found in logarithmic space, as well. In order to compute the residue representation of

$$|\boldsymbol{a} - \boldsymbol{b}| = |\boldsymbol{a}_1 - \boldsymbol{b}_1| + \cdots + |\boldsymbol{a}_k - \boldsymbol{b}_k|$$

it is sufficient to decide whether $\boldsymbol{a}_i \geq \boldsymbol{b}_i$ holds. As the residue representations for each $\boldsymbol{a}_i$ and $\boldsymbol{b}_i$ is known, all the decisions can be made in logspace, according to Lemma 3. The same conclusion can be made for the right hand side of (8). □

4 A Non-affine Language

As we saw in the previous section, $\mathsf{AfL}_{\mathbb{Q}} \subseteq \mathsf{L}$, and hence languages beyond L, are good candidates for non-affine languages.[1] In this section, we will however demonstrate that the border of non-affinity may lie considerably lower: There are languages in L which are not affine.

In an earlier work [6], we applied the method of Turakainen [17] to show that there are languages in L which however are not contained in BAfL. Here we will extend the previous result to show that those languages are not contained even in $\mathsf{AfL}_{\mathbb{A}}$. (We leave open whether a similar technique can be applied for AfL.)

Definition 10 (Lower density). *Let $L \subseteq a^*$ be a unary language. We call* ***lower density*** *of L the limit*

$$\underline{dens}(L) = \liminf_{n\to\infty} \frac{|\{a^k \in L \mid k \leq n\}|}{n+1}.$$

Definition 11 (Uniformly distributed sequence). *Let $(\boldsymbol{x}_n)$ be a sequence of vectors in $\mathbb{R}^k$ and $I = [a_1, b_1) \times \cdots \times [a_k, b_k)$ be an interval in $\mathbb{R}^k$. We define $C(I,n)$ as $C(I,n) = |\{\boldsymbol{x}_i \mod 1 \in I \mid 1 \leq i \leq n\}|$.*

We say that $(\boldsymbol{x}_n)$ ***is uniformly distributed mod 1*** *if and only if for any I of such type,*

$$\lim_{n\to\infty} \frac{C(I,n)}{n} = (b_1 - a_1)\cdots(b_k - a_k).$$

Theorem 6. *If $L \subseteq a^*$ satisfies the following conditions:*

1. *$\underline{dens}(L) = 0$.*
2. *For all $N \in \mathbb{N}$, there exists $r \in \mathbb{N}$ and an ascending sequence $(m_i) \in \mathbb{N}$ such that $a^{r+m_iN} \subseteq L$ and for any irrational number α, the sequence $((r + m_iN)\alpha)$ is uniformly distributed mod 1.*

Then L is not in $\mathsf{AfL}_{\mathbb{A}}$.

Proof. Let's assume for contradiction that $L \in \mathsf{AfL}_{\mathbb{A}}$. Then there exists an AfA A with s states, matrix M and initial vector $\boldsymbol{v}$ such that the acceptance value of A is

$$f_A(a^n) = \frac{|PM^n\boldsymbol{v}|}{|M^n\boldsymbol{v}|}. \tag{9}$$

Without loss of generality, we can assume that the cutpoint equals to $\frac{1}{2}$, and hence $w \in L \Leftrightarrow f_A(w) > \frac{1}{2}$.

[1] It is known that $\mathsf{L} \subsetneq \mathsf{PSPACE}$, so it is plausible that PSPACE-complete languages are not in $\mathsf{AfL}_{\mathbb{Q}}$.

Using the Jordan decomposition $M = PJP^{-1}$, one has $M^n = PJ^nP^{-1}$. So the coordinates of $M^n\boldsymbol{v}$ have the form

$$(M^n\boldsymbol{v})_j = \sum_{k=1}^{s} p_{jk}(n)\lambda_k^n, \tag{10}$$

where λ_k are the eigenvalues of M and p_{jk} are polynomials of degree less than the degree of the corresponding eigenvalue. For short, we denote $F(n) = f_A(a^n)$, and let $\lambda_k = |\lambda_k|\, e^{2i\pi\theta_k}$.

When studying expression (9), we can assume without loss of generality, that all numbers θ_k are irrational. In fact, replacing matrix M with αM, where $\alpha \neq 0$ does not change (9), since

$$\frac{|P(\alpha M)^n\boldsymbol{v}|}{|(\alpha M)^n\boldsymbol{v}|} = \frac{|\alpha^n PM^n\boldsymbol{v}|}{|\alpha^n M^n\boldsymbol{v}|} = \frac{|PM^n\boldsymbol{v}|}{|M^n\boldsymbol{v}|}.$$

Selecting now $\alpha = e^{2\pi i\theta}$ (where $\theta \in \mathbb{R}$) implies that the eigenvalues of M are $\lambda_k e^{2i\pi(\theta_k+\theta)}$. The field extension $\mathbb{Q}(\theta_1, \ldots, \theta_s)$ is finite, and hence there is always an irrational number $\theta \notin \mathbb{Q}(\theta_1, \ldots, \theta_s)$. It follows directly that all numbers $\theta_k + \theta$ are irrational. Hence we can assume that all the numbers θ_k are irrational in the first place.[2]

By restricting to an arithmetic progression $n = r + mN$ ($m \in \mathbb{N}$) we can also assume that no λ_i/λ_j is a root of unity for $i \neq j$. In fact, selecting $N = \operatorname{lcm}\{\operatorname{ord}(\lambda_i/\lambda_j) \mid i \neq j \text{ and } \lambda_i/\lambda_j \text{ is a root of unity}\}$ (10) becomes

$$(M^{r+mN}\boldsymbol{v})_j = \sum_{k=1}^{s} p_{jk}(r+mN)\lambda_k^r(\lambda_k)^{Nm} = \sum_{k=1}^{s'} q_{jk}(m)\mu_k^m, \tag{11}$$

where $\{\mu_1, \ldots, \mu_{s'}\}$ are the distinct elements of set $\{\lambda_1^N, \ldots, \lambda_s^N\}$ Now for $i \neq j$ μ_i/μ_j cannot be a root of unity, since $(\mu_i/\mu_j)^t - 1$ would imply $(\lambda_{i'}/\lambda_{j'})^{Nt} = 1$, which in turn implies $(\lambda_{i'}/\lambda_{j'})^N = 1$ and hence $\mu_i = \lambda_{i'}^N = \lambda_{j'}^N = \mu_j$, which contradicts the assumption $\mu_i \neq \mu_j$.

We can now write the acceptance condition $f_A(a^n) > \frac{1}{2}$ equivalently as

$$\begin{aligned} &f_A(a^n) > \frac{1}{2} \Leftrightarrow 2\,|PM^n\boldsymbol{v}| > |M^n\boldsymbol{v}| \\ &\Leftrightarrow 2\sum_{j\in E_a} |(M^n\boldsymbol{v})_j| > \sum_{j\in E} |(M^n\boldsymbol{v})_j| \Leftrightarrow \underbrace{\sum_{j\in E_a} |(M^n\boldsymbol{v})_j| - \sum_{j\in \overline{E_a}} |(M^n\boldsymbol{v})_j|}_{g(n)} > 0, \end{aligned}$$

[2] Note that the new matrix obtained may not be affine, so it would be wrong to assume that all AfAs have to admit an equivalent one with only irrational eigenvalues. However, this does not affect this proof, since we do not require the new matrix to be affine, we only study the values that the fraction $\frac{|P(\alpha M)^n v|}{|(\alpha M)^n v|} = \frac{|PM^n v|}{|M^n v|}$ take.

Where E is the set of states of A, $E_a \subseteq E$ its set of accepting states, and $\overline{E_a}$ the complement of E_a. According to (10), $g(n) := \sum_{j\in E_a} |(M^n\boldsymbol{v})_j| - \sum_{j\in \overline{E_a}} |(M^n\boldsymbol{v})_j|$ consists of combinations of absolute values of linear combination of functions of type $n^d\lambda^n$.

We say that $n^{d_1}\lambda_1^n$ is of *larger order* than $n^{d_2}\lambda_2^n$, if $|\lambda_1| > |\lambda_2|$; and in the case $|\lambda_1| = |\lambda_2|$, if $d_1 > d_2$. If $|\lambda_1| = |\lambda_2|$, we say that $n^d\lambda_1^n$ and $n^d\lambda_2^n$ and of the same order. It is clear that if term $t_1(n)$ is of larger order than $t_2(n)$, then $\lim_{n\to\infty} \frac{t_2(n)}{t_1(n)} = 0$.

We can organize the terms in expression (10) as

$$(M^n\boldsymbol{v})_j = \sum_{k=1}^{s} p_{jk}(n)\lambda_k^n = \Lambda_j^{(N)}(n) + \Lambda_j^{(N-1)}(n) + \cdots + \Lambda_j^{(0)}(n), \tag{12}$$

where each $\Lambda_j^{(m)}(n)$ consists of terms with equal order multiplier:

$$\Lambda_j^{(m)}(n) = \sum_{k=1}^{m_j} c_{mk} n^{d_m} \lambda_{mk}{}^n = n^{d_m}\lambda_m^n \sum_{k=1}^{m_j} c_{mk} e^{2\pi i n\theta_{mk}} \tag{13}$$

(for notational simplicity, we mostly omit the dependency on j in the right hand side of (13)). Here $\lambda_m \in \mathbb{R}_+$ is the common absolute value of all eigenvalues $\lambda_{mk} = \lambda_m e^{2\pi i\theta_{mk}}$, and expression (12) is organized in descending order: $\Lambda_j^{(N)}$ is the sum of terms of the highest order multiplier, $\Lambda_j^{(N-1)}$ contains the terms of the second highest order multiplier, etc. We say that $\Lambda_j^{(k_2)}$ is lower than $\Lambda_j^{(k_1)}$ if $k_2 < k_1$.

We will then fix a representation

$$\begin{aligned} g(n) &= \sum_{j\in E_a} \left| \sum_{k=1}^{s} p_{jk}(n)\lambda_k^n \right| - \sum_{j\in\overline{E_a}} \left| \sum_{k=1}^{s} p_{jk}(n)\lambda_k^n \right| \\ &= \sum_{j\in E_a} |A_j(n) + B_j(n) + C_j(n)| - \sum_{j\in\overline{E_a}} |A_j(n) + B_j(n) + C_j(n)|, \end{aligned} \tag{14}$$

where $A_j(n) + B_j(n) + C_j(n)$ is a grouping of all Λ-terms in (12) defined as follows:

1. $A_j(n) = \sum_{k=0}^{m} \Lambda_j^{(N-k)}(n)$, where $m \in [-1, N] \cap \mathbb{Z}$ is chosen as the maximal number so that

$$A = \sum_{j\in E_a} |A_j(n)| - \sum_{j\in\overline{E_a}} |A_j(n)| \tag{15}$$

is a constant function $\mathbb{N} \to \mathbb{R}$. Such an m exists, since for $m = -1$, the sum is regarded empty and $A_j(n) = 0$, but for $m = N$, all Λ-terms are included, and then (15) becomes $f_A(a^n)$, which is not constant (otherwise condition 1 or 2 of the theorem would be false).

2. $B_j(n)$ consists a single Λ-term immediately lower than those in $A_j(n)$, and
3. $C_j(n)$ contains the rest of the Λ-terms, lower than $B_j(n)$.

Lemma 4. *If $A \neq 0$, then $\forall z \in \mathbb{C}, |A+z| = |A| + \text{Re}\,\frac{|A|}{A}z + O(\frac{z^2}{A})$.*

Proof. Denote $z = x + iy$. Because $|\text{Re}\, z| \leq |z|$, we have

$$\begin{aligned}|1+z| &= |1+x+iy| = \sqrt{(1+x)^2+y^2} = \sqrt{1+2\,\text{Re}\, z + |z|^2} \\ &= 1 + \text{Re}\, z + O(z^2).\end{aligned}$$

Now

$$|A+z| = |A|\left|1+\frac{z}{A}\right| = |A|\left(1+\text{Re}\,\frac{z}{A} + O((\frac{z}{A})^2)\right) = |A| + \text{Re}\,\frac{|A|}{A}z + O(\frac{z^2}{A}).$$

□

We choose $\lambda \in \mathbb{R}_+$ and d so that the highest Λ-term in $B(n)$ is of order $n^d\lambda^n$ and define $A_j'(n) = n^{-d}\lambda^{-n}A_j(n)$, $B_j'(n) = n^{-d}\lambda^{-n}B_j(n)$, $g'(n) = g(n)n^{-d}\lambda^{-n}$. Then clearly $g'(n) > 0$ if and only if $g(n) > 0$ and each $B_j(n)$ remains bounded as $n \to \infty$. To simplify the notations, we omit the primes and recycle the notations to have a new version of $g(n)$ of (14) where A_j-terms may tend to infinity but B_j-terms remain bounded.

Recall that we may assume (by restricting to an arithmetic progression) that no λ_i/λ_j is a root of unity. By Skolem-Mahler-Lech theorem [5], this implies that functions A_j can have only a finite number of zeros, and in the continuation we assume that n is chosen so large that no function A_j becomes zero. Furthermore, by the main theorem of [4], then $|A_j(n)| = \Omega(n^d\lambda^{n-\epsilon})$ for each $\epsilon > 0$.[3] As each B_j remains bounded, we find that B_j^2/A_j tend to zero as $n \to \infty$, and hence by Lemma 4, defining

$$\begin{aligned}&g_1(n) = \\ &\sum_{j\in E_a}\left(|A_j(n)| + \text{Re}(\frac{|A_j(n)|}{A_j(n)}B_j(n))\right) - \sum_{j\in \overline{E_a}}\left(|A_j(n)| + \text{Re}(\frac{|A_j(n)|}{A_j(n)}B_j(n))\right) \\ &= \underbrace{\sum_{j\in E_a}|A_j(n)| - \sum_{j\in \overline{E_a}}|A_j(n)|}_{h(n)} + \sum_{j\in E_a}\text{Re}(\frac{|A_j(n)|}{A_j(n)}B_j(n)) + \sum_{j\in \overline{E_a}}\text{Re}(\frac{|A_j(n)|}{A_j(n)}B_j(n))\end{aligned}$$

we have a function $g_1(n)$ with the property $g_1(n) - g(n) \to 0$ (C-terms are lower than B-terms, so they can be dropped without violating this property), when $n \to \infty$. Also by the construction it is clear that $h(n) = C \cdot n^d\lambda^n$, where C is a constant, and by the conditions of the theorem, this is possible only if $C = 0$.

[3] This is the only point we need the assumption that the matrix entries are algebraic.

Notice tat $g_1(n)$ is not a constant function by construction. Also, each B_j is a linear combination of functions of form $e^{2\pi i\theta_k n}$, each θ_k can be assumed irrational, and $||A_j(n)| / A_j(n) = 1|$, so we can conclude that $g_1(n)$ is a continuous function formed of terms of form $ce^{i\theta_k n}$ and of ratios $|A_j| / A_j$. In these terms, however the behaviour is asymptotically determined by the highest Λ-terms, so the conclusion remains even if we drop the lower terms.

By assumption, for all k, the sequence $(r + mN)\theta_k$ is uniformly distributed modulo 1. It follows that the values $e^{2i\pi(r+mN)\theta_k}$ are dense in the unit circle. If for some m, $g_1(r + mN) < 0$, then $g_1(r + Nm) \leq -\varepsilon$ for some $\epsilon > 0$. Then, because of the density argument, there are arbitrarily large values of i for which $g_1(r + m_i N) \leq 0$ contradicting condition 2 of the statement. Hence $g_1(r + mN) \geq 0$ for each m large enough. As g_1 is not a constant, there must be some m_0 so that $g_1(m_0) \geq \epsilon > 0$.

Next, let $R(x_1, \ldots, x_s)$ be a function obtained from g_1 by replacing each occurrence of $e^{i\theta_k n}$ by a variable x_k, hence each x_k will assume its value in the unit circle. Moreover, by the assumptions of the theorem, the values of x_k will be uniformly distributed in the unit circle.

Note that $g_1(n) = R((e^{2i\pi(r+m_i N)\theta_k})_{k\in A})$. Then, because the sequences $((r + m_i N)\theta_k)_i$ are uniformly distributed modulo 1, it follows that any value obtained by the function $R((e^{2i\pi y_k})_{k\in A})$ can be approximated by some $g_1(r + m_i M)$ with arbitrary precision. The function R is continuous, therefore there exists an interval $I = (x_1, y_1, ...) = ((x_k, y_k))_{k\in A}$ on which $R((x_k)) > \frac{\varepsilon}{2}$. So, if m_i is large enough and satisfies

$$((r + m_i N)\theta_1 \mod 1, \ldots) = ((r + m_i M)\theta_k \mod 1)_{k\in A} \in I,$$

then $g_1(r + m_i N) > \frac{\varepsilon}{2}$, which implies $f_A(r + m_i N) > 0$ and hence $a^{r+m_i N} \in L$. Now we just have to prove that the sequence $(r + m_i N)$ is "dense enough" to have $\underline{dens}(L) > 0$, contradicting again condition 1.
Then, because of uniform distribution imposed by condition 2, one has

$$d = \lim_{i\to\infty} \frac{C(I, r + mN)}{r + mN} = \prod_{k\in A} (y_k - x_k)$$

And so for i large enough, $\frac{C(I,r+m_i N)}{r+m_i N} \geq \frac{d}{2}$, with $a^{h+n_i Q} \in L$, implying $\underline{dens}(L) > 0$, a contradiction. □

Corollary 1. *Let P be any polynomial with nonnegative coefficients and* $\deg(P) > 2$. *The language* $\{a^{P(n)} \mid n \in \mathbb{N}\}$ *is not in* $\mathsf{AfL}_{\mathbb{A}}$.

Corollary 2. *The language* $\{a^p \mid p \text{ prime}\}$ *is not in* $\mathsf{AfL}_{\mathbb{A}}$.

Proof (Proof of Corollary 1 *and Corollary* 2*).* Turakainen proved that these two languages satisfies the two conditions of Theorem 6 [17]. Therefore, these two languages not in $\mathsf{AfL}_{\mathbb{A}}$. □

Acknowledgments. Yakaryılmaz was partially supported by Akadēmiskā personāla atjaunotne un kompetenču pilnveide Latvijaspilnveide Latvijas Universitātē līg Nr. 8.2.2.0/18/A/010 LU reģistrācijas Nr. ESS2018/289 and ERC Advanced Grant MQC. Hirvensalo was partially supported by the Väisälä Foundation and Moutot by ANR project CoCoGro (ANR-16-CE40-0005).

References

1. Ambainis, A., Beaudry, M., Golovkins, M., Ķikusts, A., Mercer, M., Thérien, D.: Algebraic results on quantum automata. Theory Comput. Syst. **39**(1), 165–188 (2006)
2. Ambainis, A., Yakaryılmaz, A.: Automata and quantum computing. CoRR, abs/1507.0:1–32 (2015)
3. Díaz-Caro, A., Yakaryılmaz, A.: Affine computation and affine automaton. In: Kulikov, A.S., Woeginger, G.J. (eds.) CSR 2016. LNCS, vol. 9691, pp. 146–160. Springer, Cham (2016). https://doi.org/10.1007/978-3-319-34171-2_11
4. Evertse, J.-H.: On sums of S-units and linear recurrences. Compositio Math. **53**(2), 225–244 (1984)
5. Hansel, G.: A simple proof of the Skolem-Mahler-Lech theorem. Theoret. Comput. Sci. **43**(1), 91–98 (1986)
6. Hirvensalo, M., Moutot, E., Yakaryılmaz, A.: On the computational power of affine automata. In: Drewes, F., Martín-Vide, C., Truthe, B. (eds.) LATA 2017. LNCS, vol. 10168, pp. 405–417. Springer, Cham (2017). https://doi.org/10.1007/978-3-319-53733-7_30
7. Ibrahimov, R., Khadiev, K., Prūsis, K., Yakaryılmaz, A.: Error-free affine, unitary, and probabilistic OBDDs. In: Konstantinidis, S., Pighizzini, G. (eds.) DCFS 2018. LNCS, vol. 10952, pp. 175–187. Springer, Cham (2018). https://doi.org/10.1007/978-3-319-94631-3_15
8. Jeandel, E.: Topological automata. Theory Comput. Syst. **40**(4), 397–407 (2007)
9. Kondacs, A., Watrous, J.: On the power of quantum finite state automata. In: FOCS, pp. 66–75. IEEE (1997)
10. Macarie, I.I.: Space-efficient deterministic simulation of probabilistic automata. SIAM J. Comput. **27**(2), 448–465 (1998)
11. Yakaryılmaz, A., Cem Say, A.C.: Languages recognized by nondeterministic quantum finite automata. Quantum Inf. Comput. **10**(9&10), 747–770 (2010)
12. Paz, A.: Introduction to Probabilistic Automata (Computer Science and Applied Mathematics). Academic Press Inc., Orlando (1971)
13. Rabin, M.O.: Probabilistic automata. Inf. Control **6**, 230–243 (1963)
14. Sipser, M.: Introduction to the Theory of Computation, 1st edn. International Thomson Publishing, Stamford (1996)
15. Turakainen, P.: On Probabilistic Automata and their Generalizations. Annales Academiae Scientiarum Fennicae. Series A 429 (1969)
16. Turakainen, P.: Generalized automata and stochastic languages. Proc. Am. Math. Soc. **21**(2), 303–309 (1969)
17. Turakainen, P.: On nonstochastic languages and homomorphic images of stochastic languages. Inf. Sci. **24**(3), 229–253 (1981)
18. Villagra, M., Yakaryılmaz, A.: Language recognition power and succinctness of affine automata. In: Amos, M., Condon, A. (eds.) UCNC 2016. LNCS, vol. 9726, pp. 116–129. Springer, Cham (2016). https://doi.org/10.1007/978-3-319-41312-9_10

An Exponentially Growing Nubot System Without State Changes

Chun-Ying Hou and Ho-Lin Chen(✉)

National Taiwan University, Taipei, Taiwan
{b02401053,holinchen}@ntu.edu.tw

Abstract. Self-assembly is the process in which simple units assemble into complex structures without explicit external control. The nubot model was proposed to describe self-assembly systems involving active components capable of changing internal states and moving relative to each other. A major difference between the nubot model and many previous self-assembly models is its ability to perform exponential growth. Several previous works focused on restricting the nubot model while preserving exponential growth. In this work, we construct a nubot system which performs exponential growth without any state changes. All nubots stay in fixed internal states throughout the growth process. This construction not only improves the previous optimal construction, but also demonstrates new trade-offs between different types of rules in the nubot system.

1 Introduction

Self-assembly is the process in which simple units assemble into complex structures without explicit external control. There are many self-assembly systems in nature. For example, mineral crystallization and the embryo developing process can both be viewed as self-assembly processes. In this paper, we study algorithmic self-assembly, a type of self-assembly system whose components and assembly rules are programmable. In nanotechnology, as the size of the system gets smaller and smaller, precise external control becomes prohibitively costly. Algorithmic self-assembly has become an important tool for nano-scale fabrication and nano machines. DNA is a popular substrate for molecular implementations of algorithmic self-assembly due to its combinatorial nature and predictable geometric structure. DNA self-assembly has been used for many different applications including performing computation [2,22,30], constructing molecular patterns [8,13,20,21,24,37], and building nano-scale machines [3,9,14,23,25,35].

In many studies of algorithmic self-assembly using DNA molecules, DNA strands first assemble into small programmable structures called "tiles". These tiles then assemble into large two-dimensional and three-dimensional structures using simple rules. The first algorithmic tile assembly model, the abstract Tile Assembly Model (aTAM), was proposed by Rothemund and Winfree [19,30].

Research supported in part by MOST grant number 107-2221-E-002-031-MY3.

I. McQuillan and S. Seki (Eds.): UCNC 2019, LNCS 11493, pp. 122–135, 2019.
https://doi.org/10.1007/978-3-030-19311-9_11

Tile systems in aTAM can perform Turing-universal computation [30] and produce arbitrary shapes [16,26] (if arbitrary scaling is allowed). Many studies also focused on constructing structures of fixed sizes such as counters [1,6] and squares [11,15,19]. It is also known that aTAM is intrinsically universal [12]. Many aTAM systems have been successfully implemented by DNA tiles in lab experiments [2,5,21,31].

In contrast to tile assembly systems, the self-assembly systems in nature are often equipped with more active nano-components: these components can sense and process environmental cues; upon interaction, their internal structures may change; they can both passively diffuse and actively move. Previously, many active molecular components have been implemented and tested in the lab. Examples include autonomous walkers [14,17,18,23,25,27,34], logic and catalytic circuits [22,29,33,36], and triggered assembly of linear [10,28] and dendritic structures [33]. It seems possible to use these active components to create active algorithmic self-assembly systems.

The nubot model [32] was proposed to describe these active assembly systems. This model generalizes aTAM to include active components by introducing molecular units named nubots which, after attaching, are able to carry out active and passive movements, undergo state changes, alter the bond types and detach from assembled structures. Similar to aTAM, all these different reaction rules can be applied asynchronously and in parallel in the nubot model. This allows self-assembled structures to grow exponentially in size, which is not possible in most other (passive) self-assembly models such as aTAM. It has been shown that there exists a set of nubot systems that grow a line of length n in time $O(\log n)$ using $O(\log n)$ states [32]. Furthermore, squares and arbitrary shapes can also be assembled in poly-logarithmic time without any scaling [32].

The nubot model is an extremely powerful model originally aimed to describe the behavior of various active self-assembly systems. It is very difficult to implement all the features of the nubot model using molecular components. Several previous works focused on finding a minimal set of features in the nubot model which allows exponential growth to happen [4,7]. Previously, it has been shown that nubot systems can still perform exponential growth even if the active movement rules are disabled [4]. However, the construction relies heavily on the rigidity of structures. Chin et al. [7] studied the problem of finding the required minimal number of state changes applied to each nubot to reach exponential growth. The main result is that, under strong constraints on the structure of the assembly, two state changes per nubot is required to achieve exponential growth; one state change per nubot is sufficient for exponential growth if the structural constraints are slightly relaxed.

Our result: In this paper, we construct a nubot system capable of exponential growth without any state changes. A line of length $\Theta(2^k)$ is built in $O(k)$ expected time with all the nubots staying in fixed states throughout the whole process. We achieve our goal by iterating a subroutine that doubles the number of insertion sites. To create an insertion site within a connected line, bond breakage is a crucial step, same as it was for previous models. Intuitively, only after the presence of an alternative connected structure, so-called a "bridge", can the bond break; the structure becomes disconnected otherwise. The main difference between our construction and the previous ones is the method for signaling the proper timing for bond breakage. In all previous constructions, the decisions were made depending on the internal states of the nubots. Two nubots cut the bonds between them only when they are in certain states indicating the existance of alternative connections. Under the restriction of no state change, we use the relative position of the two nubots to encode this information. Since a local movement propagates through the system in a non-local fashion [32], we utilize the movement rule as a way for signal transduction in a sense that the bond breakage is triggered by a change in the relative position of neighboring monomers rather than by internal state changes. In order for our construction to work properly, random passive movements (called agitations) must be disabled from the nubot model. Also, our construction employs flexible (soft) bonds, which are not required in all previous exponential-growth constructions. Although there is no formal proof, we believe that disabling agitations and utilizing flexible bonds are necessary in order to achieve exponential growth without changing internal states.

2 Model

In this section, we provide a simplified description of the nubot model proposed in [32]. The model is the same as the original definition except for the following three modifications. First, each monomer has a fixed state which cannot be changed by any rule. Second, agitations are removed from the model. Third, any components disconnected from the main structure (the structure containing the initial seed) immediately disappear. The first modification is the constraint we aim to satisfy while achieving exponential growth. The second and third modifications are required for our construction to work properly. We believe that the second condition is necessary while the third one prevents abusing the second modification.

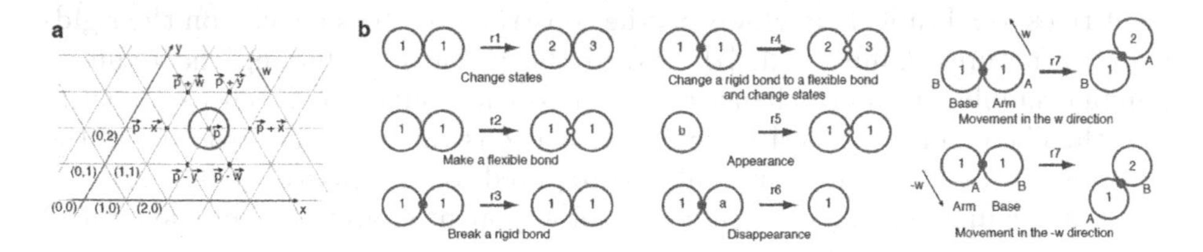

Fig. 1. (a) A triangular grid. The circle represents a monomer. (b) Examples of interaction rules. Figure obtained from [32].

The nubot model is a two-dimensional self-assembly model. As shown in Fig. 1, the basic assembly units are placed on a triangular grid. The *axial directions* $\mathcal{D} = \{\boldsymbol{x}, -\boldsymbol{x}, \boldsymbol{y}, -\boldsymbol{y}, \boldsymbol{w}, -\boldsymbol{w}\}$ are the units vectors along the grid axes. Let Σ

be the set of possible states. A *monomer* (also called a nubot), the basic assembly unit, is a pair $X = (s_X, \boldsymbol{p}(X))$ where $s_X \in \Sigma$ is the (fixed) state and $\boldsymbol{p}(X) \in \mathbb{Z}^2$ is a grid point which is the current location of the monomer. Throughout this paper, each monomer is depicted by either a disk with the fixed state written in its center or a solid circle (if the state is not important). A *configuration* $\mathcal{C}$ is a finite set of monomers along with all the bonds between them and is assumed to define the state of an entire grid at some time instance.

An *interaction rule*, $r = (s_1, s_2, b, \boldsymbol{u}) \rightarrow (s'_1, s'_2, b', \boldsymbol{u}')$, represents how two neighboring monomers interact. $s_1, s_2, s'_1, s'_2 \in \Sigma \cup \{empty\}$, where *empty* denotes lack of a monomer. In each reaction, at most one of s_1, s_2 is *empty*. $b, b' \in \{null, rigid, flexible\}$ are the bond types in between the reacting monomers before and after the reaction. The direction of a *rigid* bond cannot be changed unless by a direct movement rule, whereas the direction of a *flexible* bond can be altered by nudges from other nubots as long as the two connected monomers stay in each others' neighborhood. $\boldsymbol{u}, \boldsymbol{u}' \in \mathcal{D}$ are the relative positions of s_2 to s_1. If either of s_1, s_2 is *empty* then b is *null*, also if either or both of s'_1, s'_2 is *empty* then b' is *null*. $s_1 = s'_1$ and $s_2 = s'_2$ if both states are non-empty.

In our modified nubot model, only the following four types of rules are allowed. In the first three rules, the monomers do not make relative movements, i.e. $\boldsymbol{u} = \boldsymbol{u}'$. An *appearance rule* is a rule in which s_2 is *empty*, and s'_2 is not. A *disappearance rule* is a rule in which one or both of s_1, s_2 that are non-empty become *empty*. A *bond change* is a rule where none of s_1, s_2, s'_1, s'_2 is *empty* (and thus only the bond type b can be changed).

The last rule allows adjacent monomers to move relative to each other: a *movement rule* is an interaction rule where $\boldsymbol{u} \neq \boldsymbol{u}'$, and none of s_1, s_2, s'_1, s'_2 is *empty*. When a movement rule happens, one of the two monomers is nondeterministically chosen as the base, which remains stationary; the other monomer is the arm which moves one unit distance. The *agitation set*, denoted by $\mathcal{A}(\mathcal{C}, A, \boldsymbol{v})$, is the minimal set of monomers in a configuration $\mathcal{C}$ that must be moved together with monomer A by a unit vector $\boldsymbol{v}$ in order to satisfy the following conditions: any existing bonds remain undisturbed, and the grid points one unit distance in the $\boldsymbol{v}$ direction of the set is empty. The *movable set*, denoted by $\mathcal{M}(\mathcal{C}, A, B, \boldsymbol{v})$ where A is the arm and B is the base, is the agitation set $\mathcal{A}(\mathcal{C}, A, \boldsymbol{v})$ if the agitation set does not contain B; the movable set is empty otherwise. A formal definition of the movable set, including how to efficiently compute the set, can be found in [32].

A *nubot system* is a pair $\mathcal{N} = (\mathcal{C}_0, \mathcal{R})$ where $\mathcal{C}_0$ is the initial configuration, and $\mathcal{R}$ is the set of rules.

3 System Construction

In this section, we describe the detail of a nubot system that grows a line of length $\Theta(2^k)$ is built in expected time $O(k)$. Since our construction does not require any state change, "nubot in state A" will be referred to as "nubot A" or simply "A" hereinafter.

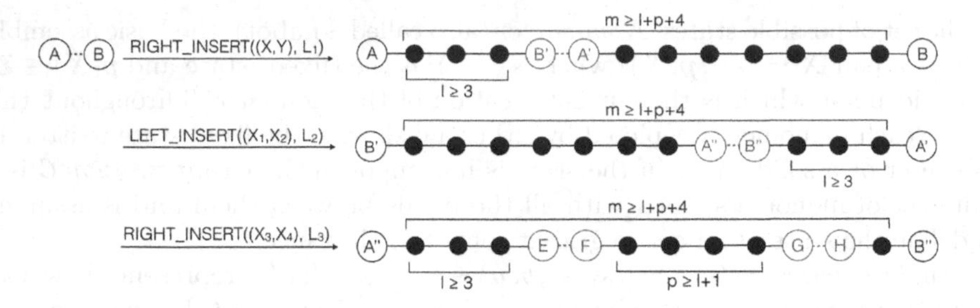

Fig. 2. The process of generating two insertion sites (E, F) and (G, H) from one insertion site (A, B). Solid lines denote rigid bonds; dashed lines denote flexible bonds.

Construction overview: In our construction, the nubots form a line in the $\boldsymbol{x}$ direction within which pairs of adjacent nubots in the $\boldsymbol{x}$-direction joined by flexible bonds are called insertion sites, being named after their function. When an insertion site is present, there exists a set of rules that instructs other nubots to attach on top of the site to create an alternative connection, so-called a "bridge", between the two nubots at the insertion site. Upon the completion of the bridge, a movement rule is then applied and moves one of the nubots at the insertion site, changing the relative position of the insertion pair. Sensing a change in relative position, the insertion site cuts the flexible bond in between, allowing a sequence of nubots to be "inserted" between the two nubots at the insertion site.

Our construction relies on two types of insertion. RIGHT_INSERT$((X, Y), L)$ breaks the insertion site (X, Y) and inserts a sequence L of nubots from left to right between X and Y. Its mirrored version LEFT_INSERT$((X, Y), L)$ breaks the insertion site (X, Y) and inserts a sequence L of nubots from right to left.

There are four different main insertion sites: (A_i, B_i), (C_i, D_i), (E_i, F_i) and (G_i, H_i), $i = 1, 2, \ldots, k-1$ where all pairs are written from the negative to the positive x-direction. Starting from an insertion site (A_i, B_i) or (C_i, D_i), a subroutine consisting of three rounds of insertion is executed and generates both (E_{i+1}, F_{i+1}) and (G_{i+1}, H_{i+1}), as shown in Fig. 2. Similarly, each insertion site (E_i, F_i) or (G_i, H_i) generates both (A_{i+1}, B_{i+1}) and (C_{i+1}, D_{i+1}). The pairs of nubots (A_k, B_k), (C_k, D_k), (E_k, F_k) and (G_k, H_k) all have rigid bonds between them and do not allow further insertions. The seed structure consists of seven nubots forming a single line in the x-direction: one single insertion site (A_1, B_1) and five nubots connected to the x-direction of B_1 using rigid bonds. These extra five nubots do not participate in any reaction rule, merely occupying the position to prevent some undesired reactions. Starting from a seed insertion site (A_1, B_1), the number of insertion sites doubles after each subroutine. Our construction allows all possible insertions to operate in parallel with at most a constant slow-down to each process, thus ensures that the size of the line grows exponentially. The following subsections describe each part of this construction in more detail and provide a formal analysis of the running time.

3.1 RIGHT_INSERT($(A, B), L$)

Figure 3 illustrates the process RIGHT_INSERT($(A, B), L$), mainly comprising three parts: bridge construction (i)–(iii), line construction (iv) and movement of B, where $L = (a_1, a_2, ..., a_m)$ is listed from the negative to the positive x-direction (in the figure, $m = 4$).

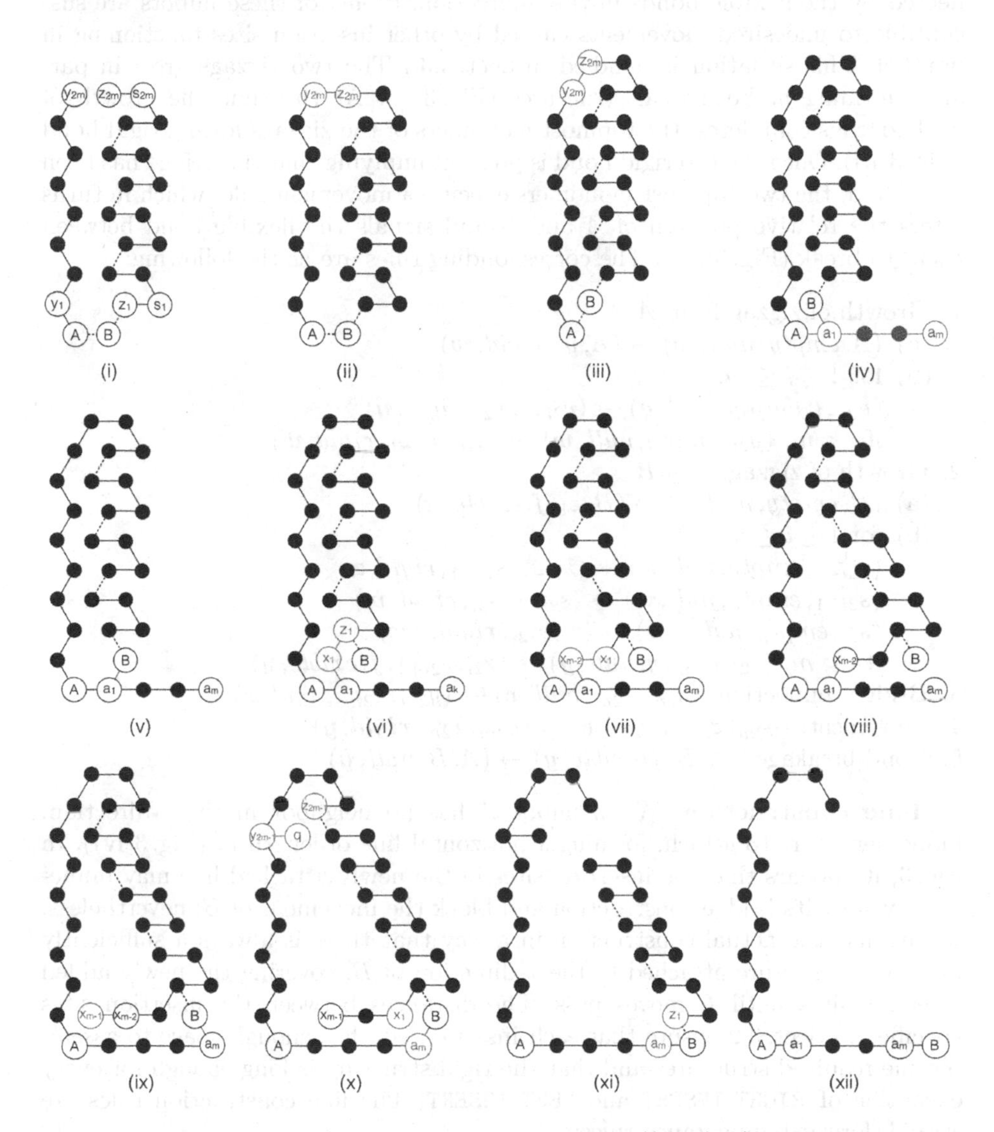

Fig. 3. The desired reaction sequence for RIGHT_INSERT($(A, B), L$). Solid lines denote rigid bonds; dashed lines denote flexible bonds.

LEFT_INSERT($(B, A), L'$) is the mirrored version of RIGHT_INSERT($(A, B), L$), where the sequence L' is listed from the positive to the negative x-direction.

Bridge construction: Nubot A grows a zigzag of length 2 m entirely connected with rigid bond, and nubot B grows a variant zigzag substituting each rigid bond in the y-direction with a flexible bond and each bond in the w-direction with two additional monomers forming a bulge connected with rigid bonds. The bulges are designed to limit the relative movement of nubots connected by the flexible bonds in the y-direction. (Some of these nubots are susceptible to undesired movements caused by other insertion sites functioning in parallel. This situation is handled in Sect. 3.3.) The two zigzags grow in parallel, forming m hexagonal structures (Fig. 3(i)). At the time the growth of both zigzags completes, the topmost monomers of the zigzags form a rigid bond (Fig. 3(ii)). Once the top rigid bond is present, implying that the bridge has been connected, the two topmost monomers execute a movement rule, which in turns alters the relative position of A and B and signals the flexible bond between them to break (Fig. 3(iii)). The corresponding rules are as the following:

1. Growth of zigzag from A
   (a) $(A, empty, null, \boldsymbol{w}) \rightarrow (A, y_1, rigid, \boldsymbol{w})$
   (b) for $1 \leq i \leq m$:
   $(y_{2i-1}, empty, null, \boldsymbol{y}) \rightarrow (y_{2i-1}, y_{2i}, rigid, \boldsymbol{y})$
   if $i < m$: $(y_{2i}, empty, null, \boldsymbol{w}) \rightarrow (y_{2i}, y_{2i+1}, rigid, \boldsymbol{w})$
2. Growth of zigzag from B
   (a) $(B, empty, null, \boldsymbol{y}) \rightarrow (B, z_1, flexible, \boldsymbol{y})$
   (b) for $1 \leq i \leq m$:
   $(z_{2i-1}, empty, null, \boldsymbol{x}) \rightarrow (z_{2i-1}, s_{2i-1}, rigid, \boldsymbol{x})$
   $(s_{2i-1}, empty, null, \boldsymbol{w}) \rightarrow (s_{2i-1}, s_{2i}, rigid, \boldsymbol{w})$
   $(s_{2i}, empty, null, -\boldsymbol{x}) \rightarrow (s_{2i}, z_{2i}, rigid, -\boldsymbol{x})$
   if $i < m$: $(z_{2i}, empty, null, \boldsymbol{y}) \rightarrow (z_{2i}, z_{2i+1}, flexible, \boldsymbol{y})$
3. Bridge connection: $(y_{2m}, z_{2m}, null, \boldsymbol{x}) \rightarrow (y_{2m}, z_{2m}, rigid, \boldsymbol{x})$
4. Movement: $(y_{2m}, z_{2m}, rigid, \boldsymbol{x}) \rightarrow (y_{2m}, z_{2m}, rigid, \boldsymbol{y})$
5. Bond breakage: $(A, B, flexible, \boldsymbol{y}) \rightarrow (A, B, null, \boldsymbol{y})$

Line construction: When nubot A has no neighbor in the x-direction, monomers start to attach, forming a horizontal line of length m (Fig. 3(iv)). In Fig. 3, it appears that the insertion sites in the newly attached line may immediately start its bridge construction and block the movement of B; nevertheless, we arrange the actual construction in a way that there is always a sufficiently long rigid structure attached to the x-direction of B, covering the newly added insertion sites until B moves pass. The distances between the insertion sites specified in Sect. 3.2 ensure that each insertion site has enough space to assemble the required structures and that the rigid structure is long enough for every execution of `RIGHT_INSERT` and `LEFT_INSERT`. The line construction rules are straightforward appearance rules:

1. $(A, empty, null, \boldsymbol{x}) \rightarrow (A, a_1, rigid, \boldsymbol{x})$
2. For $1 \leq i \leq m-1$: $(a_i, empty, null, \boldsymbol{x}) \rightarrow (a_i, a_{i+1}, b_i, \boldsymbol{x})$
   The bond b_i is *flexible* if (a_i, a_{i+1}) is an insertion site and is *rigid* otherwise.

Movement of B: Monomers in the new grown line move B towards the x-direction through interactions with B and x_i's while line construction operates simultaneously (Fig. 3(iv)–(x)). After B is moved m distance to the right, the bridge structure connecting A and B breaks, allowing B to move downward to the x-direction of a_m (Fig. 3(xi)), and the insertion process is completed as m monomers are inserted between A and B (Fig. 3(xii)). The corresponding rules are as the following:

1. Fig. 3(iv)–(vi)
   $(a_1, B, null, \boldsymbol{w}) \rightarrow (a_1, B, rigid, \boldsymbol{w})$
   $(a_1, B, rigid, \boldsymbol{w}) \rightarrow (a_1, B, rigid, \boldsymbol{y})$
   $(a_1, B, rigid, \boldsymbol{y}) \rightarrow (a_1, B, null, \boldsymbol{y})$
   $(z_1, empty, null, -\boldsymbol{y}) \rightarrow (z_1, x_1, rigid, -\boldsymbol{y})$
2. Fig. 3(vii)–(ix)
   for $1 \leq i \leq k-2$:
   $(a_1, x_i, null, \boldsymbol{w}) \rightarrow (a_1, x_i, rigid, \boldsymbol{w})$
   $(a_1, x_i, rigid, \boldsymbol{w}) \rightarrow (a_1, x_i, rigid, \boldsymbol{y})$
   $(a_1, x_i, rigid, \boldsymbol{y}) \rightarrow (a_1, x_i, null, \boldsymbol{y})$
   $(x_i, empty, null, -\boldsymbol{x}) \rightarrow (x_i, x_{i+1}, rigid, -\boldsymbol{x})$
3. Fig. 3(ix)–(xii)
   $(a_m, B, null, \boldsymbol{w}) \rightarrow (a_m, B, rigid, \boldsymbol{w})$
   $(a_m, B, rigid, \boldsymbol{w}) \rightarrow (a_m, B, rigid, \boldsymbol{y})$
   $(a_m, x_1, null, \boldsymbol{w}) \rightarrow (a_m, empty, null, \boldsymbol{w})$
   $(y_{2m-1}, empty, null, \boldsymbol{x}) \rightarrow (y_{2m-1}, q, null, \boldsymbol{x})$
   $(q, z_{2m-1}, null, \boldsymbol{y}) \rightarrow (q, empty, null, \boldsymbol{y})$
   $(a_m, B, rigid, \boldsymbol{y}) \rightarrow (a_m, B, rigid, \boldsymbol{x})$
   $(a_m, z_1, null, \boldsymbol{y}) \rightarrow (a_m, empty, null, \boldsymbol{y})$

Note that the appearance rule, $(y_{2m-1}, empty, null, \boldsymbol{x}) \rightarrow (y_{2m-1}, q, null, \boldsymbol{x})$, could happen during the bridge construction stage and in turns block the movement rule, $(y_{2m}, z_{2m}, rigid, \boldsymbol{x}) \rightarrow (y_{2m}, z_{2m}, rigid, \boldsymbol{y})$. Accordingly, we add a disappearance rule $(z_{2m}, q, null, -\boldsymbol{y}) \rightarrow (z_{2m}, empty, null, -\boldsymbol{y})$ to amend.

In the current step, some movements may be pushed back by the movement rules applied to other insertion sites. This undesired situation is also handled in Sect. 3.3.

3.2 Doubling Insertion Sites

Figure 2 depicts the subroutine applicable to (A_i, B_i)'s and (C_i, D_i)'s. Starting from an insertion site (A, B), the subroutine follows:

1. `RIGHT_INSERT`$((A, B), L)$: $L = (c_0, c_1, \ldots, c_m)$, $c_{l+1} = B'$, $c_{l+2} = A'$, (B', A') is an auxiliary insertion site.
2. `LEFT_INSERT`$((B', A'), L')$: $L' = (c'_0, c'_1, \ldots, c'_m)$, $c'_{l+1} = A''$, $c'_{l+2} = B''$, (A'', B'') is an auxiliary insertion site.
3. `RIGHT_INSERT`$((A'', B''), L'')$: $L'' = (c''_0, c''_1, \ldots, c''_m)$, $c''_{l+1} = E$, $c''_{l+2} = F$, $c''_{l+p+3} = G$, $c''_{l+p+4} = H$), (E, F) and (G, H) are two main insertion sites to which the mirrored subroutine is applicable.

Except the ones being specifically labled, all c_i, c_i', c_i'' are unique states not involved in any rules after their round of insertion is completed. The distance constraints labeled in Fig. 2 (in the figure, $m = 12, l = 3, p = 4$) ensure that each insertion site has enough space to assemble the required structures and that no insertion site starts to grow prematurely before being connected to the line. (m, l, p) remain constants in our construction, yet any three consants are viable as long as they satisfy the constraints. Interchanging `LEFT` and `RIGHT` gives the mirrored version which can be applied to (E_i, F_i)'s and (G_i, H_i)'s. Each doubling subroutines initiated by insertion sites with different internal states utilizes a distinct set of nubot states.

3.3 Handling Undesired Conditions

When multiple insertion sites operate in parallel, `RIGHT_INSERT` and `LEFT_INSERT` executed at different locations may interfere with each other, potentially leading to the system's deviation from the desired trajectory. To deal with the anomalies, we focus on a particular insertion site, (A, B), and discuss the possible aberrations caused by interactions outside of `RIGHT_INSERT`$((A, B), L)$ on the line. The following analyses are based on the ideas that from every desired configuration $\mathcal{C}_i$, there is a always a constant probability of its transition into the next desired configuration $\mathcal{C}_{i+1}$, and that for any undesired configuration $\mathcal{C}_{i+1}'$, it returns to its previous desired configuration $\mathcal{C}_i$ with probability one. Figure 3 demonstrates the site of our concern.

Case 1: Let $\mathcal{C}$ be any configuration in which the rigid bond between y_{2m} and z_{2m} is still absent. If there exists a rule r, other than the rules in `RIGHT_INSERT`-$((A, B), L)$, that would move z_{2i+1} by $-\boldsymbol{x}$, then once r is applied, z_{2i+1} will be shifted to its $-x$-direction, leading to a distorted configuration. To resolve the condition, we add several rules to the `RIGHT_INSERT` process: (s_0 denotes the x-direction neighboring monomer of B, which is not shown in Fig. 3)

1. $r_{a_i} = (s_{2i}, s_{2i+1}, null, \boldsymbol{w}) \rightarrow (s_{2i}, s_{2i+1}, rigid, \boldsymbol{w})$, $i = 0, 1, \ldots, m-1$
   This rule adds an additional rigid bond b to the structure.
2. $r_{b_i} = (s_{2i}, s_{2i+1}, rigid, \boldsymbol{w}) \rightarrow (s_{2i}, s_{2i+1}, rigid, \boldsymbol{y})$, $i = 0, 1, \ldots, m-1$
   This rule utilizes the additional bond to translate the affected nubots back to their intended positions.
3. $r_{c_i} = (s_{2i}, s_{2i+1}, rigid, \boldsymbol{y}) \rightarrow (s_{2i}, s_{2i+1}, null, \boldsymbol{y})$, $i = 0, 1, \ldots, m-1$
   This rule removes b, and the configuration returns to that before the undesired movement.

As the additional bond b prevents the undesired movement of z_{2i+1}, after r_{c_i} is applied, z_{2i+1} is again susceptible to the undesired shifting. However, once y_{2i+2} and z_{2i+2} are both attached, z_{2i+1} can no longer be pushed to its $-x$-direction. In other words, if the attachments of y_{2i+2} and z_{2i+2} is completed before r_{c_i} is applied, the structure containing monomers up to height $2i+2$ is secured. The three rule applications (attachment of y_{2i+2}, attachment of z_{2i+2} and r_{c_i}) are equally likely, hence a constant probability of the rules applying in the desired order.

The above adjustment entails a nuisance: the transition from (x) to (xi) is stymied given that s_{2m-1} and s_{2m-2} have remained bonded. Adding two rules

$$r_d = (z_{2m}, \mathit{empty}, \mathit{null}, -\boldsymbol{w}) \rightarrow (z_{2m}, m, \mathit{rigid}, -\boldsymbol{w})$$

and

$$r_e = (m, s_{2m-1}, \mathit{null}, \boldsymbol{x}) \rightarrow (m, \mathit{empty}, \mathit{null}, -\boldsymbol{x})$$

removes s_{2m-1} and thus enables the transition.

Case 2: Since every insertion site except the initial one is grown from a previous insertion site, in the configuration in Fig. 3(v), once the $-x$-direction neighbor of nubot B is empty, two rules are applicable:

$$r_1 : (z_1, \mathit{empty}, \mathit{null}, -\boldsymbol{y}) \rightarrow (z_1, x_1, \mathit{rigid}, -\boldsymbol{y})$$

in the normal insertion process and

$$r_2 : (B, \mathit{empty}, \mathit{null}, -\boldsymbol{x}) \rightarrow (B, A, \mathit{flexible}, -\boldsymbol{x})$$

resulting from the rules in the previous subroutine `LEFT_INSERT`$((E'', F''), L'')$ or `LEFT_INSERT`$((G'', H''), L'')$. Rule r_1 has a 1/2 chance of being applied first, leading onto a desired construction, but if the opposite happens, the construction is hampered. To resolve this, we introduce a rule: $r'_f = (a_1, A, \mathit{null}, \boldsymbol{w}) \rightarrow (a_1, \mathit{empty}, \mathit{null}, \boldsymbol{w})$, restoring the configuration in Fig. 3(v).

Case 3: Let $\mathcal{C}'$ be any configuration in which B has been shifted to the x-direction i times, and x_i has not yet attached. Having $(a_1, x_{i-1}, \mathit{rigid}, \boldsymbol{y}) \rightarrow (a_1, x_{i-1}, \mathit{null}, \boldsymbol{y})$ $((a_1, B, \mathit{rigid}, \boldsymbol{y}) \rightarrow (a_1, B, \mathit{null}, \boldsymbol{y})$ when $i = 1)$ being applied after the attachment of x_i delivers a proper construction and has a constant probability of taking place. In the converse case, and if either of the following is satisfied:

1. there exists a rule r, other than the rules in `RIGHT_INSERT`$((A, B), L)$, that would move B by $-\boldsymbol{x}$
2. r_{b_i} (defined in **case 1**) is applicable to $\mathcal{C}'$

then as the rule applies, B will be moved back to its $-x$-direction, blocking the attachment of x_i, yet this won't cause problem since B will again be shifted to its x-direction due to standard insertion rules.

Case 4: In the configuration in Fig. 3(x), after the disappearance of x_1, $r_3 = (a_m, B, \mathit{rigid}, \boldsymbol{y}) \rightarrow (a_m, B, \mathit{rigid}, \boldsymbol{x})$ has a constant probability of being applied before r_2 (defined above in **case 2**), leading to the transition from configuration (x) to (xi). However, if r_2 is applied first, the transition is hindered. By adding $r'_g = (a_m, A, \mathit{null}, \boldsymbol{w}) \rightarrow (a_m, \mathit{empty}, \mathit{null}, \boldsymbol{w})$, we reverse the undesired rule application.

Though we take (A, B) as an example, the above analysis applies to any insertion site operating `RIGHT_INSERT`, and mirrored conditions occur regarding the insertion sites operating `LEFT_INSERT`.

3.4 Expected Time

In the previous subsections, we show that our construction uses $\Theta(k)$ total states (since each doubling subroutine initiated by insertion sites with different internal states uses a constant number of states) and always produce an assembly structure containing a line in the x-direction of length $\Theta(2^k)$ (plus some extra structure remaining on top of the line). In this section, we will prove that the expected time for the line to complete is $O(k)$. More formally, we have the following theorem.

Theorem 1. *There exists a nubot system without state changes such that the expected time for the system to generate a line of length $\Theta(2^k)$ is $O(k)$. Furthermore, the system only uses $O(k)$ total states.*

In order to prove this theorem, we use a similar technique as [7]. In this paper, we make a more formal definition of the regular systems defined in [7].

Definition 1. *A nubot system is* regular *if the following conditions hold:*

1. *The system starts with one insertion site.*
2. *Every insertion site can and must be cut and replaced by two insertion sites (called a doubling process).*
3. *Starting from any intermediate configuration in the doubling process, the expected time to finish the doubling process is a constant.*
4. *All doubling processes can happen in parallel.*
5. *The system reaches a desired configuration as soon as k rounds of doubling process finishes.*

When a nubot system is regular, after k rounds of doubling process, 2^k different insertion sites will be generated. Let $X_1, X_2, \ldots, X_{2^k}$ denote the actual time at which these insertion sites are generated, the following lemma proved in [7] shows that the system reaches a desired configuration in time $O(k)$.

Lemma 1. *For any regular system, $E[maxX_i] = O(k)$.*

Therefore, the only remaining task is to prove that our system is regular. It suffices to check the subroutine RIGHT_INSERT$((A, B), L)$ in detail. First, in the bridge building process, the attachment of each nubot in the two (constant height) zigzag structures are irreversible. The zigzag structure attaching to B is never blocked and will attach in expected constant time. The construction in Sect. 3.3 guarantees that every nubot in the zigzag structure attaching to A will attach in expected constant time. After both zigzag structures attach, the rigid bond formation at the top and the bond breakage at the insertion site happen in expected constant time. Second, the line construction finishes in expected constant time. Third, in the movement of B nubot, each move becomes irreversible as soon as the corresponding nubot x_i attaches. Again, the analyses in Sect. 3.3 guarantees that each of these steps happen in constant expected time. Combining all the arguments above, we know that our system is regular and thus finishes in expected $O(k)$ time.

4 Conclusion and Discussion

We have proposed a nubot system with no state change that grows a line structure of length $\Theta(2^k)$ in expected time $O(k)$. The construction involves $O(k)$ states, meanwhile, each nubot monomers stays in a fixed internal state throughout the entire process. It is shown in [4] that under a nubot system without movement rules and with agitation, a length n line is assembled in sublinear expected time, and an $n \times n$ square in polylogarithmic time. Chin et al. proved that at least one state change is required for exponential growth in a 2-layered nubot system [7]. Our work suggests that an active self-assembly system without state change is nevertheless capable of exponential growth, which is the optimal construction with respect to the number of state change.

Our construction relies chiefly on the pliability of flexible bonds. The direction of flexible bonds could be altered by distant movement rules; this property serves as a signal transduction mechanism vital to our construction. Following the dependence on flexible bonds, a key assumption about our nubot system is that agitation is removed. By agitation, we mean the unintended or uncontrollable movements in the system, such as diffusion and Brownian motion. The purpose of the no-agitation condition is to prevent unexpected changes in the direction of a flexible bond. Therefore, the assumption is equivalent to that the flexible bonds could withstand any unintentional force and is only affected by rule applications. However, whether exploiting flexible bonds and forbidding agitation are necessary for exponential growth under a zero-state-change system remains unanswered.

References

1. Adleman, L., Cheng, Q., Goel, A., Huang, M.-D.: Running time and program size for self-assembled squares. In: Proceedings of the 33rd Annual ACM Symposium on Theory of Computing, pp. 740–748 (2001)
2. Barish, R.D., Rothemund, P.W.K., Winfree, E.: Two computational primitives for algorithmic self-assembly: copying and counting. Nano Lett. **5**(12), 2586–2592 (2005)
3. Bishop, J., Klavins, E.: An improved autonomous DNA nanomotor. Nano Lett. **7**(9), 2574–2577 (2007)
4. Chen, H.-L., Doty, D., Holden, D., Thachuk, C., Woods, D., Yang, C.-T.: Fast algorithmic self-assembly of simple shapes using random agitation. In: Murata, S., Kobayashi, S. (eds.) DNA 2014. LNCS, vol. 8727, pp. 20–36. Springer, Cham (2014). https://doi.org/10.1007/978-3-319-11295-4_2
5. Chen, H.-L., Schulman, R., Goel, A., Winfree, E.: Error correction for DNA self-assembly: preventing facet nucleation. Nano Lett. **7**, 2913–2919 (2007)
6. Cheng, Q., Goel, A., Moisset, P.: Optimal self-assembly of counters at temperature two. In: Proceedings of the 1st Conference on Foundations of Nanoscience: Self-Assembled Architectures and Devices, pp. 62–75 (2004)
7. Chin, Y.-R., Tsai, J.-T., Chen, H.-L.: A minimal requirement for self-assembly of lines in polylogarithmic time. In: Brijder, R., Qian, L. (eds.) DNA 2017. LNCS, vol. 10467, pp. 139–154. Springer, Cham (2017). https://doi.org/10.1007/978-3-319-66799-7_10

8. Dietz, H., Douglas, S., Shih, W.: Folding DNA into twisted and curved nanoscale shapes. Science **325**, 725–730 (2009)
9. Ding, B., Seeman, N.: Operation of a DNA robot arm inserted into a 2D DNA crystalline substrate. Science **384**, 1583–1585 (2006)
10. Dirks, R.M., Pierce, N.A.: Triggered amplification by hybridization chain reaction. Proc. Nat. Acad. Sci. **101**(43), 15275–15278 (2004)
11. Doty, D.: Randomized self-assembly for exact shapes. SIAM J. Comput. **39**(8), 3521–3552 (2010)
12. Doty, D., Lutz, J.H., Patitz, M.J., Schweller, R.T., Summers, S.M., Woods, D.: The tile assembly model is intrinsically universal. In: Proceedings of the 53rd Annual IEEE Symposium on Foundations of Computer Science (2012)
13. Douglas, S., Dietz, H., Liedl, T., Hogberg, B., Graf, F., Shih, W.: Self-assembly of DNA into nanoscale three-dimensional shapes. Nature **459**(7245), 414–418 (2009)
14. Green, S., Bath, J., Turberfield, A.: Coordinated chemomechanical cycles: a mechanism for autonomous molecular motion. Phys. Rev. Lett. **101**(23), 238101 (2008)
15. Kao, M.-Y., Schweller, R.: Reducing tile complexity for self-assembly through temperature programming. In: Proceedings of the 17th Annual ACM-SIAM Symposium on Discrete Algorithms, pp. 571–580 (2006)
16. Lagoudakis, M., LaBean, T.: 2D DNA self-assembly for satisfiability. In: Proceedings of the 5th DIMACS Workshop on DNA Based Computers in DIMACS Series in Discrete Mathematics and Theoretical Computer Science, vol. 54, pp. 141–154 (1999)
17. Pei, R., Taylor, S., Stojanovic, M.: Coupling computing, movement, and drug release (2007)
18. Reif, J.H., Sahu, S.: Autonomous programmable DNA nanorobotic devices using DNAzymes. In: Proceedings of the Thirteenth International Meeting on DNA Based Computers, Memphis, TN, June 2007
19. Rothemund, P., Winfree, E.: The program-size complexity of self-assembled squares (extended abstract). In: Proceedings of the 32nd Annual ACM Symposium on Theory of Computing, pp. 459–468 (2000)
20. Rothemund, P.W.K.: Folding DNA to create nanoscale shapes and patterns. Nature **440**(7082), 297–302 (2006)
21. Rothemund, P.W.K., Papadakis, N., Winfree, E.: Algorithmic self-assembly of DNA Sierpinski triangles. PLOS Biol. **2**, 424–436 (2004)
22. Seelig, G., Soloveichik, D., Zhang, D., Winfree, E.: Enzyme-free nucleic acid logic circuits. Science **314**, 1585–1588 (2006)
23. Sherman, W.B., Seeman, N.C.: A precisely controlled DNA bipedal walking device. Nano Lett. **4**, 1203–1207 (2004)
24. Shih, W.M., Quispe, J.D., Joyce, G.F.A.: A 1.7-kilobase single-stranded DNA that folds into a nanoscale octahedron. Nature **427**(6975), 618–621 (2004)
25. Shin, J.-S., Pierce, N.A.: A synthetic DNA walker for molecular transport. J. Am. Chem. Soc. **126**, 10834–10835 (2004)
26. Soloveichik, D., Winfree, E.: Complexity of self-assembled shapes. SIAM J. Comput. **36**, 1544–1569 (2007)
27. Tian, Y., He, Y., Chen, Y., Yin, P., Mao, C.: A DNAzyme that walks processively and autonomously along a one-dimensional track. Angewandte Chemie **44**, 4355–4358 (2005)
28. Venkataraman, S., Dirks, R.M., Rothemund, P.W.K., Winfree, E., Pierce, N.A.: An autonomous polymerization motor powered by DNA hybridization. Nature Nanotechnol. **2**, 490–494 (2007)

29. Win, M.N., Smolke, C.D.: Higher-order cellular information processing with synthetic RNA devices. Science **322**(5900), 456 (2008)
30. Winfree, E.: Algorithmic self-assembly of DNA. Ph.D. thesis, California Institute of Technology, Pasadena (1998)
31. Winfree, E., Liu, F., Wenzler, L., Seeman, N.: Design and self-assembly of two-dimensional DNA crystals. Nature **394**(6693), 539–544 (1998)
32. Woods, D., Chen, H.-L., Goodfriend, S., Dabby, N., Winfree, E., Yin, P.: Active self-assembly of algorithmic shapes and patterns in polylogarithmic time. In: Proceedings of the 4th Conference on Innovations in Theoretical Computer Science, ITCS 2013, pp. 353–354 (2013)
33. Yin, P., Choi, H.M.T., Calvert, C.R., Pierce, N.A.: Programming biomolecular self-assembly pathways. Nature **451**, 318–322 (2008)
34. Yin, P., Turberfield, A.J., Reif, J.H.: Designs of autonomous unidirectional walking DNA devices. In: Ferretti, C., Mauri, G., Zandron, C. (eds.) DNA 2004. LNCS, vol. 3384, pp. 410–425. Springer, Heidelberg (2005). https://doi.org/10.1007/11493785_36
35. Yurke, B., Turberfield, A., Mills Jr., A., Simmel, F., Neumann, J.: A DNA-fuelled molecular machine made of DNA. Nature **406**(6796), 605–608 (2000)
36. Zhang, D.Y., Turberfield, A.J., Yurke, B., Winfree, E.: Engineering entropy-driven reactions and networks catalyzed by DNA. Science **318**, 1121–1125 (2007)
37. Zhang, Y., Seeman, N.: Construction of a DNA-truncated octahedron. J. Am. Chem. Soc. **116**(5), 1661 (1994)

Impossibility of Sufficiently Simple Chemical Reaction Network Implementations in DNA Strand Displacement

Robert F. Johnson(✉)

Bioengineering, California Institute of Technology, Pasadena, CA, USA
rfjohnso@caltech.edu

Abstract. DNA strand displacement (DSD) has recently become a common technology for constructing molecular devices, with a number of useful systems experimentally demonstrated. To help with DSD system design, various researchers are developing formal definitions to model DNA strand displacement systems. With these models a DSD system can be defined, described by a Chemical Reaction Network, simulated, and otherwise analyzed. Meanwhile, the research community is trying to use DSD to do increasingly complex tasks, while also trying to make DSD systems simpler and more robust. I suggest that formal modeling of DSD systems can be used not only to analyze DSD systems, but to guide their design. For instance, one might prove that a DSD system that implements a certain function must use a certain mechanism. As an example, I show that a physically reversible DSD system with no pseudoknots, no effectively trimolecular reactions, and using 4-way but not 3-way branch migration, cannot be a systematic implementation of reactions of the form $A \rightleftharpoons B$ that uses a constant number of toehold domains and does not crosstalk when multiple reactions of that type are combined. This result is a tight lower bound in the sense that, for most of those conditions, removing just that one condition makes the desired DSD system possible. I conjecture that a system with the same restrictions using both 3-way and 4-way branch migration still cannot systematically implement the reaction $A + B \rightleftharpoons C$.

1 Introduction

DNA strand displacement (DSD) has become a common method of designing programmable *in vitro* molecular systems. DSD systems have been designed and experimentally shown to simulate arbitrary Chemical Reaction Networks (CRNs) [2,4,19,20] and some polymer systems [15], implement large logic circuits [16,22], conditionally release a cargo [7], sort objects [21], perform computation on a surface [17], and a number of other tasks [24]. Most DSD systems are based on the 3-way strand displacement mechanism, but a number of interesting devices based on a 4-way strand exchange mechanism have been shown

I. McQuillan and S. Seki (Eds.): UCNC 2019, LNCS 11493, pp. 136–149, 2019.
https://doi.org/10.1007/978-3-030-19311-9_12

[3,5,8,23]. More complex tasks may require combinations of these two mechanisms, such as a 3-way-initiated 4-way strand exchange mechanism used in Surface CRNs [17]. Meanwhile, simple DSD mechanisms such as seesaw circuits [16] have been found to function with more robustness to uncontrollable conditions, compared to more complex mechanisms [22]. The ideal case is, of course, to find as simple a DSD mechanism as possible to accomplish the desired task.

To help with design and analysis of DSD systems, a number of researchers are developing techniques to formally define and analyze the behavior of DSD systems. Reaction enumerators formally define a set of reaction types that DNA strands are assumed to do [9,13,14]. Given that, a reaction enumerator will take a set of strands and complexes and enumerate a CRN, which may have finite or countably infinite species and reactions, describing the behavior of a DSD system initialized with the given complexes. Formal verification methods of CRN equivalence define whether a given implementation CRN is a correct implementation of a given formal CRN [10,12,18]. Thus all the tools necessary are available to ask, given a formal CRN and a DSD implementation, is it correct? The Nuskell compiler, for example, combines all of the above tools to answer exactly that question [1].

I suspect that this level of formal analysis can prove that certain tasks in DSD cannot be done without certain mechanisms, or certain tasks require a certain minimum level of complexity. As an example of this, I chose general CRN implementation schemes as the task, using CRN bisimulation [10] as the measure of correctness. Current CRN implementation schemes require high initial concentrations of "fuel" complexes using 3 or more strands [2,15,19], while the seesaw gates that showed such robustness to uncertainty use only single strands and 2-stranded complexes [16,22]. Thus, I investigated whether arbitrary CRN implementations could be made using only 2-stranded signal and fuel complexes. To further probe the limits of DNA strand displacement, I investigated DSD systems with additional restrictions: the system should implement multiple CRN reactions in a systematic way (as existing CRN implementation schemes do [2,15,19]); the number of "short" or "toehold" domains (which is bounded by thermodynamics) should not increase as the number of reactions increases; the system should work under Peppercorn's condensed semantics [9] (excluding trimolecular mechanisms such as cooperative hybridization); and reversible formal reactions should be implemented by physically reversible DSD mechanisms. Under those restrictions I prove that, using only 4-way strand exchange (excluding 3-way strand displacement), the reactions $A \rightleftharpoons B$ and $C \rightleftharpoons D$ cannot be implemented without crosstalk. While 3-way strand displacement can easily implement those reactions (such as the seesaw gates), it has difficult implementing bimolecular reactions such as $A + B \rightleftharpoons C$ with only 2-stranded fuels. I conjecture that, allowing both 3-way and 4-way branch migration with all the restrictions above, the reaction $A + B \rightleftharpoons C$ cannot be implemented systematically without crosstalk; this result is intended to be the first part of that proof.

In the following sections, I first formalize the concept of DSD system and the reaction rules I use. With that formalization, I prove a locality theorem for reversible, non-pseudoknotted DSD systems without 3-way branch migration, which will be essential to the remaining proofs. I then introduce the concept of condensed reaction as formalized by Peppercorn [9], and show that in the above type of system with 2-stranded reactants there is in fact only one type of condensed reaction possible, the *two-toehold-mediated reversible 4-way strand exchange* or *2-r4 reaction.* Finally, I formalize the concept of *systematic CRN implementation*, which existing DSD implementations of CRNs satisfy [2,15,19], and the *information-bearing set* of domains which, in a systematic implementation, identify which species a DNA complex represents. I then show that the 2-r4 reaction cannot add an information-bearing domain to a complex that did not already have a copy of that domain. This implies that there is no way to build a pathway from a species A to another species B with distinct information-bearing domains, which completes the proof.

2 Formalizing DNA Strand Displacement

The syntax of a DSD system is defined by a set of *strands*, which are sequences of *domains*, and grouping of strands into *complexes* by making bonds between pairs of complementary domains. The semantics of the system will be defined by *reaction rules*, each of which says that complexes matching a certain pattern (and possibly some predicate) will react in a certain way, and defines the products of that reaction. Starting from a given set of complexes, enumerating species and reactions by iteratively applying reaction rules produces a possibly infinite Chemical Reaction Network, which models the behavior of the DSD system.

Definition 1 (Petersen et al. [14]). *The syntax of a DSD system is defined by the following grammar, in terms of* domain names x, y, z *and* bonds i, j, k.

$$\begin{aligned} d &::= x \quad \text{or} \quad x^* && \text{(Domain)} \\ o &::= d \quad \text{or} \quad d!i && \text{(Domain instance)} \\ S &::= o \quad \text{or} \quad o\,S && \text{(Sequence of domains)} \\ P &::= S \quad \text{or} \quad S \mid P && \text{(Multiset of strands)} \end{aligned}$$

A complex *is a multiset of strands P that is connected by bonds. Complexes are considered equal up to reordering of strands and/or renaming of bonds.*

In this paper I use a specific set of reaction rules: binding b, toehold unbinding u, 3-way strand displacement m_3, and 4-way strand exchange m_4. To define these rules, I use the following assumptions and notation. Each domain x has a complementary domain x^*, such that $(x^*)^* = x$. Each domain x is either *short* ("toehold") or *long*, and x^* is short iff x is short. Bonds are between exactly one domain instance and one instance of its complement: if $d!i$ appears in some P, then P has exactly one $d^*!i$ and no other instances of i.

A *pseudoknot* is a pair of bonds that cross over each other. Formally, a complex is *non-pseudoknotted* if, for some ordering of its strands, for every pair of bonds i, j, the two instances of i appear either both between or both outside the two instances of j. This *non-pseudoknotted ordering*, if it exists, is unique up to cyclic permutation [6]. Pseudoknots are poorly understood compared to non-pseudoknotted DSD systems; for this reason, DSD enumerators such as Peppercorn and Visual DSD often disallow them [9,13]. For the same reason, I define the reaction rules in such a way that no pseudoknot can form from initially non-pseudoknotted complexes (Fig. 1).

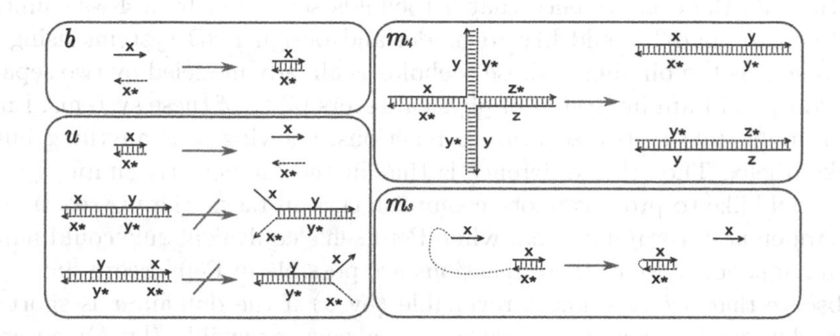

Fig. 1. The reaction rules of Definition 2. Dotted line in m_3 indicates that the reactant must be one complex; the products of m_3 and m_4 can be either one complex or two.

Definition 2 (Reaction rules). *The reactions of these DSD systems come from the following rules:*

1. *Binding (b):* $x, x^* \to x!i, x^*!i$ *if the product is non-pseudoknotted.*
2. *Toehold unbinding (u):* $x!i, x^*!i \to x, x^*$ *if* x *is short and* i *is not anchored. A bond* i *is* anchored *if it matches* $x!i\ y!j, y^*!j\ x^*!i$ *or* $y!j\ x!i, x^*!i\ y^*!j$.
3. *4-way branch migration* (m_4)*:* $x!i\ y!j, y^*!j\ z^*!k, z!k\ y!l, y^*!l\ x^*!i$ $\to x!i\ y!j, y^*!l\ z^*!k, z!k\ y!l, y^*!j\ x^*!i$.
4. *Remote-toehold 3-way strand displacement* (m_3)*:* $x, x!i, x^*!i \to x!i, x, x^*!i$ *if the reactant is one complex and the product is non-pseudoknotted.*

A Chemical Reaction Network *is a pair* $(\mathcal{S}, \mathcal{R})$ *of a set of abstract species* $\mathcal{S}$ *and a set of reactions between those species* $\mathcal{R}$*. The* DSD system enumerated from *an initial multiset of strands* P *and a set of rules is the smallest DSD system such that any complex in* P *is a species in* $\mathcal{S}$*, any application of one of the rules to reactants in* $\mathcal{S}$ *is a reaction in* $\mathcal{R}$*, and its products are species in* $\mathcal{S}$*.*

I use comma-separated sequences instead of |-separated sequences in the above definition to indicate that those sequences may each be part of a larger strand, and may or may not be on the same strand as each other. I use comma-separated sequences for the same purpose throughout the paper, which given the nature of the proofs is much more common than knowing the full sequence of a strand. Although in general checking whether a complex is non-pseudoknotted is hard due to having to find the non-pseudoknotted order, given a complex or complexes with known non-pseudoknotted order(s) checking whether an additional bond makes a pseudoknot (as in the b and m_3 conditions) is easy.

These reaction rules are similar, but not identical, to those from Petersen et al. [14]. The main difference is that Petersen's u rule counts 4-way (and n-way) junctions as "anchored" for the purpose of prohibiting u reactions. Based on Dabby's experiments and energy models of toehold-mediated 4-way branch migration [5] and the probe designed by Chen et al. using reversible 4-way branch migration [3], there is evidence that 2 toeholds separated by a 4-way junction *can* dissociate, and I would like to model and design DSD systems using this mechanism. As the binding of those toeholds is already modeled by two separate b reactions, and I am interested in physical reversibility of these systems, I modeled the unbinding as two separate u reactions, allowing u at anything but an unbroken helix. The other difference is that in the "sufficiently simple system" that I would like to prove cannot accomplish certain tasks, the m_4 reaction can only happen at a 4-way junction, while Petersen's equivalent rule could happen at larger junctions. All of these reactions are possible in Peppercorn [9].

Observe that a b reaction is reversible (by u) if the domain x is short and the bond formed is not anchored; a u is always reversible (by b); an m_3 is reversible (by m_3) if it does not separate complexes; and a m_4 is reversible (by m_4) if bonds j and l are part of a new 4-way junction. The impossibility proof I wish to present here deals with *reversible, 4-way-only DSD systems*; a DSD system is *reversible* if it contains only reversible reactions, and a DSD system is *4-way-only* if it contains no m_3 reactions. In such a system, largely because of the no-pseudoknots condition, I can prove an important locality theorem: that if a complex with a 4-way junction can, via unimolecular reactions, break the 4-way junction and eventually reach a different state with the same 4-way junction reformed, then that initial complex can reach the same final state (via unimolecular reactions) without ever breaking the 4-way junction. This, in a certain sense, one-way flow of information limits the type of reactions that can happen in such a system (Fig. 2).

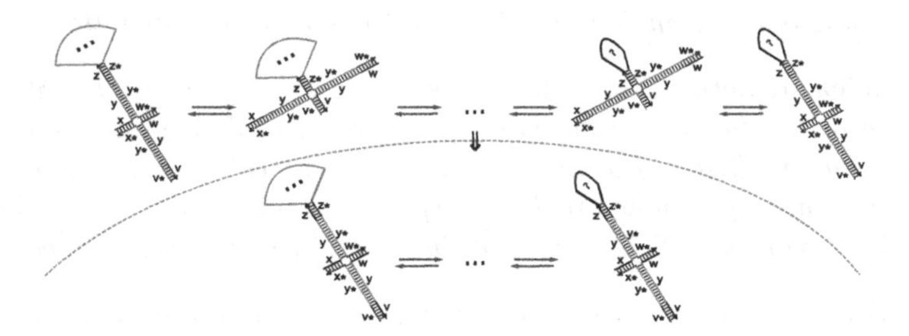

Fig. 2. Theorem 1 states roughly that if some change outside of a four-way junction can occur after which the four-way junction is reformed, then the same change can occur without breaking the four-way junction.

Theorem 1. *In a reversible, 4-way-only DSD system, consider a complex P containing a 4-way junction, where P is non-pseudoknotted. Assume $P \rightleftharpoons P'$ via a trajectory of unimolecular reactions, where P' contains the same 4-way junction (but may differ elsewhere). Then $P \rightleftharpoons P'$ via a trajectory of unimolecular reactions, not longer than the original trajectory, with no reaction changing the bonds within the 4-way junction.*

Proof. This theorem assumes an initial complex P, with no details specified except the existence of a 4-way junction:

$$x!i\ y!j, \qquad y^*!j\ z^*!k, \qquad z!k\ w!l, \qquad w^*!l\ x^*!i.$$

This theorem also assumes some sequence of reactions by which $P \rightleftharpoons P'$, again with no details specified except that the system is reversible and 4-way-only, the path contains only unimolecular reactions, and the result P' has the same 4-way junction made out of the same strand(s). I focus on three steps in the given trajectory: the first reaction that breaks the 4-way junction; the first reaction afterwards that requires that reaction to have happened; and the reaction that eventually reforms the original 4-way junction. (If the junction breaks and reforms multiple times, apply the theorem to each such sub-trajectory.) If those reactions do not exist, the result is trivial: if the junction never breaks, then the given trajectory is the desired trajectory, and if the first reaction that requires the break is the reaction that reforms the junction, then removing those two gives either the desired trajectory or (if what used to be the second reaction that breaks the 4-way junction is now the first) a shorter trajectory to which this theorem can be applied. Each of those three reactions has a limited number of possibilities, and I show that each combination of possibilities produces either a contradiction or the desired pathway.

In a reversible, 4-way-only system, a reaction that breaks a bond must be u or m_4. If the junction is first broken by u, then except for its reverse b, the only reaction that can depend on a u is a b with one of the newly opened domains:

$$x, x!i, x^*!i \xrightarrow{u} x, x, x^* \xrightarrow{b} x!i, x, x^*!i.$$

However, observe that this pair of reactions is equivalent to (and implies the possibility of) an m_3 reaction, thus the DSD system is not 4-way-only.

If the junction is first broken by m_4, to be reversible it must form a new 4-way junction with two smaller stems and two larger stems:

$$x!i\ y!j\ u!g, \qquad u^*!g\ y^*!l\ z^*!k, \qquad z!k\ y!l\ v!h, \qquad v^*!h\ y^*!j\ x^*!i.$$

To reform the junction, bonds j and l must be broken, specifically by u or m_4. If one is broken by u, then since both have domain identity y this allows an m_3 reaction with the other, contradicting 4-way-only. An m_4 reaction that is not the reverse of the original m_4, while it would break one of the bonds, would produce another bond that needs to be broken to reform the original junction, and to which this argument could be applied (formally, this is induction on the size of the remaining DNA complex). The only remaining possibility is that the original junction is reformed by the reverse m_4 of the one that broke it, implying that at that time the new (j, g, l, h) 4-way junction must be present.

I treat the above as a proof by induction on the length of the given trajectory by which $P \rightleftharpoons P'$. The base case is a trajectory of length 0, $P = P'$, which does not break the 4-way junction and thus is the desired trajectory. In a given trajectory of length $n > 0$, let P_1 be the state after the m_4 breaking the 4-way junction and P_1' the state before the reverse m_4; both have the same (j, g, l, h) 4-way junction and $P_1 \rightleftharpoons P_1'$ by a trajectory of length at most $n - 2$ satisfying the assumptions of this theorem. Thus $P_1 \rightleftharpoons P_1'$ by a trajectory not longer than the original in which no reaction breaks the (j, g, l, h) 4-way junction. But the only reactions that can require the original m_4 are a u or m_4 involving bonds j, g, l, and/or h, thus none of the new trajectory requires that m_4; and $P \rightleftharpoons P'$ by removing both m_4's and replacing the trajectory between them by the new $P_1 \rightleftharpoons P_1'$ trajectory. Because this new trajectory is valid both before and after the m_4 reaction exchanging bonds j and l ("j, l-m_4"), no reaction in it can break bonds i or k: a u reaction would be impossible post-j, l-m_4 as the pattern $x!i\ y!j, y^*!j\ x!i$ anchors i and similarly bond l anchors k, while an m_4 reaction on i or k would form a 6-way junction and thus be irreversible before the j, l-m_4.

3 The 2-r4 Condensed Reaction

Grun et al. in Peppercorn used the concept of a *condensed reaction*, which models a sequence of multiple reaction rules as one reaction [9]. They divide detailed reactions into "fast" and "slow" reactions, assume that no slow reaction can happen while any (productive sequence of) fast reactions can happen, and treat a single slow reaction followed by multiple fast reactions as a single, "condensed" reaction. The usual division, which I use, is that all unimolecular reactions are fast and all bimolecular reactions are slow.

Definition 3 (Grun et al. [9]). *Take as given a DSD system $(\mathcal{S}, \mathcal{R})$ and a division of reactions in that DSD system into "fast" unimolecular and "slow" bimolecular reactions. A* resting set *(or* resting state*) is a set of complexes connected by fast reactions, and none of which have outgoing fast reactions with products outside the resting set.*

A condensed reaction *is a reaction of resting sets, such that for some multiset of representatives of the reactant resting sets, there is a sequence of one slow reaction followed by zero or more fast reactions that produces a multiset of representatives of the product resting sets. The* condensed DSD system *corresponding to $(\mathcal{S}, \mathcal{R})$ is $(\hat{\mathcal{S}}, \hat{\mathcal{R}})$ where $\hat{\mathcal{S}}$ is the set of resting states of $\mathcal{S}$ and $\hat{\mathcal{R}}$ is the set of condensed reactions. A* reversible, condensed DSD system *is a condensed DSD system where every reaction in $\mathcal{R}$ and condensed reaction in $\hat{\mathcal{R}}$ is reversible.*

A given condensed reaction can correspond to multiple equivalent pathways of detailed reactions. To distinguish between detailed and condensed reactions, I will often use the word "step" to refer to a detailed reaction and "reaction" to

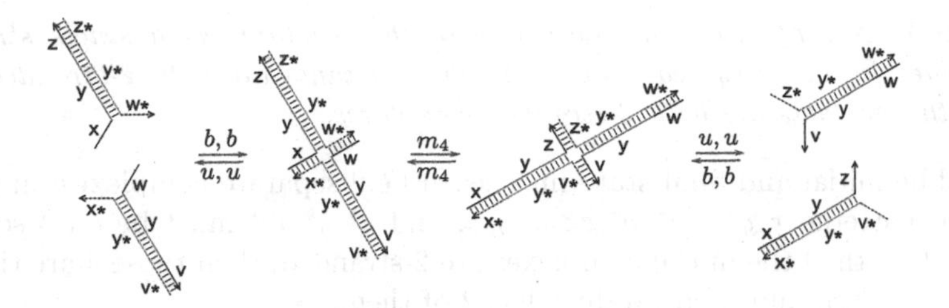

Fig. 3. The two-toehold-mediated reversible 4-way branch migration (2-r4) condensed reaction mechanism. For conciseness, the first and last two b, b and u, u detailed steps are drawn together.

refer to a condensed reaction. If all unimolecular steps are fast and all bimolecular steps slow, then a condensed reaction must be bimolecular and begin with a bimolecular b step. In a reversible system, all condensed reactions must have exactly two products and end with a u that separates complexes: one product can always reverse back to its reactants; with 3 or more products the reverse would not be a condensed reaction; and any steps after that u that do not separate complexes are reversible and thus are steps within a resting set.

An important example condensed reaction is the *two-toehold-mediated reversible 4-way strand exchange reaction*, or *2-r4* reaction, shown in Fig. 3.

Definition 4. *A* 2-r4 reaction *is a reversible condensed reaction of which one representative pathway is the following sequence of detailed reactions:*

$$x\ y!j\ z!g, z^*!g\ y^*!j\ w^*, w\ y!l\ v!h, v^*!h\ y^*!l\ x^*$$

$$\underset{u}{\overset{b}{\rightleftharpoons}} x!i\ y!j\ z!g, z^*!g\ y^*!j\ w^*, w\ y!l\ v!h, v^*!h\ y^*!l\ x^*!i$$

$$\underset{u}{\overset{b}{\rightleftharpoons}} x!i\ y!j\ z!g, z^*!g\ y^*!j\ w^*!k, w!k\ y!l\ v!h, v^*!h\ y^*!l\ x^*!i$$

$$\underset{m_4}{\overset{m_4}{\rightleftharpoons}} x!i\ y!j\ z!g, z^*!g\ y^*!l\ w^*!k, w!k\ y!l\ v!h, v^*!h\ y^*!j\ x^*!i$$

$$\underset{b}{\overset{u}{\rightleftharpoons}} x!i\ y!j\ z!g, z^*!g\ y^*!l\ w^*!k, w!k\ y!l\ v, v^*\ y^*!j\ x^*!i$$

$$\underset{b}{\overset{u}{\rightleftharpoons}} x!i\ y!j\ z, z^*\ y^*!l\ w^*!k, w!k\ y!l\ v, v^*\ y^*!j\ x^*!i$$

Any reversible condensed reaction that can be written as a sequence similar to the above, with more than one m_4 step across an unbroken sequence of migration domains, is also a 2-r4 reaction.

Recall that the comma-separated sequence notation means that the sequences above may be only part of their strands, and the complex may contain additional strands not mentioned. Note that the reverse pathway is also a 2-r4 reaction, hence the phrase "reversible condensed reaction". An important feature of the 2-r4 reaction is that if its reactants are at most 2-stranded, so are its products.

Lemma 1. *A 2-r4 reaction where one of the reactants is a single strand is impossible. In any 2-r4 reaction where the reactants are both 2-stranded complexes, the products are both 2-stranded complexes.*

Proof. The initial and final state must each be 2 separate complexes; in particular, the patterns $x\ y\ z$, $v^*\ y^*\ x$, $w\ y\ v$, and $z^*\ y^*\ w^*$ must be on 4 separate strands. If both of the initial complexes are 2-stranded, then those 4 are the only strands involved, and each product has 2 of them.

In addition to the restrictions of a reversible, condensed, 4-way-only DSD system, I would like to consider systems where all initial complexes have at most 2 strands each. In this case Lemma 1 suggests, and Theorem 2 proves, that all complexes involved will have at most 2 strands. Using Theorem 1, I show via a trace rewriting argument that any condensed reaction between 2-stranded complexes in a reversible, condensed, 4-way-only system is a 2-r4 reaction. As a first step in that proof, I observe that only b and u steps cannot make a nontrivial condensed reaction.

Lemma 2. *In a reversible, condensed, 4-way-only system, if a pathway of some condensed reaction with non-pseudoknotted reactants consists of only b and u steps, then the condensed reaction is trivial.*

Proof. Observe that any two consecutive u steps can happen in any order, and any u step that can happen after a b step can happen before it, except the reverse of that b (which can thus be removed). Then any pathway matching the assumptions of this lemma is equivalent to a pathway where all u steps happen before all b steps. Such a path is either trivial or involves one of the reactants separating unimolecularly, in which case that reactant was not in a resting state.

Theorem 2. *In a reversible, condensed, 4-way-only system, any condensed reaction between non-pseudoknotted 2-stranded complexes is a 2-r4 reaction.*

Proof. (Sketch) Given a reversible pathway representing a condensed reaction from two reactant resting states to two product resting states, I show that pathway can be "rewritten" into a pathway that represents the same condensed reaction and matches Definition 4.

Lemma 2 implies that an m_4 step, which does not eventually reverse itself, happens. That is, the first m_4 step in the reaction goes from

$$x!i\ y!j\ w!g, \qquad w^*!g\ y^*!j\ z^*!k, \qquad z!k\ y!l\ v!h, \qquad v^*!h\ y^*!l\ x^*!i$$

to

$$x!i\ y!j\ w!g, \qquad w^*!g\ y^*!l\ z^*!k, \qquad z!k\ y!l\ v!h, \qquad v^*!h\ y^*!j\ x^*!i.$$

If this m_4 was not possible in a resting state (and neither was an m_3 between what are now bonds j and l), it must be that bonds i and k were formed by inter-reactant b steps. Since these three steps cannot depend on any other b or u steps, there is an equivalent pathway where those bbm_4 are the first three steps.

(The case where y is a sequence of more than one domain and this m_4 is more than one m_4 is still covered by this pattern.)

Looking ahead to the u step that separates complexes, separating bonds g and h by two u steps (if w and v are short) completes the 2-r4 pattern. Any other possible separation can be eliminated. If the entire (j, g, l, h) 4-way junction is on one product, with the other product coming from the g or h stems, then the appropriate reactant must have been either pseudoknotted or 3-stranded. If the other product comes from the j or l stems *without* breaking bonds j or l, then it could have separated without the j, l-m_4 step. If either of bonds j or l is broken, then this will either allow an m_3 reaction, allow an irreversible reaction involving 6-way junctions, or be equivalent to the j, l-m_4 step never happening.

4 Chemical Reaction Network Implementations

Previous work on formal verification of CRNs allows one to define whether one CRN correctly implements another [10,18], and combine that work with the above definitions of modeling a DSD system as a CRN to verify that a DSD system correctly implements a CRN [1]. The definition of "systematically correct" in Definition 6 is based on (modular) CRN bisimulation [10].

Most existing DSD-based CRN implementations are not intended to be an implementation of one specific CRN, but rather a general translation scheme to implement arbitrary (or at least a wide class of) CRNs [2,15,19]. A systematic implementation is one where (a) each species has a "signal complex", and the signal complexes corresponding to two different species are identical except for the identities of certain domains; and (b) similarly, the pathways by which formal reactions of the same "type" are implemented, are identical to each other except for the identities of some domains unique to the species involved and some domains unique to the reaction. Figure 4 shows an example of this definition.

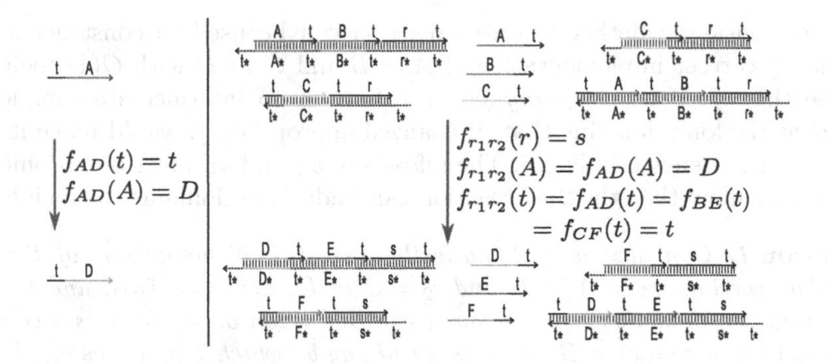

Fig. 4. An example $O(1)$ toeholds systematic DSD implementation (Cardelli et al. [2]). Left: signal species s_A and s_D, with domain isomorphism f_{AD}. Right: fuels F_{r_1} and F_{r_2} for $r_1 = A + B \rightleftharpoons C$ and $r_2 = D + E \rightleftharpoons F$, and domain isomorphism $f_{r_1r_2}$.

Definition 5 (Systematic implementation). *A* domain isomorphism *is an injective partial function on domains f with $f(x^*) = f(x)^*$. Where P is a multiset of strands and f a domain isomorphism, let $P\{f\}$ be the multiset of strands obtained from P by replacing each d with $f(d)$ whenever $f(d)$ is defined.*

A DSD implementation *of a formal CRN $(\mathcal{S}, \mathcal{R})$ is, for each species $A \in \mathcal{S}$, a "signal" DSD complex s_A, and for each reversible or irreversible reaction $r \in \mathcal{R}$, a set of "fuel" complexes F_r. A DSD implementation is* systematic *if:*

1. *Given species A, B there is a domain isomorphism f_{AB} where $s_B = s_A\{f_{AB}\}$. If $A \neq B$ there is at least one domain d appearing in s_A with $f_{AB}(d) \neq d$.*
2. *If a domain d is in both s_A and s_B for $A \neq B$, then $f_{CD}(d) = d$ for all C, D.*
3. *Given reactions r_1, r_2, if there is a bijection ϕ on species such that $r_2 = r_1\{\phi\}$, then there is a domain isomorphism $f_{r_1r_2}$ where $F_{r_2} = F_{r_1}\{f_{r_1r_2}\}$. For one such ϕ, for each A in r_1, $f_{r_1r_2} = f_{A\phi(A)}$ wherever $f_{A\phi(A)}$ is defined.*

A systematic DSD implementation is O(1) toeholds *if, whenever d is a short domain, all $f_{AB}(d) = d$ and all $f_{r_1r_2}(d) = d$.*

Where R is a multiset of species, the notation s_R means s_A for each $A \in R$.

Definition 6. *Given a CRN $(\mathcal{S}, \mathcal{R})$ and a systematic DSD implementation, let $(\mathcal{S}', \mathcal{R}')$ be the (detailed or condensed) DSD system enumerated, plus reactions $\emptyset \to F_r$ for each $r \in \mathcal{R}$. The implementation is* systematically correct *if:*

1. *There is an* interpretation *m mapping species in $\mathcal{S}'$ to multisets of species in $\mathcal{S}$, with each $m(s_A) = A$ and $x \in F_r \Rightarrow m(x) = \emptyset$.*
2. *For each $r' = R' \to P' \in \mathcal{R}'$, either $m(R') = m(P')$ (r' is "trivial") or $m(R') \to m(P')$ is some reaction $r \in \mathcal{R}$.*
3. *For each $r = R \to P \in \mathcal{R}$, there is a pathway containing exactly one nontrivial reaction, from $F_r + s_R$ to some $W_r + s_P$, with $x \in W_r \Rightarrow m(x) = \emptyset$.*
4. *For each $x \in \mathcal{S}'$, there is a pathway of trivial reactions from x to $s_{m(x)}$.*

I now consider whether the 2-r4 reaction can be used to construct a systematically correct implementation of $A \rightleftharpoons B$ and $C \rightleftharpoons D$ with $O(1)$ toeholds. I define the *information-bearing set* or *infoset* of an intermediate complex in $A \rightleftharpoons B$ as the long domains that, if changed appropriately, would make it "act like" $C \rightleftharpoons D$, as shown in Fig. 5. The infosets of s_A and s_B must be disjoint and nonempty. I show that the 2-r4 reaction can't add long domains to the infoset.

Definition 7. *Consider a systematically correct implementation of the two reversible reactions $r_1 = A \rightleftharpoons B$ and $r_2 = C \rightleftharpoons D$. Let f_{AC}, f_{BD}, and $f_{r_1r_2}$ be the appropriate domain isomorphisms as in Definition 5. Where x is a complex with $m(x) = A$ or $m(x) = B$, there is a pathway by which x produces s_B. Define the* information-bearing set *or* infoset *$I(x)$ to be the smallest set of domains for which, where x' is x with each domain $d \in I(x)$ replaced by $f_{r_1r_2}(d)$, x' can do the following: follow the pathway by which x produces s_B but using fuels from F_{r_2} to produce some complex s'_D; s'_D then mimics $D \to C$ to produce some s'_C; s'_C then mimics $C \to D$ to produce s_D.*

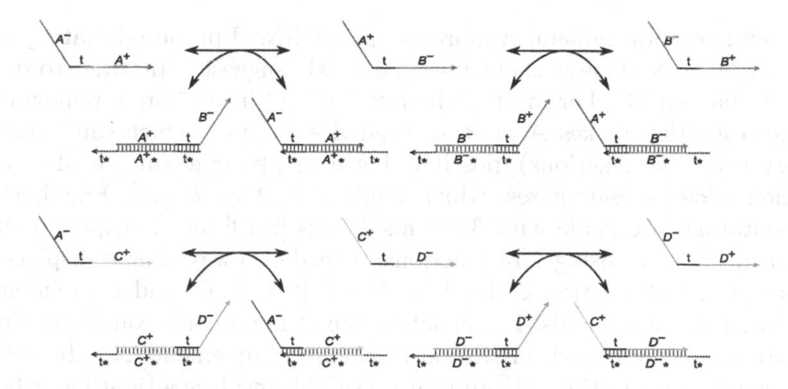

Fig. 5. In this systematic implementation of $A \rightleftharpoons B$ and $C \rightleftharpoons D$ using 3-way strand displacement, replacing A^+ with C^+ in s_A (left) lets it follow the $C \rightleftharpoons D$ pathway (bottom) instead of the $A \rightleftharpoons B$ pathway (top). Thus, the infoset $I(s_A) = \{A^+\}$.

(The use of s'_C and s'_D is a technical detail to say that domains whose identity exists but does not matter are not in the infoset.)

Lemma 3. *In a reversible, condensed, 4-way-only systematically correct $O(1)$-toehold DSD implementation of $A \rightleftharpoons B$ and $C \rightleftharpoons D$, let $x + f \rightleftharpoons x' + f'$ be a 2-r4 reaction on the pathway from s_A to s_B. Without loss of generality say $m(x)$ and $m(x')$ are each either A or B, and $m(f) = m(f') = \emptyset$. If x and f are non-pseudoknotted complexes with at most 2 strands each, then $I(x') \subset I(x)$.*

Proof. (Sketch) Observe that if x, f, x', or f' contains inter-strand long domain bonds not in the 2-r4 reaction pattern, then the 2-r4 reaction cannot happen.

Any domain $d \in I(x')$ must eventually be involved in a reaction; since $O(1)$ toeholds implies d is a long domain, the next such reaction must be an m_4 reaction. To participate in an m_4 reaction, it must be bound; for this to be a necessary reaction (i.e. not eliminated by Theorem 1; detailed proof omitted), its bound complement must be on a distinct strand. Since x' cannot have inter-strand bonds not mentioned in the 2-r4 reaction pattern, d can only be the y domain in Definition 4; thus $d \in I(x)$.

Given Theorem 2 and Lemma 3, the desired result is trivial.

Theorem 3. *No reversible, condensed, 4-way only systematically correct $O(1)$-toehold DSD implementation of $A \rightleftharpoons B$ and $C \rightleftharpoons D$ where each signal and fuel complex has at most 2 strands can exist.*

5 Discussion

The above proofs show that a reversible, condensed, 4-way-only DSD system with at most 2-stranded inputs cannot be a systematically correct with $O(1)$ toeholds implementation of multiple, independent reactions of the form $A \rightleftharpoons B$, and therefore cannot implement more complex CRNs. I proposed that a proof of "Task X cannot be done without mechanism M" suggests "In order to do task X, use mechanism M". For most of the restrictions I investigated, removing just that one restriction makes $A \rightleftharpoons B$, or even $A + B \rightleftharpoons C$ (which can implement arbitrary reversible reactions), possible. For example, removing the 4-way-only restriction allows seesaw gates, which implement $A \rightleftharpoons B$ [16]. Existing CRN implementations are made with 3-stranded fuels [2,19] and I suspect a similar mechanism can work using 2-r4 reactions instead of 3-way strand displacement reactions. The 2-r4 reaction *is* $A + B \rightleftharpoons C + D$ if A, B, C, and D are identified by combinations of toeholds [11], violating the $O(1)$ toeholds condition. Finally, a cooperative 4-way branch migration mechanism implements $A + B \rightleftharpoons C$ but is not a condensed reaction [11]. In that sense, this result is a tight lower bound.

I conjectured that a system with the same restrictions but allowing 3-way branch migration cannot implement $A + B \rightleftharpoons C$. Informally, 3-way strand displacement can easily implement $A \rightleftharpoons B$ [16], but has difficulty implementing "join" ($A + B \rightleftharpoons C$) logic without either 3-stranded fuels [2,15,19] or cooperative hybridization. Meanwhile, 4-way strand exchange can do join logic on $O(n)$ toeholds with the 2-r4 reaction [11], but is provably unable to transduce long domains with $O(1)$ toeholds. In trying to combine 3-way and 4-way mechanisms I have found numerous pitfalls. I suspect that with 2-stranded fuels, combining 3-way and 4-way mechanisms will not gain the advantages of both, and $A + B \rightleftharpoons C$ under the above restrictions will remain impossible.

Acknowledgements. I thank Chris Thachuk, Stefan Badelt, Erik Winfree, and Lulu Qian for helpful discussions on formal verification and on two-stranded DSD systems. I also thank the anonymous reviewers of a rejected previous version of this paper for their suggestions, many of which appear in this version. I thank the NSF Graduate Research Fellowship Program for financial support.

References

1. Badelt, S., Shin, S.W., Johnson, R.F., Dong, Q., Thachuk, C., Winfree, E.: A general-purpose CRN-to-DSD compiler with formal verification, optimization, and simulation capabilities. In: Brijder, R., Qian, L. (eds.) DNA 2017. LNCS, vol. 10467, pp. 232–248. Springer, Cham (2017). https://doi.org/10.1007/978-3-319-66799-7_15
2. Cardelli, L.: Two-domain DNA strand displacement. Math. Struct. Comput. Sci. **23**, 247–271 (2013)
3. Chen, S.X., Zhang, D.Y., Seelig, G.: Conditionally fluorescent molecular probes for detecting single base changes in double-stranded DNA. Nat. Chem. **5**(9), 782 (2013)
4. Chen, Y.J., et al.: Programmable chemical controllers made from DNA. Nat. Nanotechnol. **8**, 755–762 (2013)
5. Dabby, N.L.: Synthetic molecular machines for active self-assembly: prototype algorithms, designs, and experimental study. Ph.D. thesis, California Institute of Technology, February 2013

6. Dirks, R.M., Bois, J.S., Schaeffer, J.M., Winfree, E., Pierce, N.A.: Thermodynamic analysis of interacting nucleic acid strands. SIAM Rev. **49**(1), 65–88 (2007)
7. Douglas, S.M., Bachelet, I., Church, G.M.: A logic-gated nanorobot for targeted transport of molecular payloads. Science **335**(6070), 831–834 (2012)
8. Groves, B., et al.: Computing in mammalian cells with nucleic acid strand exchange. Nat. Nanotechnol. **11**(3), 287 (2016)
9. Grun, C., Sarma, K., Wolfe, B., Shin, S.W., Winfree, E.: A domain-level DNA strand displacement reaction enumerator allowing arbitrary non-pseudoknotted secondary structures. CoRR p. http://arxiv.org/abs/1505.03738 (2015)
10. Johnson, R.F., Dong, Q., Winfree, E.: Verifying chemical reaction network implementations: a bisimulation approach. Theoret. Comput. Sci. (2018). https://doi.org/10.1016/j.tcs.2018.01.002
11. Johnson, R.F., Qian, L.: Simplifying chemical reaction network implementations with two-stranded DNA building blocks, in preparation
12. Lakin, M.R., Stefanovic, D., Phillips, A.: Modular verification of chemical reaction network encodings via serializability analysis. Theoret. Comput. Sci. **632**, 21–42 (2016)
13. Lakin, M.R., Youssef, S., Polo, F., Emmott, S., Phillips, A.: Visual DSD: a design and analysis tool for DNA strand displacement systems. Bioinformatics **27**, 3211–3213 (2011)
14. Petersen, R.L., Lakin, M.R., Phillips, A.: A strand graph semantics for DNA-based computation. Theoret. Comput. Sci. **632**, 43–73 (2016)
15. Qian, L., Soloveichik, D., Winfree, E.: Efficient turing-universal computation with DNA polymers. In: Sakakibara, Y., Mi, Y. (eds.) DNA 2010. LNCS, vol. 6518, pp. 123–140. Springer, Heidelberg (2011). https://doi.org/10.1007/978-3-642-18305-8_12
16. Qian, L., Winfree, E.: Scaling up digital circuit computation with DNA strand displacement cascades. Science **332**(6034), 1196–1201 (2011)
17. Qian, L., Winfree, E.: Parallel and scalable computation and spatial dynamics with DNA-based chemical reaction networks on a surface. In: Murata, S., Kobayashi, S. (eds.) DNA 2014. LNCS, vol. 8727, pp. 114–131. Springer, Cham (2014). https://doi.org/10.1007/978-3-319-11295-4_8
18. Shin, S.W., Thachuk, C., Winfree, E.: Verifying chemical reaction network implementations: a pathway decomposition approach. Theor. Comput. Sci. (2017) https://doi.org/10.1016/j.tcs.2017.10.011
19. Soloveichik, D., Seelig, G., Winfree, E.: DNA as a universal substrate for chemical kinetics. Proc. Nat. Acad. Sci. **107**, 5393–5398 (2010)
20. Srinivas, N., Parkin, J., Seelig, G., Winfree, E., Soloveichik, D.: Enzyme-free nucleic acid dynamical systems. Science **358** (2017). https://doi.org/10.1126/science.aal2052
21. Thubagere, A.J., et al.: A cargo-sorting DNA robot. Science **357**(6356), eaan6558 (2017)
22. Thubagere, A.J., et al.: Compiler-aided systematic construction of large-scale DNA strand displacement circuits using unpurified components. Nat. Commun. **8**, 14373 (2017)
23. Venkataraman, S., Dirks, R.M., Rothemund, P.W., Winfree, E., Pierce, N.A.: An autonomous polymerization motor powered by DNA hybridization. Nat. Nanotechnol. **2**(8), 490 (2007)
24. Zhang, D.Y., Seelig, G.: Dynamic DNA nanotechnology using strand-displacement reactions. Nat. Chem. **3**(2), 103–113 (2011)

Quantum Algorithm for Dynamic Programming Approach for DAGs. Applications for Zhegalkin Polynomial Evaluation and Some Problems on DAGs

Kamil Khadiev[1,2](✉) and Liliya Safina[2]

[1] Smart Quantum Technologies Ltd., Kazan, Russia
[2] Kazan Federal University, Kazan, Russia
kamilhadi@gmail.com, liliasafina94@gmail.com

Abstract. In this paper, we present a quantum algorithm for dynamic programming approach for problems on directed acyclic graphs (DAGs). The running time of the algorithm is $O(\sqrt{\hat{n}m}\log\hat{n})$, and the running time of the best known deterministic algorithm is $O(n+m)$, where n is the number of vertices, $\hat{n}$ is the number of vertices with at least one outgoing edge; m is the number of edges. We show that we can solve problems that use OR, AND, NAND, MAX and MIN functions as the main transition steps. The approach is useful for a couple of problems. One of them is computing a Boolean formula that is represented by Zhegalkin polynomial, a Boolean circuit with shared input and non-constant depth evaluating. Another two are the single source longest paths search for weighted DAGs and the diameter search problem for unweighted DAGs.

Keywords: Quantum computation · Quantum models · Quantum algorithm · Query model · Graph · Dynamic programming · DAG · Boolean formula · Zhegalkin polynomial · DNF · AND-OR-NOT formula · NAND · Computational complexity · Classical vs. quantum · Boolean formula evaluation

1 Introduction

Quantum computing [9,32] is one of the hot topics in computer science of last decades. There are many problems where quantum algorithms outperform the best known classical algorithms [18,28]. superior of quantum over classical was shown for different computing models like query model, streaming processing models, communication models and others [1–6,27,29–31].

In this paper, we present the quantum algorithm for the class of problems on directed acyclic graphs (DAGs) that uses a dynamic programming approach. The dynamic programming approach is one of the most useful ways to solve problems in computer science [17]. The main idea of the method is to solve a problem using pre-computed solutions of the same problem, but with smaller

I. McQuillan and S. Seki (Eds.): UCNC 2019, LNCS 11493, pp. 150–163, 2019.
https://doi.org/10.1007/978-3-030-19311-9_13

parameters. Examples of such problems for DAGs that are considered in this paper are the single source longest path search problem for weighted DAGs and the diameter search problem for unweighted DAGs.

Another example is a Boolean circuit with non-constant depth and shared input evaluation. A Boolean circuit can be represented as a DAG with conjunction (AND) or disjunction (OR) in vertices, and inversion (NOT) on edges. We present it as an algorithm for computing a Zhegalkin polynomial [24,33,34]. The Zhegalkin polynomial is a way of a Boolean formula representation using "exclusive or" (XOR, $\oplus$), conjunction (AND) and the constants 0 and 1.

The best known deterministic algorithm for dynamic programming on DAGs uses depth-first search algorithm (DFS) as a subroutine [17]. Thus, this algorithm has at least depth-first search algorithm's time complexity, that is $O(n+m)$, where m is the number of edges and n is the number of vertices. The query complexity of the algorithm is at least $O(m)$.

We suggest a quantum algorithm with the running time $O(\sqrt{\hat{n}m}\log\hat{n})$, where $\hat{n}$ is the number of vertices with non-zero outgoing degree. In a case of $\hat{n}(\log\hat{n})^2 < m$, it shows speed-up comparing with deterministic algorithm. The quantum algorithm can solve problems that use a dynamic programming algorithm with OR, AND, NAND, MAX or MIN functions as transition steps. We use Grover's search [13,25] and Dürr and Høyer maximum search [23] algorithms to speed up our search. A similar approach has been applied by Dürr et al. [21,22]; Ambainis and Špalek [11]; Dörn [19,20] to several graph problems.

We apply this approach to four problems that discussed above. The first of them involves computing Boolean circuits. Such circuits can be represented as AND-OR-NOT DAGs. Sinks of such graph are associated with Boolean variables, and other vertices are associated with conjunction (AND) or a disjunction (OR); edges can be associated with inversion (NOT) function. Quantum algorithms for computing AND-OR-NOT trees were considered by Ambainis et al. [7,8,10]. Authors present an algorithm with running time $O(\sqrt{N})$, where N is the number of tree's vertices. There are other algorithms that allow as construct AND-OR-NOT DAGs of constant depth, but not a tree [15,16].

Our algorithm works with $O(\sqrt{\hat{n}m}\log\hat{n})$ running time for DAGs that can have non-constant depth.

It is known that any Boolean function can be represented as a Zhegalkin polynomial [24,33,34]. The computing of a Zhegalkin polynomial is the second problem. Such formula can be represented as an AND-OR-NOT DAG. Suppose an original Zhegalkin polynomial has t_c conjunctions and t_x exclusive-or operations. Then the corresponding AND-OR-NOT DAG have $\hat{n} = O(t_x)$ non-sink vertices and $m = O(t_c + t_x)$ edges.

If we consider AND-OR-NOT trees representation of the formula, then it has an exponential number of vertices $N \geq 2^{O(t_x)}$. The quantum algorithm for trees [7,8,10] works in $O(\sqrt{N}) = 2^{O(t_x)}$ running time. Additionally, the DAG that corresponding to a Zhegalkin polynomial has non-constant depth. Therefore, we cannot use algorithms from [15,16] that work for circuits with shared input.

The third problem is the single source longest path search problem for a weighted DAG. The best deterministic algorithm for this problem works in $O(n+m)$ running time [17]. In the case of a general graph (not DAG), it is a NP-complete problem. Our algorithm for DAGs works in $O(\sqrt{\hat{n}m}\log\hat{n})$ running time. The fourth problem is the diameter search problem for an unweighted DAG. The best deterministic algorithms for this problem works in $O(\hat{n}(n+m))$ expected running time [17]. Our algorithm for DAGs works in expected $O(\hat{n}(n+\sqrt{nm})\log n)$ running time.

The paper is organized as follows. We present definitions in Sect. 2. Section 3 contains a general description of the algorithm. The application to an AND-OR-NOT DAG evaluation and Zhegalkin polynomial evaluation is in Sect. 4. Section 5 contains a solution for the single source longest path search problem for a weighted DAG and the diameter search problem for an unweighted DAG.

2 Preliminaries

Let us present definitions and notations that we use.

A graph G is a pair $G=(V,E)$ where $V=\{v_1,\dots,v_n\}$ is a set of vertices, $E=\{e_1,\dots,e_m\}$ is a set of edges, an edge $e\in E$ is a pair of vertices $e=(v,u)$, for $u,v\in V$. A graph G is directed if all edges $e=(v,u)$ are ordered pairs, and if $(v,u)\in E$, then $(u,v)\notin E$. In that case, an edge e leads from vertex v to vertex u. A graph G is acyclic if there is no path that starts and finishes in the same vertex. We consider only directed acyclic graphs (DAGs) in the paper.

Let $D_i=(v:\exists e=(v_i,v)\in E)$ be a list of v_i vertex's out-neighbors. Let $d_i=|D_i|$ be the out-degree of the vertex v_i. Let L be a set of indices of sinks. Formally, $L=\{i:d_i=0, 1\le i\le n\}$. Let $\hat{n}=n-|L|$.

Let $D'_i=(v:\exists e=(v,v_i)\in E)$ be a list of vertex whose out-neighbor is v.

We consider only DAGs with two additional properties:

- topological sorted: if there is an edge $e=(v_i,v_j)\in E$, then $i<j$;
- last $|L|$ vertices belong to L, formally $d_i=0$, for $i>\hat{n}$.

Our algorithms use some quantum algorithms as a subroutine, and the rest is classical. As quantum algorithms, we use query model algorithms. These algorithms can do a query to black box that has access to the graph structure and stored data. As a running time of an algorithm we mean a number of queries to black box. We use an *adjacency list model* as a model for a graph representation. The input is specified by n arrays D_i, for $i\in\{1\dots n\}$. We suggest [32] as a good book on quantum computing.

3 Quantum Dynamic Programming Algorithm for DAGs

Let us describe an algorithm in a general case.

Let us consider some problem P on a directed acyclic graph $G=(V,E)$. Suppose that we have a dynamic programming algorithm for P or we can say that there is a solution of the problem P that is equivalent to computing a function f for each vertex. As a function f we consider only functions from a set $\mathcal{F}$ with following properties:

- $f : V \to \Sigma$.
- The result set Σ can be the set of real numbers $\mathbb{R}$, or integers $\{0,\dots,\mathcal{Z}\}$, for some integer $\mathcal{Z} > 0$.
- if $d_i > 0$ then $f(v_i) = h_i(f(u_1),\dots,f(u_{d_i}))$, where functions h_i are such that $h_i : \Sigma^{[1,n]} \to \Sigma$; $\Sigma^{[1,n]} = \{(r_1,\dots,r_k) : r_j \in \Sigma, 1 \le j \le k, 1 \le k \le n\}$ is a set of vectors of at most n elements from Σ; $(u_1,\dots,u_{d_i}) = D_i$.
- if $d_i = 0$ then $f(v_i)$ is classically computable in constant time.

Suppose there is a quantum algorithm Q_i that computes function h_i with running time $T(k)$, where k is a length of the argument for the function h_i. Then we can suggest the following algorithm:

Algorithm 1. Quantum Algorithm for Dynamic programming approach on DAGs. Let $t = (t[1],\dots,t[\hat{n}])$ be an array which stores results of the function f. Let $t_f(j)$ be a function such that $t_f(j) = t[j]$, if $j \le \hat{n}$; $t_f(j) = f(j)$, if $j > \hat{n}$. Note that $j > \hat{n}$ means $v_j \in L$.

---

for $i = \hat{n} \dots 1$ **do**
  $t[i] \leftarrow Q_i(t_f(j_1),\dots,t_f(j_{d_i}))$, where $(v_{j_1},\dots,v_{j_{d_i}}) = D_i$
end for
return $t[1]$

Let us discuss the running time of Algorithm 1. The proof is simple, but we present it for completeness.

Lemma 1. *Suppose that the quantum algorithms Q_i works in $T_i(k)$ running time, where k is a length of an argument, $i \in \{1,\dots,n\}$. Then Algorithm 1 works in $T^1 = \sum_{i\in\{1,\dots,n\}\backslash L} T_i(d_i)$.*

Proof. Note, that when we compute $t[i]$, we already have computed $t_f(j_1),\dots,t_f(j_{d_i})$ or can compute them in constant time because for all $e = (v_i, v_j) \in E$ we have $i < j$.

A complexity of a processing a vertex v_i is $T_i(d_i)$, where $i \in \{1,\dots,n\}\backslash L$. The algorithm process vertices one by one. Therefore

$$T^1 = \sum_{i\in\{1,\dots,n\}\backslash L} T_i(d_i).$$

□

Note, that quantum algorithms have a probabilistic behavior. Let us compute an error probability for Algorithm 1.

Lemma 2. *Suppose the quantum algorithm Q_i for the function h_i has an error probability $\varepsilon(n)$, where $i \in \{1, \ldots, n\} \backslash L$. Then the error probability of Algorithm 1 is at most* $1 - (1 - \varepsilon(n))^{\hat{n}}$.

Proof. Let us compute a success probability for Algorithm 1. Suppose that all vertices are computed with no error. The probability of this event is $(1-\varepsilon(n))^{\hat{n}}$, because error of each invocation is independent event.

Therefore, the error probability for Algorithm 1 is at most $1 - (1 - \varepsilon(n))^{\hat{n}}$, for $\hat{n} = n - |L|$. □

For some functions and algorithms, we do not have a requirement that all arguments of h should be computed with no error. In that case, we will get better error probability. This situation is discussed in Lemma 7.

3.1 Functions for Vertices Processing

We can choose the following functions as a function h

- Conjunction (AND function). For computing this function, we can use Grover's search algorithm [13,25] for searching 0 among arguments. If the element that we found is 0, then the result is 0. If the element is 1, then there is no 0s, and the result is 1.
- Disjunction (OR function). We can use the same approach, but here we search 1s.
- Sheffer stroke (Not AND or $NAND$ function). We can use the same approach as for AND function, but here we search 1s. If we found 0 then the result is 1; and 0, otherwise.
- Maximum function (MAX). We can use the Dürr and Høyer maximum search algorithm [23].
- Minimum function (MIN). We can use the same algorithm as for maximum.
- Other functions that have quantum algorithms.

As we discussed before, AND, OR and $NAND$ functions can be computed using Grover search algorithm. Therefore algorithm for these functions on vertex v_i has an error $\varepsilon_i \leq 0.5$ and running time is $T(d_i) = O(\sqrt{d_i})$, for $i \in \{1, \ldots, n\} \backslash L$. These results follow from [13,14,25,26] We have the similar situation for computing maximum and minimum functions [23].

If we use these algorithms in Algorithm 1 then we obtain the error probability $1 - (0.5)^{\hat{n}}$ due to Lemma 2.

At the same time, the error is one side. That is why we can apply the boosting technique to reduce the error probability. The modification is presented in Algorithm 2.

Let us consider MAX and MIN functions. Suppose, we have a quantum algorithm Q and a deterministic algorithm A for a function $h \in \{MAX, MIN\}$. Let k be number of algorithm's invoking according to the boosting technique for reducing the error probability. The number k is integer and $k \geq 1$. Let $(x_1, \ldots, x_d)$ be arguments (input data) of size d.

If we analyze the algorithm, then we can see that it has the following property:

Algorithm 2. The application of the boosting technique to Quantum Algorithm. Let us denote it as $\hat{Q}^k(x_1, \dots, x_d)$. Suppose, we have a temporary array $b = (b[1], \dots, b[k])$.

---

for $z = 1 \dots k$ **do**
  $b[z] \leftarrow Q(x_1, \dots, x_d)$
end for
$result \leftarrow A_i\,(b[1], \dots, b[k])$
return $result$

Lemma 3. *Let $(x_1, \dots, x_d)$ be an argument (input data) of size d, for a function $h(x_1, \dots, x_d) \in \{MAX, MIN\}$. Let k be a number of algorithm's invoking. The number k is integer and $k \geq 1$. Then the expected running time of the boosted version $\hat{Q}^k(x_1, \dots, x_d)$ of the quantum algorithm $Q(x_1, \dots, x_d)$ is $O\left(k \cdot \sqrt{d}\right)$ and the error probability is $O(1/2^k)$.*

Proof. Due to [23], the expected running time of the algorithm Q is $O\left(\sqrt{d}\right)$ and the error probability is at most 0.5. We repeat the algorithm k times and then apply A (function MAX or MIN) that has running time $O(k)$. Therefore, the expected running time is $O\left(k\sqrt{d}\right)$. The algorithm is successful if at least one invocation is successful because we apply A. Therefore, the error probability is $O\left(1/2^k\right)$. □

At the same time, if an algorithm Q measures only in the end, for example Grover Algorithm; then we can apply the amplitude amplification algorithm [14] that boosts the quantum algorithm in running time $O(\sqrt{k})$. The amplitude amplification algorithm is the generalization of the Grover's search algorithm. So, in the case of AND, OR and $NAND$ functions, we have the following property:

Lemma 4. *Let $(x_1, \dots, x_d)$ be an argument (input data) of size d for some function $h(x_1, \dots, x_d) \in \{AND, OR, NAND\}$. Suppose, we have a quantum algorithm Q for h. Let k be planned number of algorithm's invoking in a classical case. The number k is integer and $k \geq 1$. Then the running time of the boosted version $\hat{Q}^k(x_1, \dots, x_d)$ of the quantum algorithm $Q(x_1, \dots, x_d)$ is $O\left(\sqrt{k \cdot d}\right)$ and the error probability is $O\left(1/2^k\right)$.*

Proof. Due to [13,14,25,26], the running time of the algorithm Q is at most $O\left(\sqrt{d}\right)$ and the error probability is at most 0.5. We apply amplitude amplification algorithm [14] to the algorithm and gets the claimed running time and the error probability □

Let us apply the previous two lemmas to Algorithm 1 and functions from the set $\{AND, OR, NAND, MAX, MIN\}$.

Lemma 5. *Suppose that a problem P on a DAG $G = (V, E)$ has a dynamic programming algorithm such that functions $h_i \in \{AND, OR, NAND, MAX, MIN\}$, for $i \in \{1, \ldots, \hat{n}\}$. Then there is a quantum dynamic programming algorithm A for the problem P that has running time $O(\sqrt{\hat{n}m}\log\hat{n}) = O(\sqrt{nm}\log n)$ and error probability $O(1/\hat{n})$. Here $m = |E|, n = |V|, \hat{n} = n - |L|$, L is the set of sinks.*

Proof. Let us choose $k = 2\log_2\hat{n}$ in Lemmas 3 and 4. Then the error probabilities for the algorithms $Q^{2\log_2\hat{n}}$ are $O\left(0.5^{2\log_2\hat{n}}\right) = O\left(1/n^2\right)$. The running time is $O(\sqrt{d_i \log\hat{n}}) = O(\sqrt{d_i}\log\hat{n})$ for $h_i \in \{AND, OR, NAND\}$ and $O(\sqrt{d_i}\log\hat{n})$ for $h_i \in \{MAX, MIN\}$.

Due to Lemma 2, the probability of error is at most $\varepsilon(\hat{n}) = 1 - \left(1 - \frac{1}{\hat{n}^2}\right)^{\hat{n}}$. Note that

$$\lim_{\hat{n}\to\infty}\frac{\varepsilon(\hat{n})}{1/\hat{n}} = \lim_{\hat{n}\to\infty}\frac{1 - \left(1 - \frac{1}{\hat{n}^2}\right)^{\hat{n}}}{1/\hat{n}} = 1;$$

Hence, $\varepsilon(\hat{n}) = O(1/\hat{n})$.

Due to Lemma 1, the running time is

$$T^1 = \sum_{i\in\{1,\ldots,n\}\backslash L} T_i(d_i) \leq \sum_{i\in\{1,\ldots,n\}\backslash L} O\left(\sqrt{d_i}\log\hat{n}\right) = O\left((\log_2\hat{n})\cdot\sum_{i\in\{1,\ldots,n\}\backslash L}\sqrt{d_i}\right).$$

Due to the Cauchy-Bunyakovsky-Schwarz inequality, we have

$$\sum_{i\in\{1,\ldots,n\}\backslash L}\sqrt{d_i} \leq \sqrt{\hat{n}\sum_{i\in\{1,\ldots,n\}\backslash L} d_i}$$

Note that $d_i = 0$, for $i \in L$. Therefore, $\sum_{i\in\{1,\ldots,n\}\backslash L} d_i = \sum_{i\in\{1,\ldots,n\}} d_i = m$, because $m = |E|$ is the total number of edges. Hence,

$$\sqrt{\hat{n}\sum_{i\in\{1,\ldots,n\}\backslash L} d_i} = \sqrt{\hat{n}\sum_{i\in\{1,\ldots,n\}} d_i} = \sqrt{\hat{n}m}.$$

Therefore, $T_1 \leq O(\sqrt{\hat{n}m}\log\hat{n}) = O(\sqrt{nm}\log n)$. □

If $h_i \in \{AND, OR, NAND\}$, then we can do a better estimation of the running time.

Lemma 6. *Suppose that a problem P on a DAG $G = (V, E)$ has a dynamic programming algorithm such that functions $h_i \in \{AND, OR, NAND\}$, for $i \in \{1, \ldots, \hat{n}\}$. Then there is a quantum dynamic programming algorithm A for the problem P that has running time $O(\sqrt{\hat{n}m\log\hat{n}}) = O(\sqrt{nm\log n})$ and error probability $O(1/\hat{n})$. Here $m = |E|, n = |V|, \hat{n} = n - |L|$, L is the set of sinks.*

Proof. The proof is similar to the proof of Lemma 5, but we use $O(\sqrt{d_i}\log\hat{n})$ as time complexity of processing a vertex. □

If $h_i \in \{MAX, MIN\}$, then we can do a better estimation of the running time

Lemma 7. *Suppose that a problem P on a DAG $G = (V,E)$ has a dynamic programming algorithm such that functions $h_i \in \{MAX, MIN\}$, for $i \in \{1,\dots,\hat{n}\}$ and the solution is $f(v_a)$ for some $v_a \in V$. Then there is a quantum dynamic programming algorithm A for the problem P that has expected running time $O(\sqrt{\hat{n}m}\log q) = O(\sqrt{nm}\log n)$ and error probability $O(1/q)$, where q is the length of path to the farthest vertex from the vertex v_a. Here $m = |E|, n = |V|, \hat{n} = n - |L|$, L is the set of sinks.*

Proof. Let Q be the Dürr-Høyer quantum algorithm for MAX or MIN function. Let $\hat{Q}^q$ from Algorithm 2 be the boosted version of Q. Let us analyze the algorithm.

Let us consider a vertex v_i for $i \in \{1,\dots,n\}\backslash L$. When we process v_i, we should compute MAX or MIN among $t_f(j_1),\dots,t_f(j_{d_i})$. Without limit of generalization we can say that we compute MAX function. Let r be an index of maximal element. It is required to have no error for computing $t[j_r]$. At the same time, if we have an error on processing v_{j_w}, $w \in \{1,\dots,d_i\}\backslash\{r\}$; then we get value $t[j_w] < f(v_{j_w})$. In that case, we still have $t[j_r] > t[j_w]$. Therefore, an error can be on processing of any vertex v_{j_w}.

Let us focus on the vertex v_a. For computing $f(v_a)$ with no error, we should compute $f(v_{a_1})$ with no error. Here $v_{a_1} \in D_a$ such that maximum is reached on v_{a_1}. For computing $f(v_{a_1})$ with no error, we should compute $f(v_{a_2})$ with no error. Here $v_{a_2} \in D_{a_1}$ such that maximum is reached on v_{a_2} and so on. Hence, for solving problem with no error, we should process only at most q vertices with no error.

Therefore, the probability of error for the algorithm is

$$1-\left(1-\left(\frac{1}{2}\right)^{2\log q}\right)^{q} = O\left(\frac{1}{q}\right) \text{ because } \lim_{q\to\infty}\frac{1-\left(1-\frac{1}{q^2}\right)^{q}}{1/q} = 1.$$

□

4 Quantum Algorithms for Evolution of Boolean Circuits with Shared Inputs and Zhegalkin Polynomial

Let us apply ideas of quantum dynamic programming algorithms on DAGs to AND-OR-NOT DAGs.

It is known that any Boolean function can be represented as a Boolean circuit with AND, OR and NOT gates [12,33]. Any such circuit can be represented as a DAG with the following properties:

- sinks are labeled with variables. We call these vertices "variable-vertices".
- There is no vertices v_i such that $d_i = 1$.
- If a vertex v_i such that $d_i \geq 2$; then the vertex labeled with Conjunction or Disjunction. We call these vertices "function-vertices".
- Any edge is labeled with 0 or 1.
- There is one particular root vertex v_s.

The graph represents a Boolean function that can be evaluated in the following way. We associate a value $r_i \in \{0, 1\}$ with a vertex v_i, for $i \in \{1, \ldots, n\}$. If v_i is a variable-vertex, then r_i is a value of a corresponding variable. If v_i is a function-vertex labeled by a function $h_i \in \{AND, OR\}$, then $r_i = h_i\left(r_{j_1}^{\sigma(i,j_1)}, \ldots, r_{j_w}^{\sigma(i,j_w)}\right)$, where $w = d_i$, $(v_{j_1}, \ldots, v_{j_w}) = D_i$, $\sigma(i,j)$ is a label of an edge $e = (i,j)$. Here, we say that $x^1 = x$ and $x^0 = \neg x$ for any Boolean variable x. The result of the evolution is r_s.

An AND-OR-NOT DAG can be evaluated using the following algorithm that is a modification of Algorithm 1:

Algorithm 3. Quantum Algorithm for AND-OR-NOT DAGs evaluation. Let $r = (r_1, \ldots, r_n)$ be an array which stores results of functions h_i. Let a variable-vertex v_i be labeled by $x(v_i)$, for all $i \in L$. Let Q_i be a quantum algorithm for h_i; and $\hat{Q}_i^{2\log_2 \hat{n}}$ be a boosted version of Q_i using amplitude amplification (Lemma 4). Let $t_f(j)$ be a function such that $t_f(j) = r_j$, if $j \leq \hat{n}$; $t_f(j) = x(v_j)$, if $j > \hat{n}$.

---

for $i = \hat{n} \ldots s$ **do**
  $t[i] \leftarrow \hat{Q}_i^{2\log_2 \hat{n}}(t_f(j_1)^{\sigma(i,j_1)}, \ldots, t_f(j_w)^{\sigma(i,j_w)})$, where $w = d_i$, $(v_{j_1}, \ldots, v_{j_w}) = D_i$.
end for
return $t[s]$

Algorithm 3 has the following property:

Theorem 1. *Algorithm 3 evaluates a AND-OR-NOT DAG $G = (V, E)$ with running time $O(\sqrt{\hat{n}m}\log\hat{n}) = O(\sqrt{nm}\log n)$ and error probability $O(1/\hat{n})$. Here $m = |E|, n = |V|, \hat{n} = n - |L|$, L is the set of sinks.*

Proof. Algorithm 3 evaluates the AND-OR-NOT DAG G by the definition of AND-OR-NOT DAG for the Boolean function F. Algorithm 3 is almost the same as Algorithm 1. The difference is labels of edges. At the same time, the Oracle gets information on edge in constant time. Therefore, the running time and the error probability of $\hat{Q}_i^{2\log_2 \hat{n}}$ does not change. Hence, using the proof similar to the proof of Lemma 4 we obtain the claimed running time and error probability. □

Another way of a Boolean function representation is a NAND DAG or Boolean circuit with NAND gates [12,33]. We can present NAND formula as a DAG with similar properties as AND-OR-NOT DAG, but function-vertices has only NAND labels. At the same time, if we want to use more operations, then we can consider NAND-NOT DAGs and NAND-AND-OR-NOT DAGs:

Theorem 2. *Algorithm 3 evaluates a NAND-AND-OR-NOT DAG and a NAND-NOT DAG. If we consider a DAG* $G = (V, E)$*, then these algorithms work with running time* $O(\sqrt{\hat{n}m}\log\hat{n}) = O(\sqrt{nm}\log n)$ *and error probability* $O(1/\hat{n})$*. Here* $m = |E|, n = |V|, \hat{n} = n - |L|$*,* L *is the set of sinks.*

Proof. The proof is similar to proofs of Lemma 5 and Theorem 1.

Theorems 1 and 2 present us quantum algorithms for Boolean circuits with shared input and non-constant depth. At the same time, existing algorithms [7,8,10,15,16] are not applicable in a case of shared input and non-constant depth.

The third way of representation of Boolean function is Zhegalkin polynomial that is representation using AND, XOR functions and the 0, 1 constants [24,33, 34]: for some integers $k, t_1, \dots, t_k$,

$$F(x) = ZP(x) = a \oplus \bigoplus_{i=1}^{k} C_i, \text{ where } a \in \{0,1\}, C_i = \bigwedge_{z=1}^{t_i} x_{j_z}$$

At the same time, it can be represented as an AND-OR-NOT DAG with a logarithmic depth and shared input or an AND-OR-NOT tree with an exponential number of vertices and separated input. That is why the existing algorithms from [7,8,10,15,16] cannot be used or work in exponential running time.

Theorem 3. *Algorithm 3 evaluates the XOR-AND DAG* $G = (V, E)$ *with running time* $O(\sqrt{\hat{n}m}\log\hat{n}) = O(\sqrt{nm}\log n)$ *and error probability* $O(1/\hat{n})$*. Here* $m = |E|, n = |V|, \hat{n} = n - |L|$*,* L *is the set of sinks.*

Proof. XOR operation is replaced by two AND, one OR vertex and 6 edges because for any Boolean a and b we have $a \oplus b = a \wedge \neg b \vee \neg a \wedge b$. So, we can represent the original DAG as an AND-OR-NOT DAG using $\hat{n}' \leq 3 \cdot \hat{n} = O(\hat{n})$ vertices. The number of edges is $m' \leq 6 \cdot m = O(m)$. Due to Theorem 1, we can construct a quantum algorithm with running time $O(\sqrt{\hat{n}m}\log\hat{n})$ and error probability $O(1/\hat{n})$. □

The previous theorem shows us the existence of a quantum algorithm for Boolean circuits with XOR, NAND, AND, OR and NOT gates. Let us present the result for Zhegalkin polynomial.

Corollary 1. *Suppose that Boolean function* $F(x)$ *can be represented as Zhegalkin polynomial for some integers* $k, t_1, \dots, t_k$*:* $F(x) = ZP(x) = a \oplus \bigoplus_{i=1}^{k} C_i$*, where* $a \in \{0,1\}$*,* $C_i = \bigwedge_{z=1}^{t_i} x_{j_z}$*. Then, there is a quantum algorithm for* F *with running time* $O\left(\sqrt{k\log k(k + t_1 + \dots + t_k)}\right)$ *and error probability* $O(1/k)$*.*

Proof. Let us present C_i as one AND vertex with t_i outgoing edges. XOR operation is replaced by two AND, one OR vertex and 6 edges. So, $m = 6 \cdot (k-1) + t_1 + \cdots + t_k = O(k + t_1 + \cdots + t_k)$, $n = 3 \cdot (k-1) + k = O(k)$. Due to Theorem 3, we obtain claimed properties. □

5 The Quantum Algorithm for the Single Source Longest Path Problem for a Weighted DAG and the Diameter Search Problem for Weighted DAG

In this section, we consider two problems for DAGs.

5.1 The Single Source Longest Path Problem for Weighted DAG

Let us apply the approach to the Single Source Longest Path problem.

Let us consider a weighted DAG $G = (V, E)$ and the weight of an edge $e = (v_i, v_j)$ is $w(i, j)$, for $i, j \in \{1, \ldots, n\}, e \in E$.

Let we have a vertex v_s then we should compute compute $t[1], \ldots, t[n]$. Here $t[i]$ is the length of the longest path from v_s to v_i. If a vertex v_i is not reachable from v_s then $t[i] = -\infty$.

Let us preset the algorithm for longest paths lengths computing.

Algorithm 4. Quantum Algorithm for the Single Source Longest Path Search problem. Let $t = (t[1], \ldots, t[n])$ be an array which stores results for vertices. Let Q be the Dürr-Høyer quantum algorithm for MAX function. Let $\hat{Q}^{2\log_2(n)}$ be a boosted version of Q (Algorithm 2, Lemma 3).

$t \leftarrow (-\infty, \ldots, -\infty)$
$t[s] \leftarrow 0$
for $i = s+1 \ldots n$ **do**
  $t[i] \leftarrow \hat{Q}^{2\log_2 n}(t[j_1] + w(i, j_1), \ldots, t[j_w] + w(i, j_w))$, where $w = |D'_i|$, $(v_{j_1}, \ldots, v_{j_w}) = D'_i$.
end for
return t

Algorithm 4 has the following property:

Theorem 4. *Algorithm 4 solves the Single Source Longest Path Search problem with expected running time $O(\sqrt{nm}\log n)$ and error probability $O(1/n)$.*

Proof. Let us prove the correctness of the algorithm. In fact, the algorithm computes $t[i] = \max(t[j_1] + w(i, j_1), \ldots, t[j_w] + w(i, j_w))$. Assume that $t[i]$ is less than the length of the longest path. Then there is $v_z \in D'_i$ that precedes v_i in the longest text. Therefore, the length of the longest path is $t[z] + w(i, z) > t[i]$. This claim contradicts with definition of $t[i]$ as maximum. The bounds for the running time and the error probability follows from Lemmas 1 and 2.

5.2 The Diameter Search Problem for a Weighted DAG

Let us consider an unweighted DAG $G = (V, E)$. Let $len(i, j)$ be the length of the shortest path between v_i and v_j. If the path does not exist, then $len(i, j) = -1$. The diameter of the graph G is $diam(G) = \max_{i,j\in\{1,\dots,|V|\}} len(i, j)$. For a given graph $G = (V, E)$, we should find the diameter of the graph.

It is easy to see that the diameter is the length of a path between a non-sink vertex and some other vertex. If this fact is false, then the diameter is 0.

Using this fact, we can present the algorithm.

Algorithm 5. Quantum Algorithm for the Diameter Search problem. Let $t^z = (t^z[1], \dots, t^z[n])$ be an array which stores shortest paths from vertices to vertex $v_z \in V\backslash L$. Let Q be the Dürr-Høyer quantum algorithm for the MIN function. Let $\hat{Q}^{2\log_2(n)}$ be a boosted version of Q (Algorithm 2, Lemma 3). Let Q_{max} and $\hat{Q}^{\log_2(n)}_{max}$ be the quantum algorithm and the boosted version of the algorithm for the MAX function that ignore $+\infty$ values.

---

for $z = \hat{n} \dots 1$ **do**
  $t^z \leftarrow (+\infty, \dots, +\infty)$
  $t^z[z] \leftarrow 0$
  for $i = s + 1 \dots n$ **do**
    $t^z[i] \leftarrow \hat{Q}^{2\log_2 n}(t^z[j_1] + w(i, j_1), \dots, t^z[j_w] + w(i, j_w))$, where $w = |D'_i|$, $(v_{j_1}, \dots, v_{j_w}) = D'_i$.
  end for
end for
$diam(G) = \hat{Q}^{\log_2(n)}_{max}(0, t^1[2], \dots, t^1[n], t^2[3], \dots, t^2[n], \dots, t^{\hat{n}}[\hat{n} + 1], \dots, t^{\hat{n}}[n])$
return $diam(G)$

Algorithm 5 has the following property:

Theorem 5. *Algorithm 5 solves the Diameter Search problem with expected running time $O(\hat{n}(n + \sqrt{nm}) \log n)$ and error probability $O(1/n)$.*

Proof. The correctness of the algorithm can be proven similar to the proof of Theorem 4. The bounds for the running time and the error probability follows from Lemmas 1 and 2.

Acknowledgements. This work was supported by Russian Science Foundation Grant 17-71-10152.

A part of work was done when K. Khadiev visited University of Latvia. We thank Andris Ambainis, Alexander Rivosh and Aliya Khadieva for help and useful discussions.

References

1. Ablayev, F., Ablayev, M., Khadiev, K., Vasiliev, A.: Classical and quantum computations with restricted memory. In: Böckenhauer, H.-J., Komm, D., Unger, W. (eds.) Adventures Between Lower Bounds and Higher Altitudes. LNCS, vol. 11011, pp. 129–155. Springer, Cham (2018). https://doi.org/10.1007/978-3-319-98355-4_9
2. Ablayev, F., Ambainis, A., Khadiev, K., Khadieva, A.: Lower bounds and hierarchies for quantum memoryless communication protocols and quantum ordered binary decision diagrams with repeated test. In: Tjoa, A.M., Bellatreche, L., Biffl, S., van Leeuwen, J., Wiedermann, J. (eds.) SOFSEM 2018. LNCS, vol. 10706, pp. 197–211. Springer, Cham (2018). https://doi.org/10.1007/978-3-319-73117-9_14
3. Ablayev, F., Gainutdinova, A., Khadiev, K., Yakaryılmaz, A.: Very narrow quantum OBDDs and width hierarchies for classical OBDDs. Lobachevskii J. Math. **37**(6), 670–682 (2016)
4. Ablayev, F., Gainutdinova, A., Khadiev, K., Yakaryılmaz, A.: Very narrow quantum OBDDs and width hierarchies for classical OBDDs. In: Jürgensen, H., Karhumäki, J., Okhotin, A. (eds.) DCFS 2014. LNCS, vol. 8614, pp. 53–64. Springer, Cham (2014). https://doi.org/10.1007/978-3-319-09704-6_6
5. Ablayev, F., Vasilyev, A.: On quantum realisation of boolean functions by the fingerprinting technique. Discrete Math. Appl. **19**(6), 555–572 (2009)
6. Ambainis, A., Nahimovs, N.: Improved constructions of quantum automata. Theoret. Comput. Sci. **410**(20), 1916–1922 (2009)
7. Ambainis, A.: A nearly optimal discrete query quantum algorithm for evaluating NAND formulas. arXiv preprint arXiv:0704.3628 (2007)
8. Ambainis, A.: Quantum algorithms for formula evaluation. https://arxiv.org/abs/1006.3651 (2010)
9. Ambainis, A.: Understanding quantum algorithms via query complexity. arXiv preprint arXiv:1712.06349 (2017)
10. Ambainis, A., Childs, A.M., Reichardt, B.W., Špalek, R., Zhang, S.: Any and-or formula of size N can be evaluated in time $N^{1/2} + o(1)$ on a quantum computer. SIAM J. Comput. **39**(6), 2513–2530 (2010)
11. Ambainis, A., Špalek, R.: Quantum algorithms for matching and network flows. In: Durand, B., Thomas, W. (eds.) STACS 2006. LNCS, vol. 3884, pp. 172–183. Springer, Heidelberg (2006). https://doi.org/10.1007/11672142_13
12. Arora, S., Barak, B.: Computational Complexity: A Modern Approach. Cambridge University Press, New York (2009)
13. Boyer, M., Brassard, G., Høyer, P., Tapp, A.: Tight bounds on quantum searching. Fortschritte der Physik **46**(4–5), 493–505 (1998)
14. Brassard, G., Høyer, P., Mosca, M., Tapp, A.: Quantum amplitude amplification and estimation. Contemp. Math. **305**, 53–74 (2002)
15. Bun, M., Kothari, R., Thaler, J.: Quantum algorithms and approximating polynomials for composed functions with shared inputs. arXiv preprint arXiv:1809.02254 (2018)
16. Childs, A.M., Kimmel, S., Kothari, R.: The quantum query complexity of read-many formulas. In: Epstein, L., Ferragina, P. (eds.) ESA 2012. LNCS, vol. 7501, pp. 337–348. Springer, Heidelberg (2012). https://doi.org/10.1007/978-3-642-33090-2_30
17. Cormen, T.H., Leiserson, C.E., Rivest, R.L., Stein, C.: Introduction to Algorithms, 2nd edn. McGraw-Hill, New York (2001)
18. De Wolf, R.: Quantum computing and communication complexity (2001)

19. Dörn, S.: Quantum complexity of graph and algebraic problems. Ph.D. thesis, Universität Ulm (2008)
20. Dörn, S.: Quantum algorithms for matching problems. Theory Comput. Syst. **45**(3), 613–628 (2009)
21. Dürr, C., Heiligman, M., Høyer, P., Mhalla, M.: Quantum query complexity of some graph problems. In: Díaz, J., Karhumäki, J., Lepistö, A., Sannella, D. (eds.) ICALP 2004. LNCS, vol. 3142, pp. 481–493. Springer, Heidelberg (2004). https://doi.org/10.1007/978-3-540-27836-8_42
22. Dürr, C., Heiligman, M., Høyer, P., Mhalla, M.: Quantum query complexity of some graph problems. SIAM J. Comput. **35**(6), 1310–1328 (2006)
23. Durr, C., Høyer, P.: A quantum algorithm for finding the minimum. arXiv preprint arXiv:quant-ph/9607014 (1996)
24. Gindikin, S.G., Gindikin, S., Gindikin, S.G.: Algebraic Logic. Springer, New York (1985)
25. Grover, L.K.: A fast quantum mechanical algorithm for database search. In: Proceedings of the Twenty-Eighth Annual ACM Symposium on Theory of Computing, pp. 212–219. ACM (1996)
26. Grover, L.K., Radhakrishnan, J.: Is partial quantum search of a database any easier? In: Proceedings of the Seventeenth Annual ACM Symposium on Parallelism in Algorithms and Architectures, pp. 186–194. ACM (2005)
27. Ibrahimov, R., Khadiev, K., Prūsis, K., Yakaryılmaz, A.: Error-free affine, unitary, and probabilistic OBDDs. In: Konstantinidis, S., Pighizzini, G. (eds.) DCFS 2018. LNCS, vol. 10952, pp. 175–187. Springer, Cham (2018). https://doi.org/10.1007/978-3-319-94631-3_15
28. Jordan, S.: Bounded Error Quantum Algorithms Zoo. https://math.nist.gov/quantum/zoo
29. Khadiev, K., Khadieva, A.: Reordering method and hierarchies for quantum and classical ordered binary decision diagrams. In: Weil, P. (ed.) CSR 2017. LNCS, vol. 10304, pp. 162–175. Springer, Cham (2017). https://doi.org/10.1007/978-3-319-58747-9_16
30. Khadiev, K., Khadieva, A., Mannapov, I.: Quantum online algorithms with respect to space and advice complexity. Lobachevskii J. Math. **39**(9), 1210–1220 (2018)
31. Le Gall, F.: Exponential separation of quantum and classical online space complexity. Theory Comput. Syst. **45**(2), 188–202 (2009)
32. Nielsen, M.A., Chuang, I.L.: Quantum Computation and Quantum Information. Cambridge University Press, New York (2010)
33. Yablonsky, S.V.: Introduction to Discrete Mathematics: Textbook for Higher Schools. Mir Publishers, Moscow (1989)
34. Zhegalkin, I.: On the technique of calculating propositions in symbolic logic (sur le calcul des propositions dans la logique symbolique). Matematicheskii Sbornik **34**(1), 9–28 (1927). (in Russian and French)

Viewing Rate-Based Neurons as Biophysical Conductance Outputting Models

Martinius Knudsen[1], Sverre Hendseth[1(✉)], Gunnar Tufte[2(✉)], and Axel Sandvig[3(✉)]

[1] Department of Engineering Cybernetics, NTNU, Trondheim, Norway
{martinius.knudsen,sverre.hendseth}@ntnu.no
[2] Department of Computer Science, NTNU, Trondheim, Norway
gunnar.tufte@ntnu.no
[3] Department of Neuromedicine and Movement Science, NTNU, Trondheim, Norway
axel.sandvig@ntnu.no

Abstract. In the field of computational neuroscience, spiking neural network models are generally preferred over rate-based models due to their ability to model biological dynamics. Within AI, rate-based artificial neural networks have seen success in a wide variety of applications. In simplistic spiking models, information between neurons is transferred through discrete spikes, while rate-based neurons transfer information through continuous firing-rates. Here, we argue that while the spiking neuron model, when viewed in isolation, may be more biophysically accurate than rate-based models, the roles reverse when we also consider information transfer between neurons. In particular we consider the biological importance of continuous synaptic signals. We show how synaptic conductance relates to the common rate-based model, and how this relation elevates these models in terms of their biological soundness. We shall see how this is a logical relation by investigating mechanisms known to be present in biological synapses. We coin the term 'conductance-outputting neurons' to differentiate this alternative view from the standard firing-rate perspective. Finally, we discuss how this fresh view of rate-based models can open for further neuro-AI collaboration.

Keywords: Artificial neural network · Spiking neural network · Computational neuroscience · Conductance models

1 Introduction

Progress in neuroscience research has been impressive. We now understand the central nervous system with increasing anatomical detail. Additionally, we are rapidly elucidating how the brain functions down to cellular and molecular resolution. Based on brain anatomy and cellular results, research can now also focus on understanding how the neural connectome is functionally integrated

I. McQuillan and S. Seki (Eds.): UCNC 2019, LNCS 11493, pp. 164–177, 2019.
https://doi.org/10.1007/978-3-030-19311-9_14

on macro, meso- and microscale levels. Central to future progress within this research field is the development of powerful computational tools to analyse the big data sets generated by the neural connectome. Within this area, the subfield Computational Neuroscience exists where researchers attempt to model and simulate these brain processes at different levels of abstraction; in particular, cell and network level modelling where biological detail is a prerequisite as well as a constraint. In terms of network modelling, empirically derived phenomenological models such as *spiking neural networks (SNN)* [15] are commonly employed. While these network models seem to replicate biological dynamics, SNNs have had limited success with regards to task-solving.[1] However, there's growing research investigating how SNNs *can* be applied to tasks such as feature learning and control [7,13,14] using Hebbian-like [2] spike-timing dependent plasticity (STDP) [17,25] learning rules. Research in the area has however yet to show rivalling results to that of conventional feed forward *Artificial Neural Networks (ANN)* employed by the machine-learning community.

AI-ANN research has achieved impressive success in a wide variety of applications by use of abstracted neuron models. However, the name 'artificial neurons' implying a replica of biology, tends not to be well accepted by the neuroscience community. While AI frequently looks to biology for inspiration, researchers in the field have granted themselves freedom from biological detail. Even leading AI researchers are now stating that ANN models are more mathematical constructs than attempts of modelling the complexity of biological neurons [18]. These artificial neurons negate the characteristic spiking behaviour prevalent in biological neurons, and instead output rates of neural spikes. For this reason, these artificial neurons are commonly referred to as 'firing-rate' models. A common and long-standing debate is whether or not firing-rates sufficiently capture the information transfer between neurons [3,8,11,24].

SNNs are typically praised as being more biologically accurate, and are sometimes even referred to as the 3rd generation of neural networks [15].[2] In this paper we investigate commonalities between the commonly employed neuron models within the two fields, as well as their respective distinctions. In particular, we look at the resulting output signals of each model and see how they compare with respect to the receiving downstream neuron. We find that a fresh look at the standard rate-based model suggests that these models are more similar to their spiking counterparts than they may first appear. We also argue that simplified point spiking neurons may be biologically inferior to firing-rate models when considering synaptic transfer in network wide simulations. To support this argument, we investigate information flow between biological neurons, and compare this to spiking and firing-rate neurons. By this investigation, we further find that firing-rate neurons can be viewed as *conductance-outputting models*, and see how this alternative view increases these models' biological support and intuition. We finally discuss why this alternative view is of importance.

[1] Partly due to this not being the main priority of these models.

[2] With rate-based models being the 2nd generation and threshold perceptrons being the 1st.

2 Background

2.1 Information Transfer Between Biological Neurons

Information flow between biological neurons is a complex process involving a cascade of events. A summarized illustration of this flow from one neuron to the next is shown in Fig. 1. Here, upstream neuronal action potentials (APs) cause the release of neurotransmitters into the synaptic cleft between two neurons. In excitatory synapses, this neurotransmitter is commonly glutamate [19], and can be viewed as a carrier of synaptic conductance [26, p. 173]. The amount of glutamate released into the synaptic cleft follows the firing-rate of the presynaptic neuron [26, p. 175], where high pre firing-rates yield greater releases of neurotransmitter up to some saturation point. The increase of glutamate in the synaptic cleft results in higher synaptic conductance as glutamate pairs with postsynaptic ionotropic receptors, such as AMPA [9], which open and become permeable to ionic currents. This allows for an influx of sodium into the postsynaptic dendrite, resulting in a depolarization known as an excitatory postsynaptic potential (EPSP). The more receptors available, the higher the synaptic weight/efficiency, and the stronger effect the presynaptic activation has on the postsynaptic neuron. The EPSP further propagates down the dendrite to the soma, depolarizing the somatic membrane potential $V(t)$. If the potential rises above a certain threshold, the neuron fires an action potential (AP). This AP in turn propagates down the axon causing the release of glutamate at a synapse between it and a downstream neuron, repeating the whole process there.

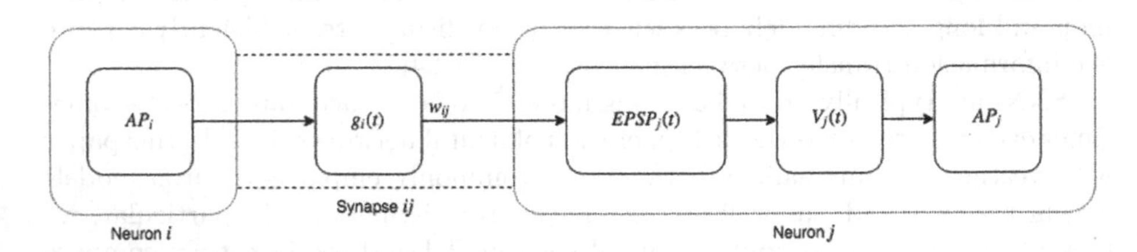

Fig. 1. Neural flow of information through an excitatory synapse between biological neurons. AP is the Dirac delta action potential, $g(t)$ is the synaptic conductance ranging from 0 to 1, w_{ij} is the synaptic efficiency, $EPSP(t)$ is the excitatory post-synaptic potential, $V(t)$ is the membrane potential at the soma.

2.2 Information Transfer in Simple Spiking Neuron Models

Many SNN simulations employ point neuron models and largely simplify the synapse in order to make these models computationally efficient [10]. These models often view discrete spike-events as the main source of information passing between neurons. Figure 2 illustrates the flow of information between spiking neurons. These discrete spikes influence postsynaptic membrane potential directly in proportion to the synaptic efficiency:

$$V_j = \sum_{i=0}^{N} \delta_i (V_i) w_{ij} \tag{1}$$

Here, N is the nr of presynaptic neurons i connected to a downstream neuron j, δ_i is the Dirac delta function which equals 1 when the membrane potential V_i of the presynaptic neuron goes above some firing threshold and 0 otherwise. In these networks the weight is often updated through STDP [17,25] which considers the timing of pre-post spiking events similarly to a Hebbian rule.

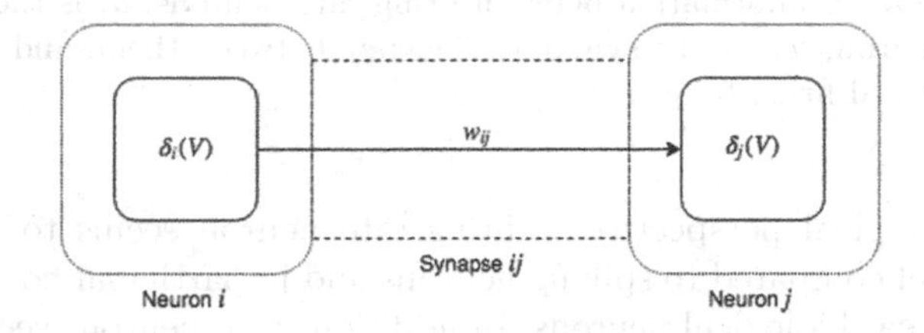

Fig. 2. Neural flow of information between simple spiking point neurons. δ is the Dirac delta function.

2.3 Information Transfer Between Firing-Rate Neuron Models

The standard ANN employs the firing-rate model [22]. These neurons do not model internal states such as spiking neurons do with their membrane potentials. They are thus typically static in nature,[3] whereas spiking neurons are highly dynamic. The function of the firing-rate neuron is modelled on the observation that biological neurons fire APs at rates proportional to the strength of their inputs [22]. By assuming that most of the information lies within these firing rates, firing-rate models simply convert upstream firing rates directly into new firing rates, foregoing the complex dynamics of membrane potentials and spikes.

Figure 3 illustrates this information exchange between firing-rate neurons. These neurons often employ some non-linear activation function on their inputs. Traditionally, this has been a squashing function which acts to bound the outputs between 0 (not firing) and 1 (firing at maximum frequency) using a sigmoid function f:[4]

$$\begin{aligned} a_j &= f\left(\sum_{i=0}^{N} a_i w_{ij} + b_j\right) \\ f(x) &= \frac{1}{1+e^{-x}} \end{aligned} \tag{2}$$

[3] Apart from a few special versions e.g. continuous ANNs.

[4] Although several successful but less biologically motivated activation functions have come about in recent years [20].

Here, a_i is the activation level (firing-rate) of the presynaptic neuron i, b_j is the neuron's natural bias (baseline activity even in the absence of input [28]), and w_{ij} is the weight (synaptic efficiency) between neuron i and j.

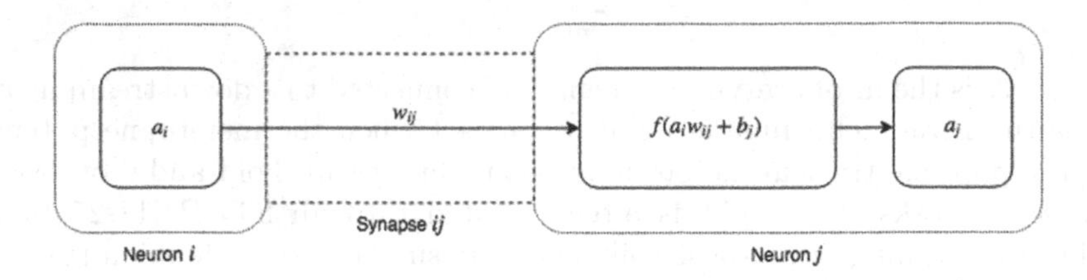

Fig. 3. Neural flow of information between firing-rate neurons. a_i is the activation of a presynaptic neuron, w_{ij} is the synaptic efficiency between them, and b_j is the bias; the neuron's natural firing frequency.

From a biological perspective, a firing-rate neuron seems to be an overly simplified model compared to spiking neurons and in particular compared to the complexity of real biological neurons. In addition, rate neurons seem to dismiss any possible information that may lie within the timing of spikes [16]. We shall in the next section examine how important the conveying of spikes really is in the view of postsynaptic neurons.

3 Portraying Rate-Based Neurons as Conductance-Outputting Models

Reviewing the information flow between biological neurons described in the previous section, it does not seem that a post-synaptic neuron actually ever receives a discrete all-or-nothing spike from its upstream connections. Instead, the postsynaptic neuron receives a continuous conductance as a *result* of presynaptic spikes. The neuron is thus oblivious to any spiking events other than the resulting conductance changes in the synaptic cleft. In this case, the passing of discrete spike events often employed in spiking networks, appears no more biologically realistic than the firing-rates used by rate-based neurons when considering network wide simulations. In fact, we state that firing-rates viewed as conductance traces are a more biologically realistic view than either of the above. In the following sections, we present three different arguments to support this claim.

3.1 Argument 1: Mathematical Equivalence

We here propose that rate-based neurons can alternatively be viewed as conductance-outputting models. To exemplify this, we investigate a mathematical model describing the total synaptic current going into a biological neuron j

at any given dendritic location [26, p. 174]. This current is the sum of synaptic currents from neurons i to j at the given location:

$$I_{syn}(t) = \sum_{i=0}^{N} g_i(t) w_{ij} (E - V_j(t)) \tag{3}$$

were $I_{syn}(t)$ is the total synaptic current, $g_i(t)$ is the conductance, E is the reversal potential of the input current, $V_j(t)$ is the post-synaptic membrane potential, and w_{ij} is the synaptic efficiency. If we allow ourselves a common simplification; that the incoming current is independent of the post-synaptic potential (PSP), we can simplify Eq. 3:

$$I_{syn}(t) = \sum_{i=0}^{N} g_i(t) w_{ij} \tag{4}$$

We now have an equation similar to (2) representing the integration of inputs employed by common ANN models. Here, the output firing-rate of the neuron $a_i(t)$ has been replaced by a conductance $g_i(t)$, thus making $g_i(t)$ essentially a function of $a_i(t)$. This conductance model effectively represents an average of the firing-rate, as seen in Fig. 5 which shows the relation between membrane potential and the output conductance. Similar to the bounded firing-rate due to the squashing function, $g(t)$ can be equally bounded between 0 and 1 with the values representing the concentration of glutamate within the synaptic cleft. Glutamate can thus be non-present (0) or at saturation (1).

In our biological model, the integrated input to neuron j further causes a rise in the EPSP and the somatic potential $V_i(t)$. $V_i(t)$ impacts whether the neuron will fire an AP. These APs in turn cause the release of glutamate into the synaptic cleft, and therefor increase the conductance at the synapse. If we were to simply bypass these intermediate steps through EPSP, $V_i(t)$ and AP, we can instead define a direct mapping function h from input current to output conductance where we obtain $g_j(t)$ (the output of neuron j) directly:

$$g_j(t) = h(I_{syn}(t)) \tag{5}$$

Figure 4 illustrates this point. $g_j(t)$ is hence obtained as a function of the total current $I_{syn}(t)$ going into neuron j, the same way $a_j(t)$ is a function of $I_{syn}(t)$ (where $I_{syn}(t) = \sum a_i w_{ij}$ in the rate-based case). As in the rate based case which often includes a bias b; $f(I_{syn}(t) + b_f)$, it is possible to include this natural and biologically relevant activation level to our conductance outputting model as well: $h(I_{syn}(t) + b_h)$.

In summary, the conductance outputting neuron model circumvents the internal voltage dynamics of the neuron as well as the firing of APs. Instead a direct mapping from input current to output conductance is performed $I_{syn}(t) \rightarrow g(t)$, and this can be viewed as similar to the mapping of the input current to the firing-rate of ANN neurons $I_{syn}(t) \rightarrow a(t)$.

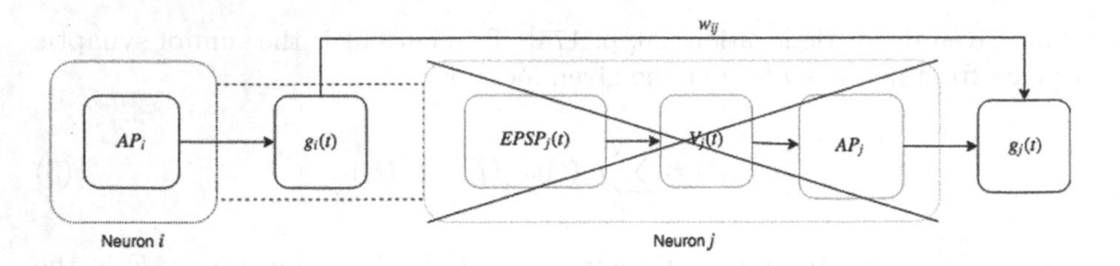

Fig. 4. Revised neural information flow demonstrating the direct mapping of conductances between neurons i and j.

3.2 Argument 2: Artificial Firing-Rate Neurons Approximate Biological Conductance Models in Simulations

To demonstrate the above point; that the synaptic conductance is equivalent to the firing rate of the neuron, we run a simulation of a biological neuron model in which we apply an external input current. This input current, which can be viewed either as input from upstream neurons or an external electrode, modulates the resulting behaviour of our simulated neuron. Here, strong input currents cause the membrane potential to depolarize faster and hence output higher AP firing rates and thus influence the synaptic conductance to post-synaptic neurons.

The neuron model we use for this experiment is the leaky integrate-and-fire (LIF) model [5]:

$$C\frac{dV(t)}{dt} = \frac{RI_{syn}(t) - V(t)}{R} \tag{6}$$

Here, $V(t)$, R and C are the membranes voltage potential, resistance and capacitance respectively. I_{syn} is the input current. The membrane potential $V(t)$ is reset when surpassing the firing threshold $\delta(AP)$. The neuron goes into a short refractory period in which $V(t)$ is held at zero.

The conductance model we use is an exponential decay function borrowed from [26], which acts as a low-pass filter:

$$\tau_g\frac{dg(t)}{dt} = \delta(AP) - g(t) \tag{7}$$

Our simulation parameters are shown in Table 1, with parameters also from [26].

Figure 5 shows the result of our simulation using a single LIF neuron and the conductance Eqs. (6) and (7) respectively. We have scaled $g(t)$ such that the its average lies between 0 and 1, as to output 1 during maximum firing frequency. This maximum frequency can be calculated using the neurons refractory period

Table 1. Simulation parameters

Parameter	Value	Unit	Description
R	10	KΩ	Membrane resistance
C	1	μF	Membrane capasitance
τ_g	10	ms	Time constant for conductance
$\delta(AP)$	20	mV	AP threshold
τ_{ref}	2	ms	Refractory period

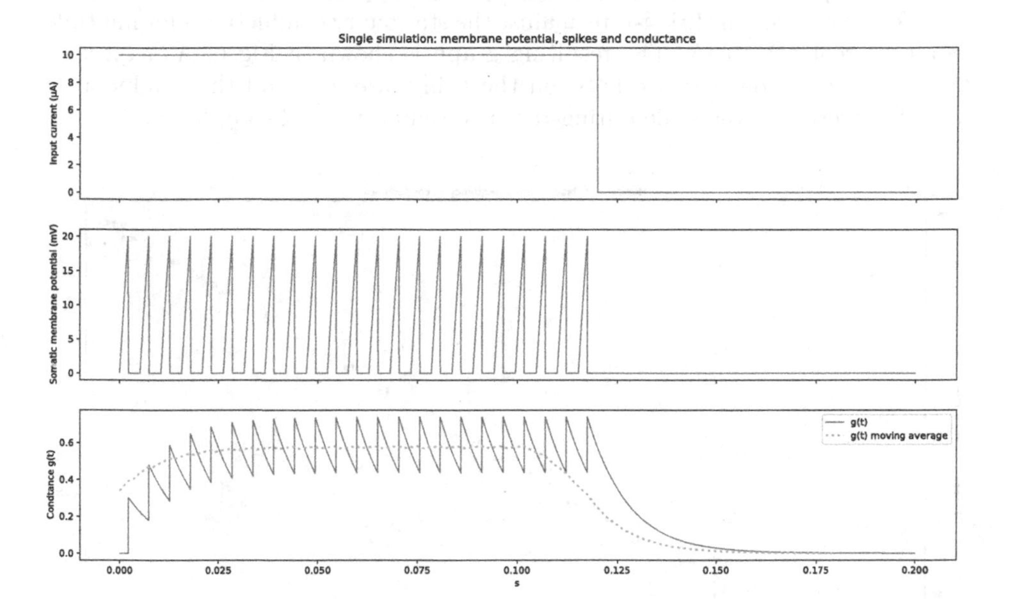

Fig. 5. *Top:* The input current of 10.6 μA, cut off at 125 ms. *Middle:* The somatic membrane potential. The spikes are APs, here firing at 191 Hz. *Bottom:* The output conductance, showing a smoothed value of about 0.6, which is equal to: $\frac{\text{firing rate}}{\text{maximum rate}} = \frac{191}{333} \approx 0.6$.

along with the duration of an AP, here set to 2 ms and 1 ms respectively [26]. For these values, the maximum firing frequency is calculated to be:[5]

$$\text{maximum rate} = \frac{1000\,\text{ms}}{2\text{ms} + 1\,\text{ms}} \approx 333\,\text{HZ} \tag{8}$$

We set a constant input current to the neuron of 10.6 μA for 125 ms before cut-off. We register that this yields a firing frequency of 191 Hz. As calculated in Eq. (8) the maximum firing frequency, given our parameters in Table 1, is 333 Hz. We find that $191/333.33 \approx 0.6$, which is the same value as our smoothed

[5] We have simplified here by disregarding depletion of neurotransmitters: i.e. we assume that neurotransmitters re-uptake is able to keep up with the pace of release.

conductance converges to. These correspondences are true for all input current values, which we have verified below.

As can be observed in Fig. 5, the conductance effectively represents the firing-rate average of the neuron, but more as a moving average due to the time constant τ_g rather than an instantaneous firing-rate. This is due to the conductance acting as a low-pass filter, which has advantages as the conductance provides a less erratic signal to the post-synaptic neuron. We can say that the function h in Eq. (5), yields an instantaneous average firing-rate similar to f in Eq. (2), while g is an low-pass filtered firing-rate. As such $g(t)$ approaches $a(t)$ for $\tau_g \to 0$.

We further plot the firing-rate against the stationary conductance for multiple input currents: 2–50 μA. The resulting graph is shown in Fig. 6. As expected, there is a linear dependence between the firing-rate $a(t)$ and the conductance $g(t)$. Furthermore, the scaled firing-rate is essentially equal to $g(t)$.

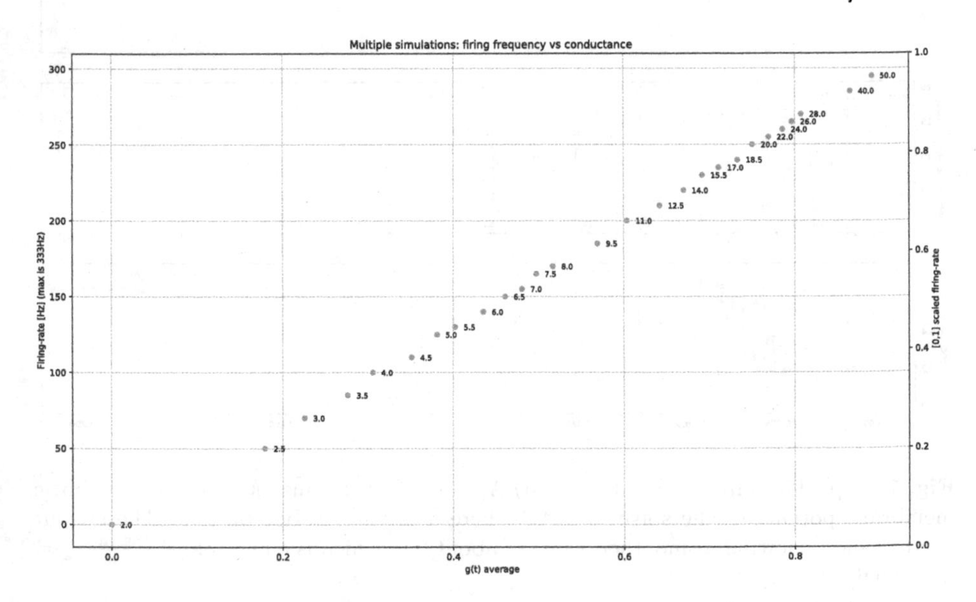

Fig. 6. The firing-rates calculated by $\frac{spikes}{200\,\mathrm{ms}}$ vs the average $g(t)$ signal. 28 simulations where run each for 200 ms real-time (20 000 timesteps at a simulation resolution of 0.01 ms). Each simulation was run with a different input current with currents ranging from 2 μA to 50 μA (higher inputs yield higher firing-rates). The y-axis to the right displays the scaled firing-rate, showing that $g(t)$ is accurately encoding the firing-rate.

3.3 Argument 3: Synaptic Plasticity Models Using Discrete Spikes Don't Work Well

When introducing plasticity into SNN simulations, it has been common practice to employ some sort of phenomenological STDP model [25]. This model takes into account the precise timing of pre- and post-synaptic action potentials in order to update the synaptic weight between two neurons. The model is motivated by experimental findings [25] as well as its computational efficiency.

Further experimental evidence however, goes against these simple STDP models, as neuroscience has discovered more and more diverse STDP phenomena [4,23] as shown in Fig. 7. This indicates that the process of synaptic tuning is governed by other factors than mere spike-times. More biophysically inspired plasticity models rely on *continuous* conductances, calcium currents and internal cell voltages rather than discrete spikes [6,23]. These models have demonstrated the ability to account for multiple STDP phenomena demonstrating the importance of continuous signals in the synapse. The dependency on continuous synaptic signals further argues against simplistic spike-based views.

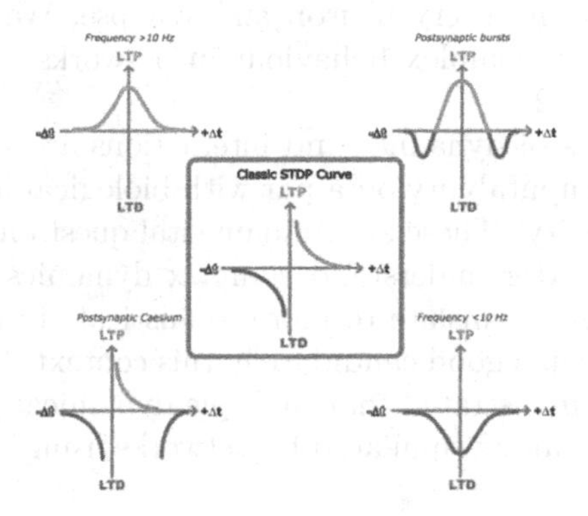

Fig. 7. The many STDP phenomena that have been experimentally observed can only be explained using continuous synaptic states. Here, we see that long-term potentiation shows varying dependence on spike-timing. The x-axis is the pre-post spiking interval, while the y-axis is the amount of positive/negative potentiation that occurs due to this timing. Courtesy of [4]

4 Discussion

4.1 Finding the Right Abstraction Level

Biological neurons transmitting spikes to one another as a means of information transfer is a simplistic view of a highly complex transaction. Within chemical synapses in the brain, transmission is in the form of continuous and bounded concentrations of neurotransmitters. Hence, utilizing spiking neuron models does not necessarily make the overall network dynamics similar to biology if the information transfer between neurons is insufficiently modelled. We argue therefore that simplistic SNN networks utilizing discrete spikes at synapses are biologically inferior compared to rate-based, or as shown in this paper 'conductance-outputting', neural network models. Only when applying conductance-based synapses in spiking networks should we expect similarity to biological dynamics.

Neurons portrayed as point conductance-outputting models is naturally a large simplification too, compared to the vast internal dynamics and structures of biological neurons. However, in our quest towards understanding neural networks and replicating some of their amazing abilities, reducing complexity and computational cost of our models is essential. We must ask which details really matter. For example, one could argue that if (a) we had a complete architectural map of the brain down to the molecular level, and (b) we had the computational resources to simulate all molecular interactions, we could essentially simulate a complete working brain. Most researchers do not think (or at least hope) that this level of detail is necessary to understand the principles governing computation within neural networks. It may be that the complex emergent behaviour can be replicated by simple local interactions without modelling the complex dynamics within each and every neuron and synapse. We know from work on Cellular Automata that complex behaviour in networks can come about from such simple local rules [27].

So how many levels of dynamics and interactions are we required to model in order to achieve computability on a par with biological networks? Is there an upper limit to complexity? These are fundamental questions we need to answer. For this we need to better understand complex dynamics in simpler networks and how these dynamics translate to more sophisticated models. Conductance-outputting neurons seem a good candidate in this context. In fact, several experiments have indeed demonstrated that multiple dynamical phenomena observed in biological networks can be simulated by networks using continuous firing-rate models [24].

4.2 Firing-Rates vs Conductance: Why the Perspective Matters

In most sensory systems, the firing rate generally increases non-linearly with increasing stimulus intensity [12]. This discovery has led to the widely discussed assumption that neurons convey information through firing rates. Such a view has been highly influential in the resulting ANN models which employ firing-rate neurons. However, it is not obvious how downstream neurons would be able to observe upstream firing-rates at any instance of time. In order for such an analogy to work, there has to be some mechanism that dynamically encodes this firing-rate. Synaptic conductance resolves this problem by effectively representing the firing-rate as g.

The alternative view is important not only with respect to biological accuracy, but also because although firing-rate models are inspired by findings in neuroscience, today AI and neuroscience speak very different languages. This is largely due to the use of seemingly incompatible models. The 'artificial neuron' is often frowned upon by neuroscientists due to it's ambitious name and discrepancy with biological principles. We believe that the gap between the two fields can be bridged through an alternative view and terminology; viewing rate-based models as conductance-outputting, which revive these models in a more biologically plausible manner. An analogy such as this makes it easier to compare the benefits and limitations of models on both sides and creates a common platform for discussion. A different view can additionally stimulate new ideas and opportunities for more cross-disciplinary research. Such research is in fact becoming increasingly more common these days as many research institutions are connecting neuroscientists and AI researchers [1,21]. Despite the impressive results of ANN research, as well as the fascinating findings in neuroscience, we are still far from understanding the computational abilities of biological neural networks. As researchers from both fields work on similar problems, a common language is both beneficial and highly necessary for collaboration and fruitful scientific progress.

5 Future Work

This paper introduces the concept of conductance-based models: a model that is simpler than SNNs, yet more biologically intuitive than firing-rate neurons. Future work will involve the comparison of larger scale network dynamics of: (a) spiking networks with discrete synapses, (b) spiking networks with conductance modelled synapses, and (c) the continuous conductance-outputting model put forth here.

6 Conclusion

The common and widely applied firing-rate neuron model employed in ANN research has been examined and presented from a fresh point of view as a conductance-outputting model. The alternative view allows for better biological appreciation of these models, and also argues that they may be more biologically accurate than simple SNNs when employed in network wide simulations. This is especially prevalent when the synapse and plasticity models of SNNs base themselves on discrete events, which they often do. The takeaways from this is that one should not naively assume spiking neuron models to be biologically superior to firing-rate neurons. Information transfer between neurons is a large part of the equation and getting this part wrong can not be compensated by employing biologically sophisticated and detailed neural models. Furthermore finding good modelling abstraction levels is essential in order to better understand the computational abilities in neural networks. This is also important for creating a common language between researchers in neuroscience and AI.

References

1. Aaser, P., et al.: Towards making a cyborg: a closed-loop reservoir-neuro system. In: Proceedings of the 14th European Conference on Artificial Life ECAL 2017, pp. 430–437. MIT Press, Cambridge (2017). https://doi.org/10.7551/ecal_a_072
2. Attneave, F., B., M., Hebb, D.O.: The organization of behavior: a neuropsychological theory. Am. J. Psychol. **63**(4), 633 (2006). https://doi.org/10.2307/1418888
3. Brette, R.: Philosophy of the spike: rate-based vs. spike-based theories of the brain. Front. Syst. Neurosci. **9**, 151 (2015). https://doi.org/10.3389/fnsys.2015.00151
4. Buchanan, K.A., Mellor, J.: The activity requirements for spike timing-dependent plasticity in the hippocampus. Front. Synaptic Neurosci. **2**, 11 (2010). https://doi.org/10.3389/fnsyn.2010.00011

5. Burkitt, A.N.: A review of the integrate-and-fire neuron model: I. homogeneous synaptic input. Biol. Cybern. **95**(1), 1–19 (2006). https://doi.org/10.1007/s00422-006-0068-6
6. Clopath, C., Gerstner, W.: Voltage and spike timing interact in STDP - a unified model. Front. Synaptic Neurosci. **2**, 25 (2010). https://doi.org/10.3389/fnsyn.2010.00025
7. Diehl, P.U., Cook, M.: Unsupervised learning of digit recognition using spike-timing-dependent plasticity. Front. Comput. Neurosci. **9**, 99 (2015). https://doi.org/10.3389/fncom.2015.00099
8. Gerstner, W., Kreiter, A.K., Markram, H., Herz, A.V.: Neural codes: firing rates and beyond. Proc. Nat. Acad. Sci. U.S.A. **94**(24), 12740–1 (1997). https://doi.org/10.1073/PNAS.94.24.12740
9. Honoré, T., Lauridsen, J., Krogsgaard-Larsen, P.: The binding of [3H]AMPA, a structural analogue of glutamic acid, to rat brain membranes. J. Neurochem. (1982). https://doi.org/10.1111/j.1471-4159.1982.tb10868.x
10. Izhikevich, E.M.: Which model to use for cortical spiking neurons? IEEE Trans. Neural Netw. **15**(5), 1063–1070 (2004). https://doi.org/10.1109/TNN.2004.832719
11. de Kamps, M., van der Velde, F.: From artificial neural networks to spiking neuron populations and back again. Neural Netw. **14**(6–7), 941–953 (2001). https://doi.org/10.1016/S0893-6080(01)00068-5
12. Kandel, E.R., Schwartz, J.H., Jessell, T.M.: Principles of Neural Science, vol. 4. McGraw-Hill Education, New York (2013). https://doi.org/10.1036/0838577016
13. Kheradpisheh, S.R., Ganjtabesh, M., Masquelier, T.: Bio-inspired unsupervised learning of visual features leads to robust invariant object recognition. Neurocomputing **205**, 382–392 (2016). https://doi.org/10.1016/j.neucom.2016.04.029
14. Kheradpisheh, S.R., Ghodrati, M., Ganjtabesh, M., Masquelier, T.: Deep network scan resemble human feed-forward vision in invariant object recognition. Sci. Rep. **6** (2016). https://doi.org/10.1038/srep32672
15. Maass, W.: Networks of spiking neurons: the third generation of neural network models. Neural Netw. **10**(9), 1659–1671 (1997). https://doi.org/10.1016/S0893-6080(97)00011-7
16. Mainen, Z.F., Seinowski, T.J.: Reliability of spike timing in neocortical neurons. Science (1995). https://doi.org/10.1126/science.7770778
17. Markram, H., Gerstner, W., Sjøstrøm, P.J.: Spike-timing-dependent plasticity: a comprehensive overview. Front. Res. Topics **4**, 2010–2012 (2012). https://doi.org/10.3389/fnsyn.2012.00002
18. Medium: Google brain's co-inventor tells why he's building Chinese neural networks: Andrew Ng on the state of deep learning at Baidu. Medium (2015)
19. Meldrum, B.S.: Glutamate as a neurotransmitter in the brain: review of physiology and pathology. J. Nutr. **130**, 1007S-15S (2000). 10736372
20. Nair, V., Hinton, G.E.: Rectified linear units improve restricted Boltzmann machines. In: Proceedings of the 27th International Conference on Machine Learning (2010). https://doi.org/10.1.1.165.6419
21. Numenta. https://numenta.com/
22. Rumelhart, D.E., Widrow, B., Lehr, M.A.: The basic ideas in neural networks. Commun. ACM (1994). https://doi.org/10.1145/175247.175256
23. Shouval, H.Z., Wang, S.S.H., Wittenberg, G.M.: Spike timing dependent plasticity: a consequence of more fundamental learning rules. Front. Comput. Neurosci. **4**, 1–13 (2010). https://doi.org/10.3389/fncom.2010.00019

24. Sompolinsky, H.: Computational neuroscience: beyond the local circuit. Current Opinion Neurobiol. **25**, xiii–xviii (2014). https://doi.org/10.1016/J.CONB.2014.02.002
25. Song, S., Miller, K.D., Abbott, L.F.: Competitive Hebbian learning through spike-timing-dependent synaptic plasticity. Nature Neurosci. **3**(9), 919–926 (2000). https://doi.org/10.1038/78829
26. Sterratt, D., Graham, B., Gillies, A., Willshaw, D.: Principles of Computational Modelling in Neuroscience. Cambridge University Press (2011). https://doi.org/10.1109/MPUL.2012.2196841
27. Wolfram, S.: Cellular automata as models of complexity. Nature **311**(5985), 419–424 (1984). https://doi.org/10.1038/311419a0
28. Wurtz, R.H.: Visual receptive fields of striate cortex neurons in awake monkeys. J. Neurophysiol. (1969). https://doi.org/10.1152/jn.1969.32.5.727

The Lyapunov Exponents of Reversible Cellular Automata Are Uncomputable

Johan Kopra(✉)

Department of Mathematics and Statistics, University of Turku, 20014 Turku, Finland
jtjkop@utu.fi

Abstract. We will show that the class of reversible cellular automata (CA) with right Lyapunov exponent 2 cannot be separated algorithmically from the class of reversible CA whose right Lyapunov exponents are at most $2-\delta$ for some absolute constant $\delta > 0$. Therefore there is no algorithm that, given as an input a description of an arbitrary reversible CA F and a positive rational number $\epsilon > 0$, outputs the Lyapunov exponents of F with accuracy ϵ.

Keywords: Cellular automata · Lyapunov exponents · Reversible computation · Computability

1 Introduction

A cellular automaton (CA) is a model of parallel computation consisting of a uniform (in our case one-dimensional) grid of finite state machines, each of which receives input from a finite number of neighbors. All the machines use the same local update rule to update their states simultaneously at discrete time steps. The following question of error propagation arises naturally: If one changes the state at some of the coordinates, then how long does it take for this change to affect the computation at other coordinates that are possibly very far away? Lyapunov exponents provide one tool to study the asymptotic speeds of error propagation in different directions. The concept of Lyapunov exponents originally comes from the theory of differentiable dynamical systems, and the discrete variant of Lyapunov exponents for CA was originally defined in [11].

The Lyapunov exponents of a cellular automaton F are interesting also when one considers F as a topological dynamical system, because they can be used to give an upper bound for the topological entropy of F [12]. In [2] a closed formula for the Lyapunov exponents of linear one-dimensional cellular automata is given, which is a first step in determining for which classes of CA the Lyapunov exponents are computable. It is previously known that the entropy of one-dimensional cellular automata is uncomputable [7] (and furthermore from [4] it follows that

The work was partially supported by the Academy of Finland grant 296018 and by the Vilho, Yrjö and Kalle Väisälä Foundation.

I. McQuillan and S. Seki (Eds.): UCNC 2019, LNCS 11493, pp. 178–190, 2019.
https://doi.org/10.1007/978-3-030-19311-9_15

there exists a single cellular automaton whose entropy is uncomputable), which gives reason to suspect that also the Lyapunov exponents are uncomputable in general.

The uncomputability of Lyapunov exponents is easy to prove for (not necessarily reversible) cellular automata by using the result from [8] which says that nilpotency of cellular automata with a spreading state is undecidable. We will prove the more specific claim that the Lyapunov exponents are uncomputable even for reversible cellular automata. In the context of proving undecidability results for reversible CA one cannot utilize undecidability of nilpotency for non-reversible CA. An analogous decision problem, the (local) immortality problem, has been used to prove undecidability results for reversible CA [10]. We will use in our proof the undecidability of a variant of the immortality problem, which in turn follows from the undecidability of the tiling problem for 2-way deterministic tile sets.

2 Preliminaries

For sets A and B we denote by B^A the collection of all functions from A to B.

A finite set A of *letters* or *symbols* is called an *alphabet*. The set $A^{\mathbb{Z}}$ is called a *configuration space* or a *full shift* and its elements are called *configurations*. An element $c \in A^{\mathbb{Z}}$ is interpreted as a bi-infinite sequence that contains the symbol $c(i)$ at position i. A *factor* of c is any finite sequence $c(i)c(i+1)\cdots c(j)$ where $i, j \in \mathbb{Z}$, and we interpret the sequence to be empty if $j < i$. Any finite sequence $w = w(1)w(2)\cdots w(n)$ (also the empty sequence, which is denoted by ϵ) where $w(i) \in A$ is a *word* over A. If $w \neq \epsilon$, we say that w *occurs* in c at position i if $c(i)\cdots c(i+n) = w(1)\cdots w(n)$ and we denote by $w^{\mathbb{Z}} \in A^{\mathbb{Z}}$ the configuration in which w occurs at all positions of the form in ($i \in \mathbb{Z}$). The set of all words over A is denoted by A^*, and the set of non-empty words is $A^+ = A^* \setminus \{\epsilon\}$. More generally, for $L, K \subseteq A^*$ we denote $LK = \{w_1w_2 \mid w_1 \in L, w_2 \in K\}$, $L^* = \{w_1\cdots w_n \mid n \geq 0, w_i \in L\}$ and $L^+ = \{w_1\cdots w_n \mid n \geq 1, w_i \in L\}$. The set of words of length n is denoted by A^n. For a word $w \in A^*$, $|w|$ denotes its length, i.e. $|w| = n \iff w \in A^n$.

If A is an alphabet and C is a countable set, then A^C becomes a compact metrizable topological space when endowed with the product topology of the discrete topology of A (in particular a set $S \subseteq A^{\mathbb{Z}}$ is compact if and only if it is closed). In our considerations $C = \mathbb{Z}$ or $C = \mathbb{Z}^2$. We define the *shift* $\sigma : A^{\mathbb{Z}} \to A^{\mathbb{Z}}$ by $\sigma(c)(i) = c(i+1)$ for $c \in A^{\mathbb{Z}}$, $i \in \mathbb{Z}$, which is a homeomorphism. We say that a closed set $X \subseteq A^{\mathbb{Z}}$ is a *subshift* if $\sigma(X) = X$.

Occasionally we consider configuration spaces $(A_1 \times A_2)^{\mathbb{Z}}$ and then we may write $(c_1, c_2) \in (A_1 \times A_2)^{\mathbb{Z}}$ where $c_i \in A_i^{\mathbb{Z}}$ using the natural bijection between the sets $A_1^{\mathbb{Z}} \times A_2^{\mathbb{Z}}$ and $(A_1 \times A_2)^{\mathbb{Z}}$. We may use the terminology that c_1 is on the upper layer or on the A_1-layer, and similarly that c_2 is on the lower layer or on the A_2-layer.

Definition 1. *A (one-dimensional)* cellular automaton *(or a CA) is a 3-tuple (A, N, f), where A is a finite* state set, $N = (n_1, \ldots, n_m) \in \mathbb{Z}^m$ *is a* neighborhood vector *(containing distinct integers) and $f : A^m \to A$ is a* local rule. *A given CA (A, N, f) is customarily identified with a corresponding* CA function *$F : A^{\mathbb{Z}} \to A^{\mathbb{Z}}$ defined by*

$$F(c)(i) = f(c(i + n_1), \ldots, c(i + n_m))$$

for every $c \in A^{\mathbb{Z}}$ and $i \in \mathbb{Z}$. F is a radius-r *CA ($r \in \mathbb{N}$) if it can be defined with the neighborhood vector $(-r, \ldots, 0, \ldots, r)$ and it is a* radius-$\frac{1}{2}$ *CA if it can be defined with the neighborhood vector $(0, 1)$. If $X \subseteq A^{\mathbb{Z}}$ is a subshift and $F(X) \subseteq X$, the restriction map $F : X \to X$ is a* cellular automaton on X. *In the case we consider only the restriction $F : X \to X$, it is sometimes sufficient that the local rule f is a partial function.*

The *space-time diagram* of $c \in X$ (with respect to a CA $F : X \to X$) is the map $\theta \in A^{\mathbb{Z}^2}$ defined by $\theta(i, j) = F^j(c)(i)$. These are occasionally represented pictorially. The CA-functions on X are characterized as those continuous maps on X that commute with the shift [5]. We say that a CA $F : X \to X$ is *reversible* if it is bijective as a CA function. Reversible CA are homeomorphisms on X. The book [9] is a standard reference for subshifts and cellular automata on them.

The definition of Lyapunov exponents is from [11,12]. For a fixed subshift $X \subseteq A^{\mathbb{Z}}$ and for $c \in X$, $s \in \mathbb{Z}$, denote $W_s^+(c) = \{e \in X \mid \forall i \geq s : e(i) = c(i)\}$ and $W_s^-(c) = \{e \in X \mid \forall i \leq s : e(i) = c(i)\}$. Then for given cellular automaton $F : X \to X$, $c \in X$, $n \in \mathbb{N}$, define

$$\Lambda_n^+(c, F) = \min\{s \geq 0 \mid \forall 1 \leq i \leq n : F^i(W_{-s}^+(c)) \subseteq W_0^+(F^i(c))\}$$
$$\Lambda_n^-(c, F) = \min\{s \geq 0 \mid \forall 1 \leq i \leq n : F^i(W_s^-(c)) \subseteq W_0^-(F^i(c))\}.$$

Finally, the quantities

$$\lambda^+(F) = \lim_{n\to\infty} \max_{c \in A^{\mathbb{Z}}} \frac{\Lambda_n^+(c, F)}{n}, \qquad \lambda^-(F) = \lim_{n\to\infty} \max_{c \in A^{\mathbb{Z}}} \frac{\Lambda_n^-(c, F)}{n}$$

are called respectively the right and left *Lyapunov exponents* of F. These limits exist by an application of Fekete's lemma (e.g. Lemma 4.1.7 in [9]). We write $\Lambda_n^+(c)$, $\Lambda_n^-(c)$, λ^+ and λ^- when F is clear by the context.

3 Tilings and Undecidability

In this section we recall the well-known connection between cellular automata and tilings on the plane. We use this connection to prove an auxiliary undecidability result for reversible cellular automata.

Definition 2. *A* Wang tile *is formally a function $t : \{N, E, S, W\} \to C$ whose value at I is denoted by t_I. Informally, a Wang tile t should be interpreted as a unit square with edges colored by elements of C. The edges are called* north, east, south *and* west *in the natural way, and the colors in these edges of t are t_N, t_E, t_S and t_W respectively. A* tile set *is a finite collection of Wang tiles.*

Definition 3. *A tiling over a tile set T is a function $\eta \in T^{\mathbb{Z}^2}$ which assigns a tile to every integer point of the plane. A tiling η is said to be valid if neighboring tiles always have matching colors in their edges, i.e. for every $(i,j) \in \mathbb{Z}^2$ we have $\eta(i,j)_N = \eta(i,j+1)_S$ and $\eta(i,j)_E = \eta(i+1,j)_W$. If there is a valid tiling over T, we say that T* admits *a valid tiling.*

We say that a tile set T is NE-deterministic if for every pair of tiles $t, s \in T$ the equalities $t_N = s_N$ and $t_E = s_E$ imply $t = s$, i.e. a tile is determined uniquely by its north and east edge. A SW-deterministic tile set is defined similarly. If T is both NE-deterministic and SW-deterministic, it is said to be *2-way deterministic*.

The *tiling problem* is the problem of determining whether a given tile set T admits a valid tiling.

Theorem 1. *[10, Theorem 4.2.1] The tiling problem is undecidable for 2-way deterministic tile sets.*

Definition 4. *Let T be a 2-way deterministic tile set and C the collection of all colors which appear in some edge of some tile of T. T is* complete *if for each pair $(a,b) \in C^2$ there exist (unique) tiles $t, s \in T$ such that $(t_N, t_E) = (a,b)$ and $(s_S, s_W) = (a,b)$.*

A 2-way deterministic tile set T can be used to construct a complete tile set. Namely, let C be the set of colors which appear in tiles of T, let $X \subseteq C \times C$ be the set of pairs of colors which do not appear in the northeast of any tile and let $Y \subseteq C \times C$ be the set of pairs of colors which do not appear in the southwest of any tile. Since T is 2-way deterministic, there is a bijection $p : X \to Y$. Let $T^{\mathtt{C}}$ be the set of tiles formed by matching the northeast corners X with the southwest corners Y via the bijection p. Then the tile set $A = T \cup T^{\mathtt{C}}$ is complete.

Every complete 2-way deterministic tile set A determines a reversible CA $(A, (0,1), f)$ with the local rule $f : A^2 \to A$ defined by $f(a,b) = c \in A$, where c is the unique tile such that $a_E = c_W$ and $b_N = c_S$. If we denote the corresponding CA function by F, the space-time diagram of a configuration $c \in A^{\mathbb{Z}}$ corresponds to a valid tiling η via $\theta(i,j) = F^j(c)(i) = \eta(i+j, -i)$, i.e. configurations $F^j(c)$ are diagonals of η going from northwest to southeast.

Definition 5. *A cellular automaton $F : A^{\mathbb{Z}} \to A^{\mathbb{Z}}$ is (x,y)-locally immortal ($x, y \in \mathbb{N}$) with respect to a subset $B \subseteq A$ if there exists a configuration $c \in A^{\mathbb{Z}}$ such that $F^{iy+j}(c)(ix) \in B$ for all $i \in \mathbb{Z}$ and $0 \leq j \leq y$. Such a configuration c is an (x,y)-witness.*

Generalizing the definition in [10], we call the following decision problem the *(x,y)-local immortality problem*: given a reversible CA $F : A^{\mathbb{Z}} \to A^{\mathbb{Z}}$ and a subset $B \subseteq A$, is F (x,y)-locally immortal with respect to B? In Theorem 5.1.5 of [10] it is shown that the $(0,1)$-local immortality problem is undecidable for reversible CA. We adapt the proof to get the following result.

Lemma 1. *The $(1,5)$-local immortality problem is undecidable for reversible radius-$\frac{1}{2}$ CA.*

Proof. We will reduce the problem of Theorem 1 to the $(1,5)$-local immortality problem. Let T be a 2-way deterministic tile set and construct a complete tile set $T \cup T^{\mathsf{C}}$ as indicated above. Then also $A_1 = (T \times T_1) \cup (T^{\mathsf{C}} \times T_2)$ (T_1 and T_2 as in Fig. 1) is a complete tile set[1]. We denote the blank tile of the set T_1 by t_b and call the elements of $R = A_1 \setminus (T \times \{t_b\})$ arrow tiles. As indicated above, the tile set A_1 determines a reversible radius-$\frac{1}{2}$ CA $G_1 : A_1^{\mathbb{Z}} \to A_1^{\mathbb{Z}}$.

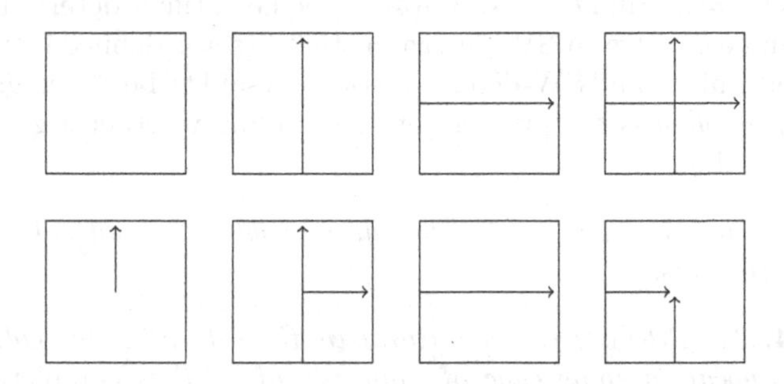

Fig. 1. The tile sets T_1 (first row) and T_2 (second row) from [10].

Let $A_2 = \{0,1,2\}$. Define $A = A_1 \times A_2$ and natural projections $\pi_i : A \to A_i$, $\pi_i(a_1,a_2) = a_i$ for $i \in \{1,2\}$. By extension we say that $a \in A$ is an arrow tile if $\pi_1(a) \in R$. Let $G : A^{\mathbb{Z}} \to A^{\mathbb{Z}}$ be defined by $G(c,e) = (G_1(c), e)$ where $c \in A_1^{\mathbb{Z}}$ and $e \in A_2^{\mathbb{Z}}$, i.e. G simulates G_1 in the upper layer. We define, using the notation of Definition 1, involutive CA J_1, J_2 and H by tuples $(A_2,(0),j_1)$, $(A_2,(0,1),j_2)$ and $(A_1 \times A_2,(0),h)$ where

$$j_1(0) = 0 \qquad j_1(1) = 2 \qquad j_1(2) = 1 \qquad j_2(a,b) = \begin{cases} 1 \text{ when } (a,b) = (0,2) \\ 0 \text{ when } (a,b) = (1,2) \\ a \text{ otherwise} \end{cases}$$

$$h((a,b)) = \begin{cases} (a,1) \text{ when } a \in R \text{ and } b = 0 \\ (a,0) \text{ when } a \in R \text{ and } b = 1 \\ (a,b) \text{ otherwise.} \end{cases}$$

If $\mathrm{id} : A_1^{\mathbb{Z}} \to A_1^{\mathbb{Z}}$ is the identity map, then $J = (\mathrm{id} \times J_2) \circ (\mathrm{id} \times J_1)$ is a CA on $A^{\mathbb{Z}} = (A_1 \times A_2)^{\mathbb{Z}}$. We define the radius-$\frac{1}{2}$ automaton $F = H \circ J \circ G : A^{\mathbb{Z}} \to A^{\mathbb{Z}}$ and select $B = (T \times \{t_b\}) \times \{0\}$. We will show that T admits a valid tiling if and only if F is $(1,5)$-locally immortal with respect to B.

Assume first that T admits a valid tiling η. Then by choosing $c \in A^{\mathbb{Z}}$ such that $c(i) = ((\eta(i,-i), t_b), 0) \in A_1 \times A_2$ for $i \in \mathbb{Z}$ it follows that $F^j(c)(i) \in B$ for all $i,j \in \mathbb{Z}$ and in particular that c is a $(1,5)$-witness.

Assume then that T does not admit any valid tiling and for a contradiction assume that c is a $(1,5)$-witness. Let θ be the space-time diagram of c with

[1] The arrow markings are used as a shorthand for some coloring such that the heads and tails of the arrows in neighboring tiles match in a valid tiling.

respect to F. Since c is a $(1,5)$-witness, it follows that $\theta(i,j) \in B$ whenever $(i,j) \in N$, where $N = \{(i,j) \in \mathbb{Z}^2 \mid 5i \leq j \leq 5(i+1)\}$. There is a valid tiling η over A_1 such that $\pi_1(\theta(i,j)) = \eta(i+j,-i)$ for $(i,j) \in \mathbb{Z}^2$, i.e. η can be recovered from the upper layer of θ by applying a suitable linear transformation on the space-time diagram. In drawing pictorial representations of θ we want that the heads and tails of all arrows remain properly matched in neighboring coordinates, so we will use tiles with "bent" labelings, see Fig. 2. Since T does not admit valid tilings, it follows by a compactness argument that $\eta(i,j) \notin T \times T_1$ for some $(i,j) \in D$ where $D = \{(i,j) \in \mathbb{Z}^2 \mid j < -\lfloor i/6 \rfloor\}$ and in particular that $\eta(i,j)$ is an arrow tile. Since θ contains a "bent" version of η, it follows that $\theta(i,j)$ is an arrow tile for some $(i,j) \in E$, where $E = \{(i,j) \in \mathbb{Z}^2 \mid j < 5i\}$ is a "bent" version of the set D. In Fig. 3 we present the space-time diagram θ with arrow markings of tiles from T_1 and T_2 replaced according to the Fig. 2. In Fig. 3 we have also marked the sets N and E. Other features of the figure become relevant in the next paragraph.

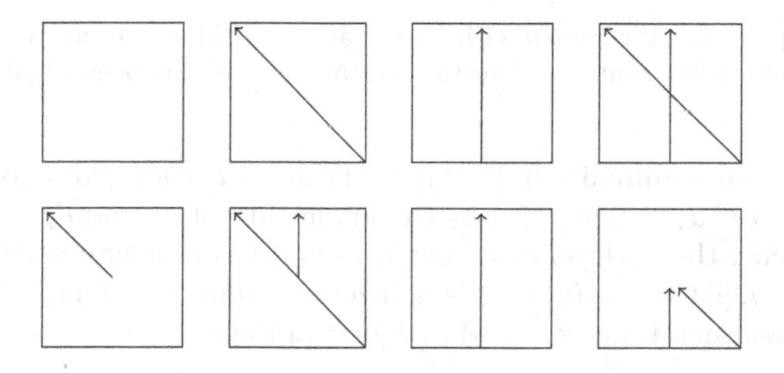

Fig. 2. The tile sets T_1 and T_2 presented in a "bent" form.

The minimal distance between a tile in N and an arrow tile in E situated on the same horizontal line in θ is denoted by $d_1 > 0$. Then, among those arrow tiles in E at horizontal distance d_1 from N, there is a tile with minimal vertical distance $d_2 > 0$ from N (see Fig. 3). Fix $x, y \in \mathbb{Z}$ so that $\theta(x, y+2)$ is one such tile and in particular $(x-d_1, y+2), (x, y+2+d_2) \in N$. Then $\theta(x, y+j)$ contains an arrow for $-2 \leq j \leq 2$, because if there is a $j \in [-2, 2)$ such that $\theta(x, y+j)$ does not contain an arrow and $\theta(x, y+j+1)$ does, then $\theta(x, y+j+1)$ must contain one of the three arrows on the left half of Fig. 2. These three arrows continue to the northwest, so then also $\theta(x-1, y+j+2)$ contains an arrow. Because $\theta(i', j') \in B$ for $(i', j') \in N$, it follows that $(x-1, y+j+2) \notin N$ and thus $(x-1, y+j+2) \in E$. Since $(x-d_1, y+2) \in N$, it follows that one of the $(x-d_1-1, y+j+2)$, $(x-d_1, y+j+2)$ and $(x-d_1+1, y+j+2)$ belong to N. Thus the horizontal distance of the tile $\theta(x-1, y+j+2)$ from the set N is at most d_1, and is actually equal to d_1 by the minimality of d_1. Since N is invariant under translation by the vector $-(1,5)$, then from $(x, y+2+d_2) \in N$ it follows that $(x-1, y-3+d_2) \in N$ and that the vertical distance of the tile $\theta(x-1, y+j+2)$ from N is at most $(y-3+d_2) - (y+j+2) \leq d_2 - 3$,

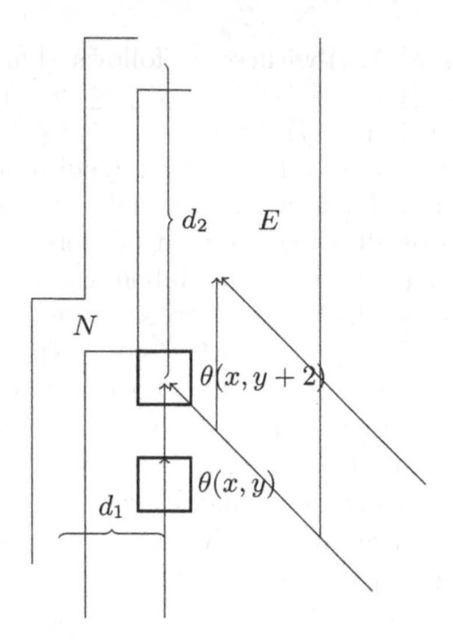

Fig. 3. The space-time diagram θ with "bent" arrow markings. An arrow tile $\theta(x, y+2)$ in E with minimal horizontal and vertical distances to N has been highlighted.

contradicting the minimality of d_2. Similarly, $\theta(x-i, y+j)$ does not contain an arrow for $0 < i \leq d_1$, $-2 \leq j \leq 2$ by the minimality of d_1 and d_2.

Now consider the A_2-layer of θ. For the rest of the proof let $e = F^y(c)$. Assume that $\pi_2(\theta(x-i, y)) = \pi_2(e(x-i))$ is non-zero for some $i \geq 0$, $(x-i, y) \in E$ and fix the greatest such i, i.e. $\pi_2(e(z)) = 0$ for z in the set

$$I_0 = \{x' \in \mathbb{Z} \mid x' < x - i, (x', y) \in N \cup E\}.$$

We start by considering the case $\pi_2(e(x-i)) = 1$. Denote

$$I_1 = \{x' \in \mathbb{Z} \mid x' < x - i, (x', y+1) \in N \cup E\} \subseteq I_0.$$

From the choice of (x, y) it follows that $\pi_1(\theta(z, y+1)) = \pi_1(G(e)(z))$ are not arrow tiles for $z \in I_1$, and therefore we can compute step by step that

$$\begin{aligned}
&\pi_2((\mathrm{id} \times J_1)(G(e))(x-i)) = 2, && \pi_2((\mathrm{id} \times J_1)(G(e))(z)) = 0 \text{ for } z \in I_0 \subseteq I_1, \\
&\pi_2(J(G(e))(x-(i+1))) = 1, && \pi_2(J(G(e))(z)) = 0 \text{ for } z \in I_1 \setminus \{x-(i+1)\}, \\
&\pi_2(F(e))(x-(i+1))) = 1, && \pi_2(F(e)(z)) = 0 \text{ for } z \in I_1 \setminus \{x-(i+1)\}
\end{aligned}$$

and $\pi_2(\theta(x-(i+1), y+1)) = 1$. By repeating this argument inductively we see that the digit 1 propagates to the upper left in the space-time diagram as indicated by Fig. 4 and eventually reaches N, a contradiction. If on the other hand $\pi_2(\theta(x-i, y)) = 2$, a similar argument shows that the digit 2 propagates to the lower left in the space-time diagram as indicated by Fig. 4 and eventually reaches N, also a contradiction.

Assume then that $\pi_2(\theta(x-i, y))$ is zero whenever $i \geq 0$, $(x-i, y) \in E$. If $\pi_2(\theta(x+1, y)) = \pi_2(e(x+1)) \neq 1$, then $\pi_2((\mathrm{id} \times J_1)(G(e))(x+1)) \neq 2$

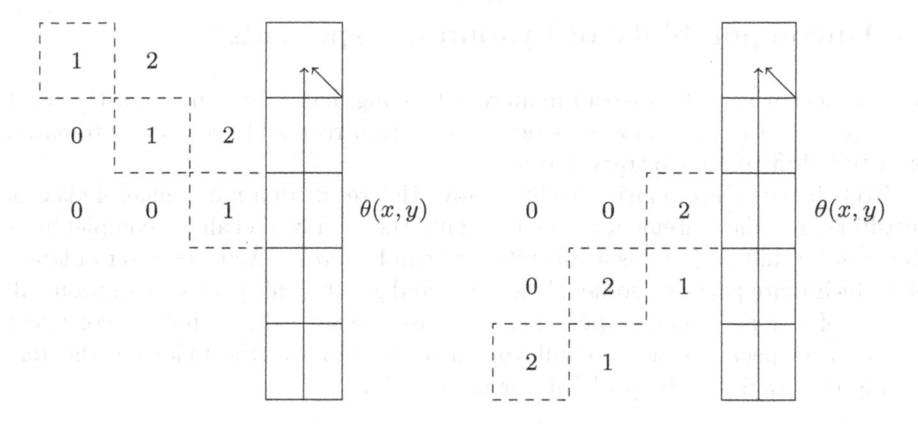

Fig. 4. Propagation of digits to the left of $\theta(x, y)$.

and $\pi_2(J(G(e))(x)) = 0$. Since $\pi_1(\theta(x, y+1))$ is an arrow tile, it follows that $\pi_2(\theta(x, y+1)) = \pi_2(H(J(G(e)))(x)) = 1$. The argument of the previous paragraph shows that the digit 1 propagates to the upper left in the space-time diagram as indicated by the left side of Fig. 5 and eventually reaches N, a contradiction.

Finally consider the case $\pi_2(\theta(x+1, y)) = \pi_2(e(x+1)) = 1$. Then

$$\pi_2(J(G(e))(x))\pi_2(J(G(e))(x+1)) = 12 \text{ and}$$
$$\pi_2(F(e))(x))\pi_2(F(e))(x+1)) = 02.$$

As in the previous paragraph we see that $\pi_2(\theta(x, y+2)) = 1$. This occurrence of the digit 1 propagates to the upper left in the space-time diagram as indicated by the right side of Fig. 5 and eventually reaches N, a contradiction. □

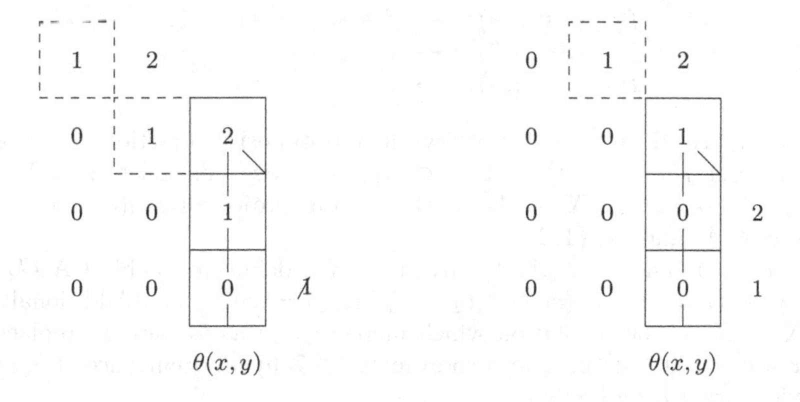

Fig. 5. Propagation of digits at $\theta(x, y)$.

4 Uncomputability of Lyapunov Exponents

In this section we will prove our main result saying that there is no algorithm that can compute the Lyapunov exponents of a given reversible cellular automaton on a full shift to an arbitrary precision.

To achieve greater clarity we first prove this result in a more general class of subshifts. For the statement of the following theorem, we recall for completeness that a sofic shift $X \subseteq A^{\mathbb{Z}}$ is a subshift that can be represented as the set of labels of all bi-infinite paths on some labeled directed graph. This precise definition will not be of any particular importance, because the sofic shifts that we construct are of very specific form. We will appeal to the proof of the following theorem during the course of the proof of our main result.

Theorem 2. *For reversible CA $F : X \to X$ on sofic shifts such that $\lambda^+(F) \in [0, \frac{5}{3}] \cup \{2\}$ it is undecidable whether $\lambda^+(F) \leq \frac{5}{3}$ or $\lambda^+(F) = 2$.*

Proof. We will reduce the decision problem of Lemma 1 to the present problem. Let $G : A_2^{\mathbb{Z}} \to A_2^{\mathbb{Z}}$ be a given reversible radius-$\frac{1}{2}$ cellular automaton and $B \subseteq A_2$ some given set. Let $A_1 = \{0, \|, \leftarrow, \rightarrow, \nwarrow, \nearrow\}$ and define a sofic shift $Y \subseteq A_1^{\mathbb{Z}}$ as the set of those configurations containing a symbol from $Q = \{\leftarrow, \rightarrow, \nwarrow, \nearrow\}$ in at most one position. We will interpret elements of Q as particles going in different directions at different speeds and which bounce between walls denoted by $\|$. Let $S : Y \to Y$ be the reversible radius-2 CA which does not move occurrences of $\|$ and which moves $\leftarrow$ (resp. $\rightarrow$, $\nwarrow$, $\nearrow$) to the left at speed 2 (resp. to the right at speed 2, to the left at speed 1, to the right at speed 1) with the additional condition that when an arrow meets a wall, it changes into the arrow with the same speed and opposing direction. More precisely, S is determined by the tuple $(A_1, \{-2, -1, 0, 1, 2\}, f)$ where the local rule $f : A_1^5 \to A_1$ is defined as follows (where $*$ denotes arbitrary symbols):

$$\begin{array}{ll} f(\rightarrow, 0, 0, *, *) = \rightarrow & f(*, \nearrow, 0, *, *) = \nearrow \\ f(*, \rightarrow, 0, \|, *) = \leftarrow & f(*, *, \nearrow, \|, *) = \nwarrow, \\ f(*, *, 0, \rightarrow, \|) = \leftarrow & \\ f(*, \|, \rightarrow, \|, *) = \rightarrow & \end{array}$$

with symmetric definitions for arrows in the opposite directions at reflected positions and $f(*, *, a, *, *) = a$ $(a \in A_1)$ otherwise. Then let $X = Y \times A_2^{\mathbb{Z}}$ and $\pi_1 : X \to Y$, $\pi_2 : X \to A_2^{\mathbb{Z}}$ be the natural projections $\pi_i(c_1, c_2) = c_i$ for $c_1 \in Y, c_2 \in A_2^{\mathbb{Z}}$ and $i \in \{1, 2\}$.

Let $c_1 \in Y$ and $c_2 \in A_2^{\mathbb{Z}}$ be arbitrary. We define reversible CA $G_2, F_1 : X \to X$ by $G_2(c_1, c_2) = (c_1, G^{10}(c_2))$, $F_1(c_1, c_2) = (S(c_1), c_2)$. Additionally, let $F_2 : X \to X$ be the involution which maps (c_1, c_2) as follows: F_2 replaces an occurrence of $\rightarrow 0 \in A_1^2$ in c_1 at a coordinate $i \in \mathbb{Z}$ by an occurrence of $\nwarrow\| \in A_1^2$ (and vice versa) *if and only if*

$$G^j(c_2)(i) \notin B \text{ for some } 0 \leq j \leq 5 \quad \text{or} \quad G^j(c_2)(i+1) \notin B \text{ for some } 5 \leq j \leq 10,$$

and otherwise F_2 makes no changes. Finally, define $F = F_1 \circ G_2 \circ F_2 : X \to X$. The reversible CA F works as follows. Typically particles from Q move in the upper layer in the intuitive manner indicated by the map S and the lower layer is transformed according to the map G^{10}. There are some exceptions to the usual particle movements: If there is a particle $\rightarrow$ which does not have a wall immediately at the front and c_2 does not satisfy a local immortality condition in the next 10 time steps, then $\rightarrow$ changes into $\nwarrow$ and at the same time leaves behind a wall segment $\|$. Conversely, if there is a particle $\nwarrow$ to the left of the wall $\|$ and c_2 does not satisfy a local immortality condition, $\nwarrow$ changes into $\rightarrow$ and removes the wall segment.

We will show that $\lambda^+(F) = 2$ if G is $(1,5)$-locally immortal with respect to B and $\lambda^+(F) \leq \frac{5}{3}$ otherwise. Intuitively the reason for this is that if $c, e \in X$ are two configurations that differ only to the left of the origin, then the difference between $F^i(c)$ and $F^i(e)$ can propagate to the right at speed 2 only via an arrow $\rightarrow$ that travels on top of a $(1,5)$-witness. Otherwise, a signal that attempts to travel to the right at speed 2 is interrupted at bounded time intervals and forced to return at a slower speed beyond the origin before being able to continue its journey to the right. We will give more details.

Assume first that G is $(1,5)$-locally immortal with respect to B. Let $c_2 \in A_2^{\mathbb{Z}}$ be a $(1,5)$-witness and define $c_1 \in Y$ by $c_1(0) = \rightarrow$ and $c_1(i) = 0$ for $i \neq 0$. Let $c = (0^{\mathbb{Z}}, c_2) \in X$ and $e = (c_1, c_2) \in X$. It follows that $\pi_1(F^i(c))(2i) = 0$ and $\pi_1(F^i(e))(2i) = \rightarrow$ for every $i \in \mathbb{N}$, so $\lambda^+(F) \geq 2$. On the other hand, F has a neighborhood vector $(-2, -1, 0, 1, \dots, 10)$ so necessarily $\lambda^+(F) = 2$.

Assume then that there are no $(1,5)$-witnesses for G. Let us denote

$$C(n) = \{c \in A_1^{\mathbb{Z}} \mid G^{5i+j}(c)(i) \in B \text{ for } 0 \leq i \leq n, 0 \leq j \leq 5\} \text{ for } n \in \mathbb{N}.$$

Since there are no $(1,5)$-witnesses, by a compactness argument we may fix some $N \in \mathbb{N}_+$ such that $C(2N) = \emptyset$. We claim that $\lambda^+(F) \leq \frac{5}{3}$, so let us assume that $(c^{(n)})_{n\in\mathbb{N}}$ with $c^{(n)} = (c_1^{(n)}, c_2^{(n)}) \in X$ is a sequence of configurations such that $\Lambda_n^+(c^{(n)}, F) = s_n n$ where $(s_n)_{n\in\mathbb{N}}$ tends to λ^+. There exist $e^{(n)} = (e_1^{(n)}, e_2^{(n)}) \in X$ such that $c^{(n)}(i) = e^{(n)}(i)$ for $i > -s_n n$ and $F^{t_n}(c)(i_n) \neq F^{t_n}(e)(i_n)$ for some $0 \leq t_n \leq n$ and $i_n \geq 0$.

First assume that there are arbitrarily large $n \in \mathbb{N}$ for which $c_1^{(n)}(i) \in \{0, \|\}$ for $i > -s_n n$ and consider the subsequence of such configurations $c^{(n)}$ (starting with sufficiently large n). Since G is a radius-$\frac{1}{2}$ CA, it follows that $\pi_2(F^{t_n}(c^{(n)}))(j) = \pi_2(F^{t_n}(e^{(n)}))(j)$ for $j \geq 0$. Therefore the difference between $c^{(n)}$ and $e^{(n)}$ can propagate to the right only via an arrow from Q, so without loss of generality (by swapping $c^{(n)}$ and $e^{(n)}$ if necessary) $\pi_1(F^{t_n}(c^{(n)}))(j_n) \in Q$ for some $0 \leq t_n \leq n$ and $j_n \geq i_n - 1$. Fix some such t_n, j_n and let $w_n \in Q^{t_n+1}$ be such that $w_n(i)$ is the unique state from Q in the configuration $F^i(c^{(n)})$ for $0 \leq i \leq t_n$. The word w_n has a factorization of the form $w_n = u(v_1 u_1 \cdots v_k u_k)v$ ($k \in \mathbb{N}$) where $v_i \in \{\rightarrow\}^+$, $v \in \{\rightarrow\}^*$ and $u_i \in (Q \setminus \{\rightarrow\})^+$, $u \in (Q \setminus \{\rightarrow\})^*$. By the choice of N it follows that all v_i, v have length at most N and by the definition of the CA F it is easy to see that each u_i contains at least $2(|v_i|-1)+1$ occurrences of $\nwarrow$ and at least $2(|v_i|-1)+1$ occurrences of $\nearrow$ (after $\rightarrow$ turns into

$\nwarrow$, it must return to the nearest wall to the left and back and at least once more turn into $\nwarrow$ before turning back into $\rightarrow$. If $\rightarrow$ were to turn into $\leftarrow$ instead, it would signify an impassable wall on the right). If we denote by x_n the number of occurrences of $\rightarrow$ in w_n, then $x_n \leq |w_n|/3 + \mathcal{O}(1)$ (this upper bound is achieved by assuming that $|v_i| = 1$ for every i) and

$$s_n n \leq |w_n| + 2x_n \leq |w_n| + \frac{2}{3}|w_n| + \mathcal{O}(1) \leq \frac{5}{3}n + \mathcal{O}(1).$$

After dividing this inequality by n and passing to the limit we find that $\lambda^+ \leq \frac{5}{3}$.[2]

Next assume that there are arbitrarily large $n \in \mathbb{N}$ for which $c_1^{(n)}(i) \in Q$ for some $i > -s_n n$. The difference between $c^{(n)}$ and $e^{(n)}$ can propagate to the right only after the element from Q in $c^{(n)}$ reaches the coordinate $-s_n n$, so without loss of generality there are $0 < t_{n,1} < t_{n,2} \leq n$ and $i_n \geq 0$ such that $\pi_1(F^{t_{n,1}}(c^{(n)}))(-s) \in Q$ for some $s \geq s_n n$ and $\pi_1(F^{t_{n,2}}(c^{(n)}))(i_n) \in Q$. From this the contradiction follows in the same way as in the previous paragraph. □

We are ready to prove our main result.

Theorem 3. *For reversible CA $F : A^{\mathbb{Z}} \to A^{\mathbb{Z}}$ such that $\lambda^+(F) \in [0, \frac{5}{3}] \cup \{2\}$ it is undecidable whether $\lambda^+(F) \leq \frac{5}{3}$ or $\lambda^+(F) = 2$.*

Proof. Let $G : A_2^{\mathbb{Z}} \to A_2^{\mathbb{Z}}$, A_1, $F = F_1 \circ G_2 \circ F_2 : X \to X$, etc. be as in the proof of the previous theorem. We will adapt the conveyor belt construction from [3] to define a CA F' on a full shift which simulates F and has the same right Lyapunov exponent as F.

Denote $Q = \{\leftarrow, \rightarrow, \nwarrow, \nearrow\}$, $\Sigma = \{0, \|\}$, $\Delta = \{-, 0, +\}$, define the alphabets

$$\Gamma = (\Sigma^2 \times \{+,-\}) \cup (Q \times \Sigma \times \{0\}) \cup (\Sigma \times Q \times \{0\}) \subseteq A_1 \times A_1 \times \Delta$$

and $A = \Gamma \times A_2$ and let $\pi_{1,1}, \pi_{1,2} : A^{\mathbb{Z}} \to A_1^{\mathbb{Z}}$, $\pi_\Delta : A^{\mathbb{Z}} \to \Delta^{\mathbb{Z}}$ $\pi_2 : A^{\mathbb{Z}} \to A_2^{\mathbb{Z}}$ be the natural projections $\pi_{1,1}(c) = c_{1,1}$, $\pi_{1,2}(c) = c_{1,2}$, $\pi_\Delta(c) = c_\Delta$, $\pi_2(c) = c_2$ for $c = (c_{1,1}, c_{1,2}, c_\Delta, c_2) \in A^{\mathbb{Z}} \subseteq (A_1 \times A_1 \times \Delta \times A_2)^{\mathbb{Z}}$. For arbitrary $c = (c_1, c_2) \in (\Gamma \times A_2)^{\mathbb{Z}}$ define $G_2' : A^{\mathbb{Z}} \to A^{\mathbb{Z}}$ by $G_2'(c) = (c_1, G^{10}(c_2))$.

Next we define $F_1' : A^{\mathbb{Z}} \to A^{\mathbb{Z}}$. Every element $c = (c_1, c_2) \in (\Gamma \times A_2)^{\mathbb{Z}}$ has a unique decomposition of the form $(c_1, c_2) = \cdots (w_{-2}, v_{-2})(w_{-1}, v_{-1})(w_0, v_0)$ $(w_1, v_1)(w_2, v_2) \cdots$ where

$$\begin{aligned} w_i \in &(\Sigma^2 \times \{+\})^*((Q \times \Sigma \times \{0\}) \cup (\Sigma \times Q \times \{0\}))(\Sigma^2 \times \{-\})^* \\ &\cup (\Sigma^2 \times \{+\})^*(\Sigma^2 \times \{-\})^* \end{aligned}$$

with the possible exception of the leftmost w_i beginning or the rightmost w_i ending with an infinite sequence from $\Sigma^2 \times \{+,-\}$.

Let $(x_i, y_i) \in (\Sigma \times \Sigma)^*((Q \times \Sigma) \cup (\Sigma \times Q))(\Sigma \times \Sigma)^* \cup (\Sigma \times \Sigma)^*$ be the word that is derived from w_i by removing the symbols from Δ. The pair (x_i, y_i) can

[2] By performing more careful estimates it can be shown that $\lambda^+ = 1$, but we will not attempt to formalize the argument for this.

be seen as a conveyor belt by gluing the beginning of x_i to the beginning of y_i and the end of x_i to the end of y_i. The map F'_1 will shift arrows like the map F_1, and at the junction points of x_i and y_i the arrow can turn around to the opposite side of the belt. More precisely, define the permutation $\rho : A_1 \to A_1$ by

$$\rho(0) = 0 \quad \rho(\|) = \|$$
$$\rho(\leftarrow) = \rightarrow \quad \rho(\rightarrow) = \leftarrow \quad \rho(\nwarrow) = \nearrow \quad \rho(\nearrow) = \nwarrow$$

and for a word $u \in A_1^*$ let $\rho(u)$ denote the coordinatewise application of ρ. For any word $w = w(1) \cdots w(n)$ define its reversal by $w^R(i) = w(n+1-i)$ for $1 \leq i \leq n$. Then consider the periodic configuration $e = [(x_i, v_i)(\rho(y_i), v_i)^R]^{\mathbb{Z}} \in (A_1 \times A_2)^{\mathbb{Z}}$. The map $F_1 : X \to X$ extends naturally to configurations of the form e: e can contain infinitely many arrows, but they all point in the same direction and occur in identical contexts. By applying F_1 to e we get a new configuration of the form $[(x'_i, v_i)(\rho(y'_i), v_i)^R]$. From this we extract the pair (x'_i, y'_i), and by adding plusses and minuses to the left and right of the arrow (or in the same coordinates as in (x_i, y_i) if there is no occurrence of an arrow) we get a word w'_i which is of the same form as w_i. We define $F'_1 : A^{\mathbb{Z}} \to A^{\mathbb{Z}}$ by $F'_1(c) = c'$ where $c' = \cdots (w'_{-2}, v_{-2})(w'_{-1}, v_{-1})(w'_0, v_0)(w'_1, v_1)(w'_2, v_2) \cdots$. Clearly F'_1 is shift invariant, continuous and reversible.

We define the involution $F'_2 : A^{\mathbb{Z}} \to A^{\mathbb{Z}}$ as follows. For $c \in A^{\mathbb{Z}}$ and $j \in \{1, 2\}$ F'_2 replaces an occurrence of $\rightarrow 0$ in $\pi_{1,j}(c)$ at coordinate $i \in \mathbb{Z}$ by an occurrence of $\nwarrow \|$ (and vice versa) *if and only if* $\pi_\Delta(c)(i+1) = -$ and

$$G^j(\pi_2(c))(i) \in B \text{ for } 0 \leq j \leq 5 \qquad \text{and} \qquad G^j(\pi_2(c))(i+1) \in B \text{ for } 5 \leq j \leq 10,$$

and otherwise F_2 makes no changes. F'_2 simulates the map F_2 and we check the condition $\pi_\Delta(c)(i+1) = -$ to ensure that F'_2 does not transfer information between neighboring conveyor belts.

Finally, we define $F' = F'_1 \circ G'_2 \circ F'_2 : A^{\mathbb{Z}} \to A^{\mathbb{Z}}$. The reversible CA F' simulates $F : X \to X$ simultaneously on two layers and it has the same right Lyapunov exponent as F. □

The following corollary is immediate.

Corollary 1. *There is no algorithm that, given a reversible CA $F : A^{\mathbb{Z}} \to A^{\mathbb{Z}}$ and a rational number $\epsilon > 0$, returns the Lyapunov exponent $\lambda^+(F)$ within precision ϵ.*

5 Conclusions

We have shown that the Lyapunov exponents of a given reversible cellular automaton on a full shift cannot be computed to arbitrary precision. Ultimately this turned out to follow from the fact that the tiling problem for 2-way deterministic Wang tile sets reduces to the problem of computing the Lyapunov exponents of reversible CA. In our constructions we controlled only the right exponent λ^+ and let the left exponent λ^- vary freely. Controlling both Lyapunov exponents would be necessary to answer the following.

Problem 1. Is it decidable whether the equality $\lambda^+(F) + \lambda^-(F) = 0$ holds for a given reversible cellular automaton $F : A^{\mathbb{Z}} \to A^{\mathbb{Z}}$?

We mentioned in the introduction that there exists a single CA whose topological entropy is an uncomputable number. We ask whether a similar result holds also for the Lyapunov exponents.

Problem 2. Does there exist a single cellular automaton $F : A^{\mathbb{Z}} \to A^{\mathbb{Z}}$ such that $\lambda^+(F)$ is an uncomputable number?

We are not aware of a cellular automaton on a full shift that has an irrational Lyapunov exponent (see Question 5.7 in [1]), so constructing such a CA (or proving the impossibility of such a construction) should be the first step. This problem has an answer for CA $F : X \to X$ on general subshifts X, and furthermore for every real $t \geq 0$ there is a subshift X_t and a reversible CA $F_t : X_t \to X_t$ such that $\lambda^+(F_t) = \lambda^-(F_t) = t$ [6].

References

1. Cyr, V., Franks, J., Kra, B.: The spacetime of a shift endomorphism. Trans. Am. Math. Soc. **371**(1), 461–488 (2019)
2. D'Amico, M., Manzini, G., Margara, L.: On computing the entropy of cellular automata. Theoret. Comput. Sci. **290**, 1629–1646 (2003)
3. Guillon, P., Salo, V.: Distortion in one-head machines and cellular automata. In: Dennunzio, A., Formenti, E., Manzoni, L., Porreca, A.E. (eds.) AUTOMATA 2017. LNCS, vol. 10248, pp. 120–138. Springer, Cham (2017). https://doi.org/10.1007/978-3-319-58631-1_10
4. Guillon, P., Zinoviadis, Ch.: Densities and entropies in cellular automata. In: Cooper, S.B., Dawar, A., Löwe, B. (eds.) CiE 2012. LNCS, vol. 7318, pp. 253–263. Springer, Heidelberg (2012). https://doi.org/10.1007/978-3-642-30870-3_26
5. Hedlund, G.: Endomorphisms and automorphisms of the shift dynamical system. Math. Syst. Theory **3**, 320–375 (1969)
6. Hochman, M.: Non-expansive directions for $\mathbb{Z}^2$ actions. Ergodic Theory Dyn. Syst. **31**(1), 91–112 (2011)
7. Hurd, L.P., Kari, J., Culik, K.: The topological entropy of cellular automata is uncomputable. Ergodic Theory Dyn. Syst. **12**, 255–265 (1992)
8. Kari, J.: The nilpotency problem of one-dimensional cellular automata. SIAM J. Comput. **21**, 571–586 (1992)
9. Lind, D., Marcus, B.: An Introduction to Symbolic Dynamics and Coding. Cambridge University Press, Cambridge (1995)
10. Lukkarila, V.: On undecidable dynamical properties of reversible one-dimensional cellular automata (Ph.D. thesis). Turku Centre for Computer Science (2010)
11. Shereshevsky, M.A.: Lyapunov exponents for one-dimensional cellular automata. J. Nonlinear Sci. **2**(1), 1–8 (1992)
12. Tisseur, P.: Cellular automata and Lyapunov exponents. Nonlinearity **13**(5), 1547–1560 (2000)

Geometric Tiles and Powers and Limitations of Geometric Hindrance in Self-assembly

Daniel Hader and Matthew J. Patitz(✉)

Department of Computer Science and Computer Engineering,
University of Arkansas, Fayetteville, AR, USA
dhader@email.uark.edu, patitz@uark.edu

Abstract. Tile-based self-assembly systems are capable of universal computation and algorithmically-directed growth. Systems capable of such behavior typically make use of "glue cooperation" in which the glues on at least 2 sides of a tile must match and bind to those exposed on the perimeter of an assembly for that tile to attach. However, several models have been developed which utilize "weak cooperation", where only a single glue needs to bind but other preventative forces (such as geometric, or steric, hindrance) provide additional selection for which tiles may attach, and where this allows for algorithmic behavior. In this paper we first work in a model where tiles are allowed to have geometric bumps and dents on their edges. We show how such tiles can simulate systems of square tiles with complex glue functions (using asymptotically optimal sizes of bumps and dents). We also show that with only weak cooperation via geometric hindrance, no system in any model can simulate even a class of tightly constrained, deterministic cooperative systems, further defining the boundary of what is possible using this tool.

1 Introduction

Systems of tile-based self-assembly in models such as the abstract Tile Assembly Model (aTAM) [20] have been shown to be very powerful in the sense that they are computationally universal [20] and are also able to algorithmically build complex structures very efficiently [18,19]. The key to their computational and algorithmic power arises from the ability of tiles to convey information via the glues that they use to bind to growing assemblies and the preferential binding of some types of tiles over others based upon the requirement that they simultaneously match the glues of multiple tiles already in an assembly. This is called

---

The original version of this chapter was revised: The original "Section 4" was removed. The correction to this chapter is available at https://doi.org/10.1007/978-3-030-19311-9_22

This author's research was supported in part by National Science Foundation Grants CCF-1422152 and CAREER-1553166.

I. McQuillan and S. Seki (Eds.): UCNC 2019, LNCS 11493, pp. 191–204, 2019.
https://doi.org/10.1007/978-3-030-19311-9_16

(glue) cooperation, and in physical implementations it can require a difficult balance of conditions to enforce. It is conjectured that systems which do not utilize cooperation, that is, those in which tiles can bind to a growing assembly by matching only a single glue, do not have the power to perform computations or algorithmically guided growth [4,15,16]. However, several past results have shown that a middle ground, which we call weak cooperation, can be used to design systems which are capable of at least some of the power of cooperative systems. It has been shown that using geometric hindrance [5–7,10] or repulsive glue forces [9,13,17], systems with a binding threshold of 1 (a.k.a. temperature-1 systems) are capable of universal computation. This is because they are able to simulate temperature-2 *zig-zag* aTAM systems, which are in many ways the most restrictive and deterministic of aTAM systems, but which are still capable of simulating arbitrary Turing machines.

In this paper, we further explore some of the powers and limitations of self-assembly in weakly-cooperative systems. First, we investigate the abilities of so-called geometric tiles (those with bumps and dents on their edges), which were shown in [6] to be able to self-assemble $n \times n$ squares using only temperature-1 and $\Theta(\sqrt{\log n})$ unique tile types (beating the lower bound of $\log n / \log \log n$ required for square aTAM tiles), and also at temperature-1 to be able to simulate temperature-2 zig-zag systems in the aTAM, and thus arbitrary Turing machines. Here we prove their ability to simulate non-cooperative, temperature-1, aTAM systems that have complex glue functions which allow glue types to bind to arbitrary sets of other glue types. We provide a construction and then show that it uses the minimum possible number of unique glue types (which is 2), and that it is asymptotically optimal with respect to the size of the geometries used.

Our final contribution is to expose a fundamental limitation of weakly cooperative self-assembling systems which rely on geometric hindrance. As previously mentioned, they are able to simulate the behaviors of temperature-2 aTAM systems which are called zig-zag systems. These systems have the properties that at all points during assembly, there is exactly one frontier location where the next tile can attach (or zero once assembly completes), and for every tile type, every tile of that type which attaches does so by using the exact same input sides (i.e. the initial sides with which it binds), and also has the same subset of sides used as output sides (i.e. sides to which later tiles attach).[1] It has previously been shown in [10] that there exist temperature-2 aTAM systems which cannot be simulated by temperature-1 systems with duples, but that proof fundamentally utilizes a nondeterministic, undirected aTAM system with an infinite number of unique terminal assemblies. Here we try to find a "tighter" gap and so we explore aTAM temperature-2 systems which are directed and only ever have a single frontier location and whose tile types always have fixed input sides, but we make the slight change of allowing for tiles of the same type to sometimes use

[1] Note that *rectilinear* systems can also be simulated, and they have similar properties except that they may have multiple frontier locations available.

different output sides. We prove that with this minimal addition of uncertainty, no system which utilizes weak cooperation with geometric hindrance can simulate such a system, at any scale factor. Thus, while geometric hindrance is an effective tool for allowing the simulation of general computation, the dynamics which such weakly-cooperative systems can capture is severely restricted.

Due to space limitations, this version of the paper contains only high-level definitions of some of the models and terminology used, and only sketches of several of the poofs. Full versions can be found in the technical appendix which is included in the online version [8].

2 Preliminaries

Here we provide brief descriptions of models used in this paper. References are provided for more thorough definitions.

2.1 Informal Description of the Abstract Tile Assembly Model

This section gives a brief informal sketch of the abstract Tile Assembly Model (aTAM) [20] and uses notation from [18] and [12]. For more formal definitions and additional notation, see [18] and [12].

A *tile type* is a unit square with four sides, each consisting of a *glue label* which is often represented as a finite string. There is a finite set T of tile types, but an infinite number of copies of each tile type, with each copy being referred to as a *tile*. A *glue function* is a symmetric mapping from pairs of glue labels to a non-negative integer value which represents the strength of binding between those glues. An *assembly* is a positioning of tiles on the integer lattice $\mathbb{Z}^2$, described formally as a partial function $\alpha : \mathbb{Z}^2 \dashrightarrow T$. Let $\mathcal{A}^T$ denote the set of all assemblies of tiles from T, and let $\mathcal{A}^T_{<\infty}$ denote the set of finite assemblies of tiles from T. We write $\alpha \sqsubseteq \beta$ to denote that α is a *subassembly* of β, which means that dom $\alpha \subseteq$ dom β and $\alpha(p) = \beta(p)$ for all points $p \in$ dom α. Two adjacent tiles in an assembly *interact*, or are *attached*, if the glue labels on their abutting sides have positive strength between them according to the glue function. Each assembly induces a *binding graph*, a grid graph whose vertices are tiles, with an edge between two tiles if they interact. The assembly is τ*-stable* if every cut of its binding graph has strength at least τ, where the strength of a cut is the sum of all of the individual glue strengths in the cut.

A *tile assembly system* (TAS) is a 4-tuple $\mathcal{T} = (T, \sigma, G, \tau)$, where T is a finite set of tile types, $\sigma : \mathbb{Z}^2 \dashrightarrow T$ is a finite, τ-stable *seed assembly*, G is a *glue function*, and τ is the *temperature*. In the case that the glue function G is *diagonal*, meaning that each glue only has a non-zero strength with itself, G is often omitted from the definition and a TAS is defined as the triple $\mathcal{T} = (T, \sigma, \tau)$ where the strengths between identical glues are given as part of T. Glue functions which are not diagonal are often said to define *flexible* glues. Given an assembly α, the *frontier*, $\partial^\tau \alpha$, is the set of locations to which tiles can τ-stably attach. An assembly α is *producible* if either $\alpha = \sigma$ or if β is a producible assembly and α

can be obtained from β by the stable binding of a single tile to a location in $\partial^\tau\beta$. In this case we write $\beta \rightarrow_1^\mathcal{T} \alpha$ (to mean α is producible from β by the attachment of one tile), and we write $\beta \rightarrow^\mathcal{T} \alpha$ if $\beta \rightarrow_1^{\mathcal{T}*} \alpha$ (to mean α is producible from β by the attachment of zero or more tiles). We let $\mathcal{A}[\mathcal{T}]$ denote the set of producible assemblies of $\mathcal{T}$. An assembly α is *terminal* if no tile can be τ-stably attached to it, i.e. $|\partial^\tau\alpha| = 0$. We let $\mathcal{A}_\Box[\mathcal{T}] \subseteq \mathcal{A}[\mathcal{T}]$ denote the set of producible, terminal assemblies of $\mathcal{T}$. A TAS $\mathcal{T}$ is *directed* if $|\mathcal{A}_\Box[\mathcal{T}]| = 1$. Hence, although a directed system may be nondeterministic in terms of the order of tile placements, it is deterministic in the sense that exactly one terminal assembly. We say that a system $\mathcal{T}$ is a *single-assembly-sequence* system (SASS), if for every producible assembly $\alpha \in \mathcal{A}[\mathcal{T}]$, $|\partial^\tau\alpha| \leq 1$, i.e. there is never more than one location to which a new tile can bind. If a system $\mathcal{T}$ is a SASS and also directed, then it is fully deterministic. We say that a system $\mathcal{T}$ is a *zig-zag system* if it is a SASS where, for every producible assembly $\alpha \in \mathcal{A}[\mathcal{T}]$ and $\beta \sqsubseteq \alpha$, the y coordinate of $\partial^\tau\alpha$ is never smaller than the y coordinate of $\partial^\tau\beta$.

2.2 Informal Description of the Geometric Tile Assembly Model

The geometric tile assembly model (GTAM) is similar to the aTAM with the addition of geometric bumps along the sides of tiles and the restriction that the glue function be diagonal. This section will provide an informal introduction to the model, but a more complete introduction can be found in [6]. The introduction presented here differs slightly from that in [6] in that we focus only on 1-dimensional geometry and try to match the notation as closely as possible to the aTAM definition in the previous section (Fig. 1).

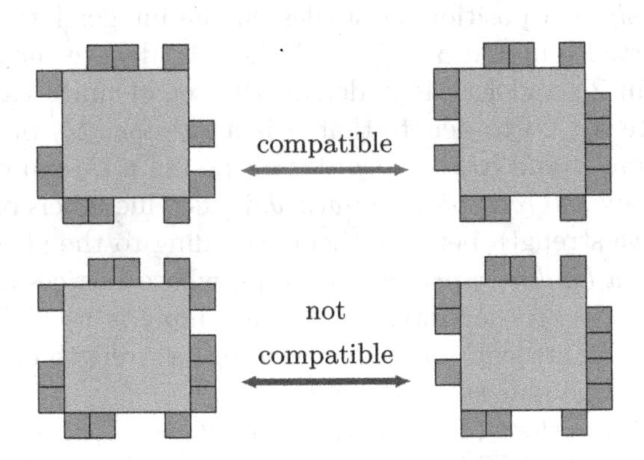

Fig. 1. Examples of compatible and incompatible geometric tiles

A *geometry* of size n is a mapping from $\{1, \ldots, n\}$ to $\{0, 1\}$. This represents n possible locations for bumps with a 1 representing a bump at that location and a 0 representing no bump. A *geometric tile type* is a unit square with a glue label and a geometry on each side. Similarly to tiles in the aTAM, two geometric tiles *interact* if the glue labels on their abutting sides have a positive strength; however, if the two abutting geometries have bumps in corresponding locations, they are called *incompatible* and cannot bind regardless of glue strength. It's important to note that opposite sides of a tile can posses the same geometry. Geometry is rotated along the sides of the tile, so in this case, the geometry on the opposite side would be reversed. Thus any geometry which has a pair of bumps symmetric about its middle would be incompatible with itself. A *geometric tile assembly system* (GTAS) is a triple $\mathcal{T} = (T, \sigma, \tau)$, where T is a finite set of geometric tile types, $\sigma : \mathbb{Z}^2 \dashrightarrow T$ is a finite, τ-stable *seed assembly*, and τ is the *temperature*. Since the glue function for a GTAM system is diagonal, it is omitted from the definition for convenience. Also note that the size of the geometries in any GTAS is fixed.

2.3 Additional Models

A wide variety of models which generalize and extend certain aspects of the aTAM have been developed. Of those, due to space constraints we will briefly mention a few and cite references which can be used to find full definitions.

The 3D aTAM [1] is the natural extension of the aTAM from 2-dimensional square tiles to 3-dimensional cubes. In [10], the Dupled abstract Tile Assembly Model (DaTAM) is defined as an extension to the aTAM which allows for not only the standard square, 1×1, tiles, but also the inclusion of "duples" (or dominoes) which are tiles of dimension 2×1 or 1×2. Allowing for more complex tile shapes, the Polyomino Tile Assembly Model (polyTAM) [5] allows for tiles composed of arbitrary numbers of unit squares which are connected along aligned faces. The polygonal TAM [2,7] allows for tiles to have arbitrary polygonal shapes, and this is the only model mentioned which doesn't have an underlying regular lattice. Finally, another extension to the aTAM which we discuss is one which includes *negative glues*, or glues which exhibit repulsive forces [17]. In systems with negative glues, two tiles may have adjacent faces with matching glues whose interaction strength is a negative integer. This is subtracted from the overall sum of binding strengths of adjacent glues to determine if the tile can attach.[2]

2.4 Cooperation

Self-assembling systems in tile assembly models contain a parameter known as the *minimum binding threshold*, often called the *temperature*. This parameter specifies the minimum biding strength, summed over all binding glues, that a tile must have with an assembly in order to attach to it. The binding strengths of glues are typically discrete, positive integer values. If the temperature parameter is set to 2 or greater, we say that the system is *strongly-cooperative*, uses *strong cooperation*, or uses *glue cooperation*, because it is possible for the attachment

[2] Note that with negative glues, more complex dynamics, which include the breaking apart of assemblies, are possible [13].

of a new tile to an assembly to require that it bind its glues, with positive affinity, to more than one tile already existing in the assembly. For example, in a temperature-2 system (i.e. one where the temperature parameter = 2), a tile may attach by binding with two glues each of strength 1, and thus two tiles already in the assembly *cooperate* to allow for the new attachment. In contrast, we say that a system is *non-cooperative* if its temperature parameter is set to 1 and in all situations where there is an empty location with an exposed incident glue, any tile with a matching glue can attach there. Finally, we say a system is *weakly-cooperative*, or uses *weak-cooperation*, if its temperature parameter is set to 1 but it is able to make use of any form of *binding hindrance*. In such a system, the binding strength of any single glue is strong enough to allow a tile to attach to an assembly; however, it is possible that some tiles are prevented from binding by another factor. Two such forms of binding hindrance we'll consider are *geometric hindrance* and *glue repulsion*. In this context, geometric (a.k.a. steric) hindrance occurs when a tile cannot bind in a location because at least some of the space that it would occupy is already occupied by some portion of another tile. This is relevant when tiles have more complex shapes than unit squares. See Fig. 2 for an example, as well as [5–7, 10]. Glue repulsion occurs when glues are able to experience negative strength interactions, i.e. when they can repel each other and their interaction subtracts from the total binding strength of a tile to an assembly. Examples can be found in [3, 14, 17]. An example of the combination of geometric hindrance and glue repulsion can be found in [9].

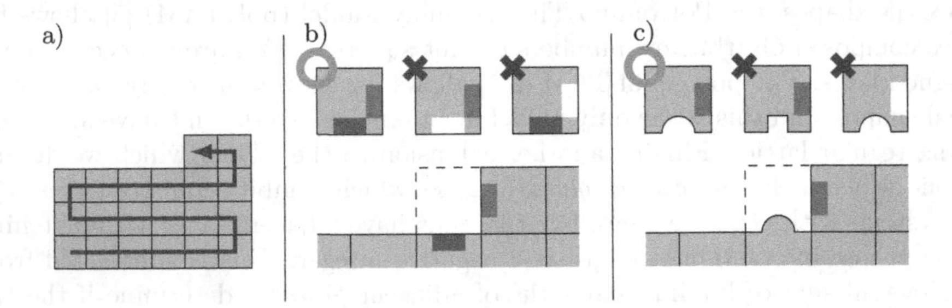

Fig. 2. (a) An illustration of how a zig-zag system propagates upward, snaking east and west. (b) Each of the new tile additions can be thought of as acting with 2 inputs. In a temperature-2 system, as in most zig-zag systems, the bottom and side inputs come from cooperating glues and only the tile that matches both can grow. (c) In temperature-1 GTAM systems there is no cooperation so the side input uses a glue while the bottom geometry is used to prevent the wrong tiles from growing.

2.5 Simulation

Here we give a very brief intuitive definition of what it means for one tile assembly system to simulate another. See the technical appendix of [8] for more technically detailed definitions related to simulation, especially as it relates to scale factors greater than 1.

Intuitively, simulation of a system $\mathcal{T}$ by a system $\mathcal{S}$ requires that there is some scale factor $m \in \mathbb{Z}^+$ such that $m \times m$ squares of tiles in $\mathcal{S}$ represent individual tiles in $\mathcal{T}$, and there is a "representation function" capable of inspecting assemblies in $\mathcal{S}$ and mapping them to assemblies in $\mathcal{T}$. A representation function R takes as input an assembly in $\mathcal{S}$ and returns an assembly in $\mathcal{T}$ to which it maps. In order for $\mathcal{S}$ to correctly simulate $\mathcal{T}$, it must be the case that for any producible assembly $\alpha \in \mathcal{A}[\mathcal{T}]$ that there is a corresponding assembly $\beta \in \mathcal{A}[\mathcal{S}]$ such that $R(\beta) = \alpha$. (Note that there may be more than one such β.) Furthermore, for any $\alpha' \in \mathcal{A}[\mathcal{T}]$ which can result from a tile addition to α, there exists $\beta' \in \mathcal{A}[\mathcal{S}]$ which can result from the addition of one or more tiles to β, and conversely, β can only grow into assemblies which can be mapped into valid assemblies of $\mathcal{T}$ into which α can grow.

3 Simulation of Temperature-1 aTAM Systems

Theorem 1. *For any temperature-1 aTAM system, $\mathcal{T} = (T, \sigma_{\mathcal{T}}, G_{\mathcal{T}}, 1)$, with arbitrary symmetric glue function (a.k.a. flexible glues), there exists a temperature-1 GTAM system $\mathcal{U} = (U, \sigma_{\mathcal{U}}, 1)$ that simulates $\mathcal{T}$ using only 2 distinct glues, tile geometries of size $4n$, where n is the number of glues in $\mathcal{T}$, and scale factor 1.*

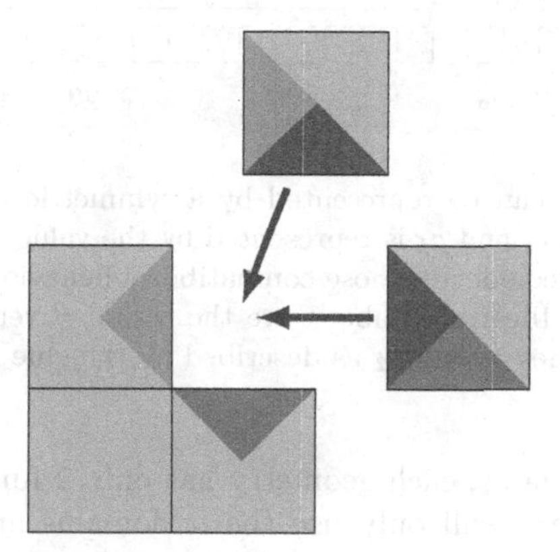

Fig. 3. An example of a situation in an aTAM system which mandates the use of two glues in the simulating GTAM system. Here the blue glue is incompatible with both the red and green glues. If only a single glue was used in the GTAM system, both of the tiles would necessarily be incompatible and could not fit. (Color figure online)

To prove this theorem (as is done fully in the appendix of [8]), we construct a GTAM system which simulates a given aTAM system. The construction seeks to implement the binding behaviour of glues using the compatibility behaviour of geometries. To represent any given glue we construct two corresponding geometries which we call the α and β versions. Both of these geometries are divided into 4 domains of size n, meaning that there are $4n$ potential bump locations on each. The domains, from left to right, are named α_1, β_1, β_2, and α_2 and examples of what these domains look like can be seen in Fig. 4. Also keep in mind that geometries are rotated in order to be placed on the various faces of a geometric tile. This means that the westernmost domain of a north geometry on a geometric tile would be α_1, whereas the westernmost domain of a south geometry would be α_2. When indexing bump locations, we count from left to right for domains α_1 and β_1 and count from right to left for domains α_2 and β_2. This means that if two geometries were on abutting faces of adjacent tiles, the location i, for $1 \leq i \leq n$, of domain α_1 on the first geometry would line up with location i of domain α_2 on the second geometry.

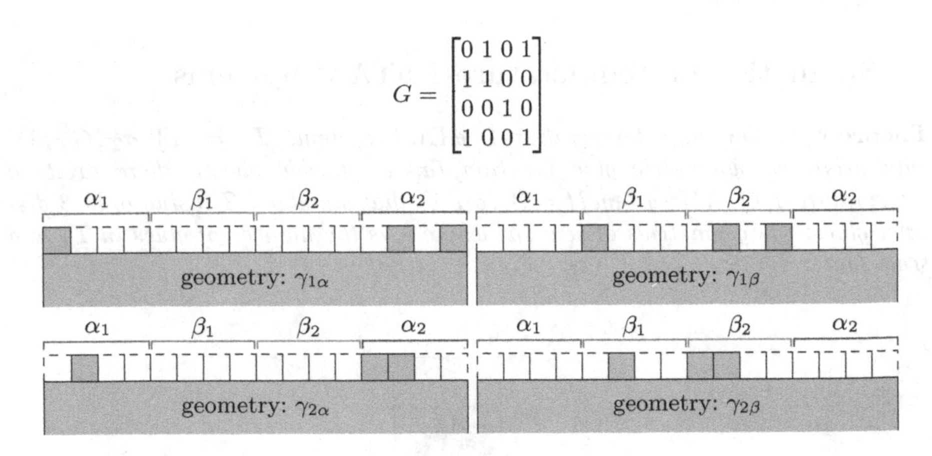

Fig. 4. A glue function can be represented by a symmetric matrix. The strength of the bond between glues g_i and g_j is represented by the value $G_{i,j}$. Given a glue function, we can construct geometries whose compatibility behaviour emulates the binding behaviour of the glues. Illustrated above are the α and β versions of the geometries corresponding to the glues g_1 and g_2 as described by the glue function.

Out of these 4 domains, each geometry has only 2 functional domains. The α version of a geometry will only use the α domains and the β version of a geometry will only use the β domains. The first functional domain in each of these geometries, the α_1 and β_1 domains respectively, encode which glue is being represented by placing a bump in the corresponding location. For example, if a geometry corresponds to glue 3, it will have a bump in location 3 of its first functional domain. The second domains in each of the two geometries, α_2 and β_2 respectively, encode the binding behaviour of the corresponding glue. This is done by placing a bump in all of the locations corresponding to glues to which the represented glue cannot interact. For example, if the represented glue does not interact with glues 1 and 3, then there will be bumps in locations 1 and 3 of the second functional domain. Consider the α versions of the geometries corresponding to two glues, say g_1 and g_2, which cannot interact. The α_2 domain of the geometry corresponding to g_1 will have a bump in position 2 indicating that it cannot interact with glue g_2. This bump will be incompatible with the bump in location 2 of the α_1 domain of the geometry corresponding to g_2. The same is true for the β domains of the β version geometries. So, whenever two glues cannot interact, the corresponding geometries of the same version will be incompatible. Figure 4 demonstrates a glue function and what some of the corresponding geometries look like.

The reason that we need 2 versions of each geometry is to accommodate mismatches. Because mismatched glues can, and often do, legitimately occur in aTAM systems, like in Fig. 3, we need two versions, α and β, of each geometry which are always compatible with geometries of the other version. Because geometries of different versions use exclusive functional domains, they will always be geometrically compatible with one another. The α domains occur on the outside of a geometry and the β domains on the inside, so when abutting, they cannot overlap. Moreover, we need 2 distinct glues in $\mathcal{U}$; since, while we do want geometries to be compatible with opposite versions, we don't want them initiating growth with tiles whose geometries are of opposite version. Thus all α version geometries will have one glue and all β version geometries another. This allows the α and β versions of glues to represent mismatches in the simulated system.

This construction demonstrates that the behaviour of temperature-1 aTAM systems with arbitrary symmetric glue functions can be simulated by temperature-1 GTAM systems with no cost in scale using a fixed number of glues, namely 2. It's important to note that the GTAM systems in our construction only use standard glues which bind only to themselves. This proof implies that two glues in a GTAM system are sufficient for simulating arbitrary aTAM systems at temperature-1; the following theorem shows that, using fewer than two glues, not all aTAM systems can be simulated by GTAM systems. This implies that two glues are necessary for allowing glue mismatches to be properly simulated.

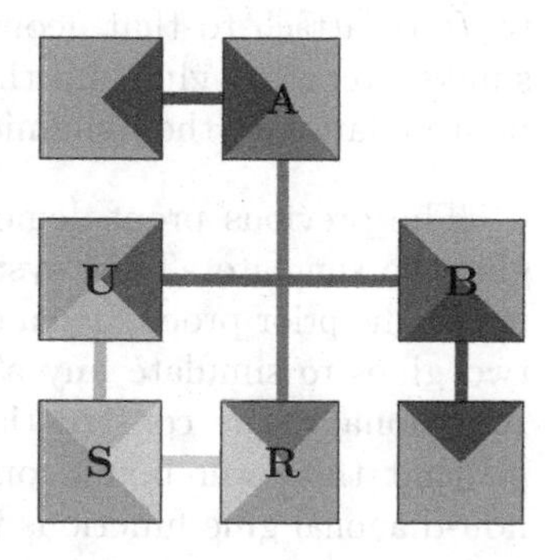

Fig. 5. The tile set used in Theorem 2. The lines between tiles represent possible attachments. Notice that, if tile S is the seed, the final configuration must be a 2 by 2 square.

Theorem 2. *There exists an aTAM system at temperature-1 which cannot be simulated by a temperature-1 GTAM system at scale factor 1 using < 2 glues, regardless of the geometry size of the tiles.*

Proof. Consider the temperature 1 aTAM system, say $\mathcal{T}$, presented in Fig. 5, wherein the tile labelled S is the seed. Notice that $\mathcal{T}$ is not directed; there are multiple final configurations, each of which are 2×2 squares. Also, let it be the case that each glue binds only to itself, so that the blue glue does not bind with any other glue, for example. Now, for contradiction, suppose that there is some temperature-1 GTAM system $\mathcal{U}$ that does simulate $\mathcal{T}$ at scale factor 1 using only a single glue. Since $\mathcal{U}$ simulates $\mathcal{T}$, it must be able to simulate the growth of $\mathcal{T}$ from any possible configuration. Consider then the configuration of $\mathcal{T}$ in which the tiles labelled U and R have grown to form an L shape. In this configuration, there are two tiles which can grow into the corner opposite to tile S: tiles A and B. Furthermore, notice that if either tile attaches, there will be some glue mismatch since the blue glue does not match with either the red or green glues.

Now imagine a corresponding, L shaped configuration in $\mathcal{U}$. Since $\mathcal{U}$ simulates $\mathcal{T}$, there must be some geometric tile corresponding to either tile A or B which can attach in the corner opposite S. If we suppose, without loss of generality, that this was a tile corresponding to A, then it must be the case that this geometric tile has a geometry on its west face which is compatible with the geometry on the east face of a tile corresponding to U despite the fact that the glues don't match in $\mathcal{T}$. Therefore, in the case where a geometric tile corresponding to tile R hasn't yet attached, there would be the possibility for the geometric tile corresponding to U to attach to that geometric tile corresponding to A since there is only a single strength-1 glue and the geometries are compatible. This, however, would be a violation of the dynamics of $\mathcal{T}$ and thus such a $\mathcal{U}$ could not simulate $\mathcal{T}$. □

The previous proof demonstrated that a GTAM system needs at least two glues to simulate aTAM systems at temperature-1 and scale factor 1. Furthermore, the prior proof gave a construction of a GTAM system which used exactly two glues to simulate any aTAM system at temperature-1 and scale factor 1. Additionally, the construction used geometries of size $4n$, and it is shown in [6] that the lower bound on the size of geometries needed to represent some non-diagonal glue functions is $\Theta(n)$.

4 Glue Cooperation Cannot Be Simulated with Geometric Hindrance

In this section, we first show that there exists a directed temperature-2 aTAM SASS (i.e. a fully deterministic temperature-2 aTAM system) which cannot be simulated by any temperature-1 GTAM system. A brief overview is given here, and the full proof can be found in the Appendix.

Theorem 3. *There exists a directed temperature-2 aTAM SASS $\mathcal{S}$ that cannot be simulated by any GTAM system at temperature 1.*

Figure 6 shows a high-level, schematic drawing of the system $\mathcal{S}$ which cannot be simulated. Essentially, it grows a "`planter`" module (similar to that of [11]) to form an infinite assembly growing to the right, which initiates an infinite series of counters which grow upward to every height ≥ 4. (Figure 7a shows the pattern of growth which allows $\mathcal{S}$ to be a SASS.) Each counter then grows an arm down which crashes into a portion of the assembly below it, but since the arms grow longer and longer, eventually they reach a point where they must "pump", or grow in a periodic manner. However, in order to correctly grow macrotiles which simulate the cooperative growth between the end of each arm and the bottom portion of the assembly, there must be path of tiles which can grow out from each arm. Since the arms must become periodic, those paths could also grow in higher locations, which leads to invalid simulation. Examples can be seen in Figs. 7b and 8.

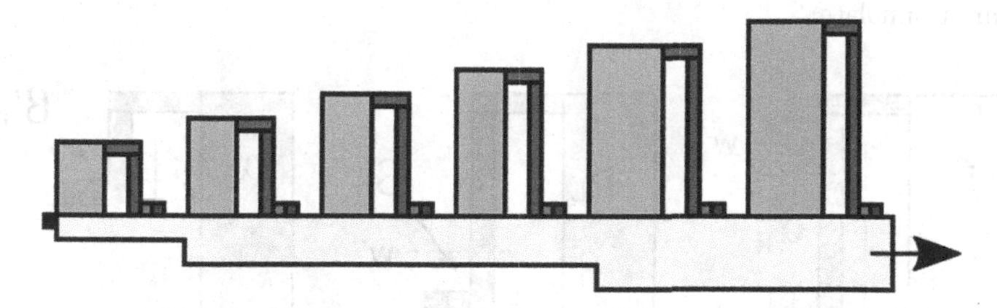

Fig. 6. Overview of a temperature-2 aTAM system which cannot be simulated by any temperature-1 GTAM system at scale factor 1.

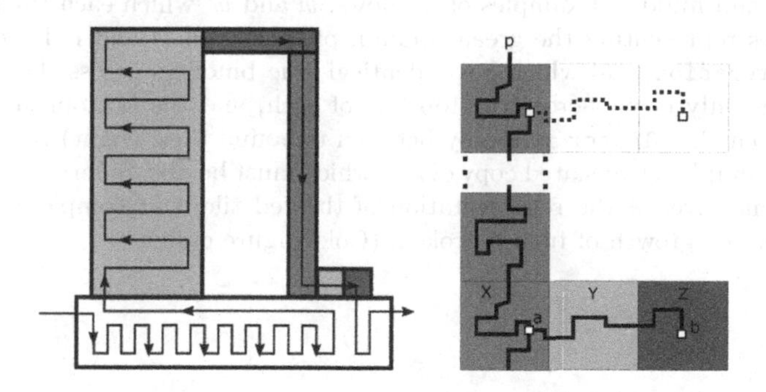

Fig. 7. (a) Depiction of one iteration of the growth of $\mathcal{T}$, with arrows showing the ordering of growth, (b) Zoomed in portion of the construction shown in Fig. 6 which shows (with a solid line) an example of a path of tiles bound by glues which must extend from a tile, a, in the supertile representing a green tile, to a tile, b, in the supertile representing the red tile. The dashed line shows how a previous copy of a could allow growth of the same path in a higher location. (Color figure online)

While Theorem 3 states that temperature-1 GTAM systems cannot even simulate the full class of directed temperature-2 aTAM single-assembly-sequence systems, the following result generalizes that to show that the same is true across all systems relying on weak cooperation across any tile assembly model.

Theorem 4. *There exists a directed temperature-2 aTAM SASS $\mathcal{S}$ that cannot be simulated by any weakly-cooperative tile assembly system that relies on geometric hindrance.*

The proof of Theorem 4 is essentially identical to that of Theorem 3.

Note that Theorem 4 is proven for weakly-cooperative systems using geometric hindrance, but that this does not include the second category of types of binding hindrance, namely systems which use glue repulsion. Although such systems make possible dynamic behavior in which portions of an assembly may break off, which may make the proof more difficult, we conjecture that they also cannot simulate $\mathcal{S}$.

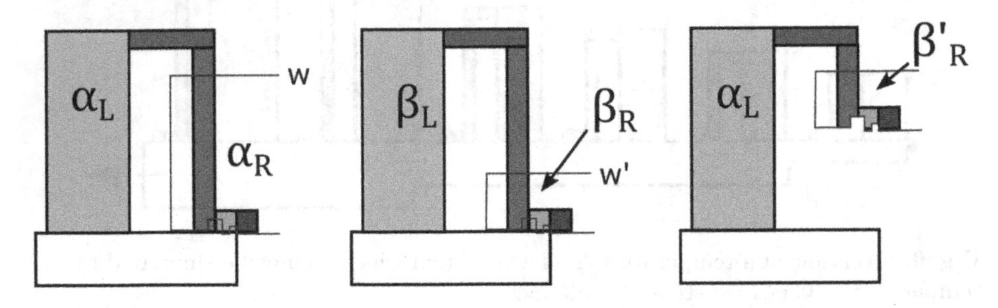

Fig. 8. (left and middle) Examples of windows, w and w' which each cut a portion of the supertiles representing the green column, plus the yellow and red tiles, from the rest of an `iteration`, and which have identical glue bindings across them. Note that glue bindings only occur across the top line of each, and the bottom line separating the inside from the `planter` goes only between unbound tiles, (right) Assembly $\alpha_L \beta'_R$ (where β'_R is simply a translated copy of β_R) which must be able to form by the Window Movie Lemma. Even if the representation of the red tile isn't complete, the allowed boundary for the growth of fuzz is broken. (Color figure online)

References

1. Cook, M., Fu, Y., Schweller, R.T.: Temperature 1 self-assembly: deterministic assembly in 3D and probabilistic assembly in 2D. In: SODA 2011: Proceedings of the 22nd Annual ACM-SIAM Symposium on Discrete Algorithms. SIAM (2011)
2. Demaine, E.D., et al.: One tile to rule them all: simulating any tile assembly system with a single universal tile. In: Esparza, J., Fraigniaud, P., Husfeldt, T., Koutsoupias, E. (eds.) ICALP 2014. LNCS, vol. 8572, pp. 368–379. Springer, Heidelberg (2014). https://doi.org/10.1007/978-3-662-43948-7_31

3. Doty, D., Kari, L., Masson, B.: Negative interactions in irreversible self-assembly. Algorithmica **66**(1), 153–172 (2013)
4. Doty, D., Patitz, M.J., Summers, S.M.: Limitations of self-assembly at temperature 1. Theor. Comput. Sci. **412**, 145–158 (2011)
5. Fekete, S.P., Hendricks, J., Patitz, M.J., Rogers, T.A., Schweller, R.T.: Universal computation with arbitrary polyomino tiles in non-cooperative self-assembly. In: Proceedings of the Twenty-Sixth Annual ACM-SIAM Symposium on Discrete Algorithms (SODA 2015), San Diego, CA, USA, 4–6 January 2015, pp. 148–167 (2015). https://doi.org/10.1137/1.9781611973730.12
6. Fu, B., Patitz, M.J., Schweller, R.T., Sheline, R.: Self-assembly with geometric tiles. In: Czumaj, A., Mehlhorn, K., Pitts, A., Wattenhofer, R. (eds.) ICALP 2012. LNCS, vol. 7391, pp. 714–725. Springer, Heidelberg (2012). https://doi.org/10.1007/978-3-642-31594-7_60
7. Gilber, O., Hendricks, J., Patitz, M.J., Rogers, T.A.: Computing in continuous space with self-assembling polygonal tiles. In: Proceedings of the Twenty-Seventh Annual ACM-SIAM Symposium on Discrete Algorithms (SODA 2016), Arlington, VA, USA, 10–12 January 2016, pp. 937–956 (2016)
8. Hader, D., Patitz, M.J.: Geometric tiles and powers and limitations of geometric hindrance in self-assembly. Technical report 1903.05774, Computing Research Repository (2019). http://arxiv.org/abs/1903.05774
9. Hendricks, J., Patitz, M.J., Rogers, T.A.: Doubles and negatives are positive (in self-assembly). Nat. Comput. **15**(1), 69–85 (2016). https://doi.org/10.1007/s11047-015-9513-6
10. Hendricks, J., Patitz, M.J., Rogers, T.A., Summers, S.M.: The power of duples (in self-assembly): it's not so hip to be square. Theor. Comput. Sci. (2015). https://doi.org/10.1016/j.tcs.2015.12.008. http://www.sciencedirect.com/science/article/pii/S030439751501169X
11. Lathrop, J.I., Lutz, J.H., Patitz, M.J., Summers, S.M.: Computability and complexity in self-assembly. Theory Comput. Syst. **48**(3), 617–647 (2011)
12. Lathrop, J.I., Lutz, J.H., Summers, S.M.: Strict self-assembly of discrete Sierpinski triangles. Theor. Comput. Sci. **410**, 384–405 (2009)
13. Luchsinger, A., Schweller, R., Wylie, T.: Self-assembly of shapes at constantscale using repulsive forces. Nat. Comput. (2018). https://doi.org/10.1007/s11047-018-9707-9
14. Luchsinger, A., Schweller, R., Wylie, T.: Self-assembly of shapes at constant scale using repulsive forces. In: Patitz, M.J., Stannett, M. (eds.) UCNC 2017. LNCS, vol. 10240, pp. 82–97. Springer, Cham (2017). https://doi.org/10.1007/978-3-319-58187-3_7
15. Meunier, P.E., Patitz, M.J., Summers, S.M., Theyssier, G., Winslow, A., Woods, D.: Intrinsic universality in tile self-assembly requires cooperation. In: Proceedings of the ACM-SIAM Symposium on Discrete Algorithms (SODA 2014), Portland, OR, USA, 5–7 January 2014, pp. 752–771 (2014)
16. Meunier, P., Woods, D.: The non-cooperative tile assembly model is not intrinsically universal or capable of bounded Turing machine simulation. In: Proceedings of the 49th Annual ACM SIGACT Symposium on Theory of Computing, STOC 2017, Montreal, QC, Canada, 19–23 June 2017, pp. 328–341 (2017). https://doi.org/10.1145/3055399.3055446
17. Patitz, M.J., Schweller, R.T., Summers, S.M.: Exact shapes and turing universality at temperature 1 with a single negative glue. In: Cardelli, L., Shih, W. (eds.) DNA 2011. LNCS, vol. 6937, pp. 175–189. Springer, Heidelberg (2011). https://doi.org/10.1007/978-3-642-23638-9_15

18. Rothemund, P.W.K., Winfree, E.: The program-size complexity of self-assembled squares (extended abstract). In: STOC 2000: Proceedings of the Thirty-second Annual ACM Symposium on Theory of Computing, pp. 459–468. ACM, Portland (2000). https://doi.org/10.1145/335305.335358
19. Soloveichik, D., Winfree, E.: Complexity of self-assembled shapes. SIAM J. Comput. **36**(6), 1544–1569 (2007)
20. Winfree, E.: Algorithmic self-assembly of DNA. Ph.D. thesis, California Institute of Technology, June 1998

DNA Computing Units Based on Fractional Coding

Sayed Ahmad Salehi(✉) and Peyton Moore

University of Kentucky, Lexington, KY 40506, USA
SayedSalehi@uky.edu, psmo224@g.uky.edu

Abstract. *Fractional encoding* has been recently proposed as a promising convention to represent information in molecular computing systems. This paper presents new 2-input molecular computing units based on unipolar fractional representation. The units calculate simple computational equations that can be used for the computation of more complex functions. The design of these molecular computing units is inspired by fan-in 2 logic gates in the field of stochastic computing. Each computing unit consists of four chemical reactions with two reactants and one product. We design the DNA reactions implementing the chemical reactions of each unit based on the *toehold-mediated DNA strand-displacement mechanism.* Every unit is designed by four input strands and eight fuel gate strands of DNA. Since DNA molecules related to the input and output of the units have the same form of *domain-toehold-domain-toehold*, output molecules of each unit can be used as input for other units and this provides the cascading of the units for designing complex circuits. The whole DNA pathway for each unit is composed of twenty DNA reactions. The simulation results by Visual DSD show that the DNA implementations follow the theoretically expected computations of each unit with the maximum of 9.33% error.

Keywords: DNA computing · Fractional coding · DNA strand-displacement

1 Introduction

The demand for computing has a long history and there have been many efforts to embody the principles of computation in a variety of physical substrates from mechanical gears to electronic transistors. Although semiconductor technology has provided a significant computational power for data crunching, researchers are striving for alternative technologies for appropriate domains of applications. Molecular computing, as an unconventional emerging technology with biological compatibility, is promising for transforming areas such as smart drugs and drug delivery. In molecular computing, variables (including inputs and outputs) are represented by molecular concentrations and molecular reactions are designed to control these concentrations.

The conventional way to represent variables in molecular computing is to assign one molecular type to each variable. This is called direct representation, where greater concentration of a molecular type means greater value for its corresponding variable. Since molecular concentrations cannot be negative, direct representation is not able to represent negative values. In another representation, so called dual-rail representation,

I. McQuillan and S. Seki (Eds.): UCNC 2019, LNCS 11493, pp. 205–218, 2019.
https://doi.org/10.1007/978-3-030-19311-9_17

two molecular types are used for each variable [1, 2]. The variable's value is the signed difference between the concentrations of the two molecular types. If the concentration of the first molecule is greater than the second one, the variable is positive; otherwise, it is negative.

Recently, a new convention called fractional encoding has been introduced for representing variables in molecular computing systems [3]. In this convention, a pair of molecular types is used to represent each variable. The value of the variable is encoded by the ratio of the concentrations for one molecular type over the total concentrations of both molecular types. For example, if the pair of (X_0, X_1) is assigned to a variable x, the value of x would be:

$$x = \frac{[X_1]}{[X_0] + [X_1]}, \tag{1}$$

where $[X_0]$ and $[X_1]$ represent concentrations of molecules X_0 and X_1, respectively. Note that the value of x is limited to the unit interval, [0, 1]. This representation is referred to as a unipolar fractional encoding. With no need of extra molecular type, the unipolar fractional encoding can be extended to bipolar fractional encoding to cover negative numbers. In bipolar fractional encoding, the value of x is defined as:

$$x = \frac{[X_1] - [X_0]}{[X_0] + [X_1]}, \tag{2}$$

where the value of x is limited to the $[-1, 1]$ interval.

The computational power of chemical reaction networks (CRNs) based on direct and dual-rail representations have been extensively investigated in the literature of the molecular computing [1–10]. However, the field lacks the extensive study of CRNs based on fractional coding while they hold the promise for expanding the computational power of CRNs beyond the current boundaries.

Fractional coding is promising because it relates molecular computing circuits to stochastic computing circuits [3, 11]. In the stochastic computing, logic gates are designed to combine two (or more) random bitstreams of 0's and 1's and produce a new random bitstream with a desired probability of seeing 1's in it [12–15]. Similarly, molecular reactions can be designed to combine two (or more) pairs of molecular types and produce a new pair of molecular types representing a desired value based on the fractional coding. This similarity has its origins in the stochasticity of each combination among individual reactants of a molecular reaction.

Prior work [11] has proposed so called Mult and NMult molecular units that compute, respectively, $c = a \times b$ and $c = 1 - a \times b$ where a and b are the inputs and c is the output, all in the fractional representation. An AND gate in stochastic computing computes multiplication $c = a \times b$, where a, b, and c are the probabilities of having 1's in input and output random bitstreams [12–15]. In fact, the design of molecular Mult unit has been inspired by AND logic gate in stochastic computing [11]. In the same way, it is easy to show that NMult unit is corresponding to NAND logic gate in stochastic computing.

Using the connection between molecular computing and stochastic computing, we investigate new molecular computing units corresponding to other stochastic gates such as OR and NOR gates in this paper. Similar to Mult and NMult, the proposed units are composed of four simple bimolecular reactions; however, they perform more complex computations compared to the multiplication performed by Mult and NMult units. First, we design the CRNs for each of these units and then, using the DNA strand-displacement mechanism [16, 17], we design DNA reactions to implement the CRNs. We simulate and verify our DNA designs by Visual DSD [18, 19].

This paper is organized as follows: Sect. 2 introduces the M-OR, M-NOR, M-XOR, and M-XNOR molecular units and their implementing CRNs; Sect. 3 explains how to map the CRNs to DNA reactions and the results for the simulation of DNA realization of the proposed units; Sect. 4 concludes the paper and its results.

2 Molecular Computing Units

In this section, CRNs for new molecular computing units are described. The design of these units is inspired by stochastic logic gates. The molecular units corresponding to AND, NAND, OR, NOR, XOR, and XNOR logic gates are called M-AND, M-NAND, M-OR, M-NOR, M-XOR, and M-XNOR, respectively.

Figure 1 shows the symbols, CRNs, and computational equations in the unipolar fractional coding for the molecular computing units. Evidently, the form of CRNs for all units is the same: four reactions with two reactants and one product. The combinations of reactants, i.e., A_0, A_1, B_0, and B_1 are the same for all of the units. However, the CRNs for units are distinguished only by the order of their products C_0 and C_1.

M-AND	M-NAND	M-OR	M-NOR	M-XOR	M-XNOR
$A_0 + B_0 \rightarrow C_0$	$A_0 + B_0 \rightarrow C_1$	$A_0 + B_0 \rightarrow C_0$	$A_0 + B_0 \rightarrow C_1$	$A_0 + B_0 \rightarrow C_0$	$A_0 + B_0 \rightarrow C_1$
$A_0 + B_1 \rightarrow C_0$	$A_0 + B_1 \rightarrow C_1$	$A_0 + B_1 \rightarrow C_1$	$A_0 + B_1 \rightarrow C_0$	$A_0 + B_1 \rightarrow C_1$	$A_0 + B_1 \rightarrow C_0$
$A_1 + B_0 \rightarrow C_0$	$A_1 + B_0 \rightarrow C_1$	$A_1 + B_0 \rightarrow C_1$	$A_1 + B_0 \rightarrow C_0$	$A_1 + B_0 \rightarrow C_1$	$A_1 + B_0 \rightarrow C_0$
$A_1 + B_1 \rightarrow C_1$	$A_1 + B_1 \rightarrow C_0$	$A_1 + B_1 \rightarrow C_1$	$A_1 + B_1 \rightarrow C_0$	$A_1 + B_1 \rightarrow C_0$	$A_1 + B_1 \rightarrow C_1$
$c = ab$	$c = 1 - ab$	$c = a + b - ab$	$c = 1 - a - b + ab$	$c = a + b - 2ab$	$c = 1 + 2ab - a - b$

Fig. 1. Each column shows molecular computing units, their symbols, molecular reactions, and the computed outputs based on unipolar fractional coding. The last row shows the computed outputs in unipolar fractional coding.

The M-AND and M-NAND units are the same as Mult and NMult units proposed in [10]. Proof for the computational equation of the M-OR computing unit is described below. Details for other computing units are similar to M-OR unit; thus, they are not discussed in this paper.

2.1 M-OR Molecular Computing Unit

M-OR is composed of the four reactions shown in the third column of Fig. 1 and it computes $c = a + b - ab$ as the function of two inputs a and b, all in the unipolar fractional representation. So, if $a = \frac{[A_1]}{[A_0]+[A_1]}$ and $b = \frac{[B_1]}{[B_0]+[B_1]}$ then $c = \frac{[C_1]}{[C_0]+[C_1]} = a + b - ab$. Based on both the stochastic model and mass-action kinetics model of CRNs, it can be shown that M-OR calculates $c = a + b - ab$ as follows.

Based on stochastic kinetics model, the concentrations of molecules are considered as discrete quantities. A simple interpretation for the set of reactions of M-OR is that the probability of generating a molecule of C_0 is the probability of reacting a molecule of A_0 and a molecule of B_0. Thus, it yields

$$\begin{aligned} P(C_0) = P(A_0) \times P(B_0) &\Rightarrow \frac{[C_0]}{[C_0]+[C_1]} = \frac{[A_0]}{[A_0]+[A_1]} \times \frac{[B_0]}{[B_0]+[B_1]} \\ &\Rightarrow (1-c) = (1-a) \times (1-b) \Rightarrow c = a + b - ab. \end{aligned} \tag{3}$$

Mass-action kinetics is a modeling scheme for chemical reaction networks which states that the rate of a chemical reaction is proportional to the product of the concentrations of the reacting chemical species. This is how the ODE of chemical reactions are made in Visual DSD, a software tool used to simulate our model later.

Based on mass-action kinetics model, the ODEs for the reactions of M-OR are given by

$$\begin{aligned} \tfrac{d[A_0]}{dt} &= -[A_0][B_0] - [A_0][B_1] & \tfrac{d[A_1]}{dt} &= -[A_1][B_0] - [A_1][B_1] \\ \tfrac{d[B_0]}{dt} &= -[A_0][B_0] - [A_1][B_0] & \tfrac{d[B_1]}{dt} &= -[A_0][B_1] - [A_1][B_1] \\ \tfrac{d[C_0]}{dt} &= [A_0][B_0] & \tfrac{d[C_1]}{dt} &= [A_1][B_1] + [A_0][B_1] + [A_1][B_1]. \end{aligned}$$

If the initial concentrations are considered as $[C_0(t=0)] = [C_1(t=0)] = 0$, $[A_0(t=0)] = a_0$, $[A_1(t=0)] = a_1$,$[B_0(t=0)] = b_0$, and $[B_1(t=0)] = b_1$, solving these ODEs yields

$$\frac{[C_1(t)]}{[C_0(t)]+[C_1(t)]} = \frac{a_0}{a_0+a_1} \times \frac{b_0}{b_0+b_1} \Rightarrow c = a + b - ab. \tag{4}$$

The computational equations for other molecular computing units can be obtained similarly to the M-OR unit.

3 Simulation Results

The design of DNA reactions and the related simulations are carried out using a form of strand displacement called *toehold-mediated branch migration* [16, 17], a method that can exchange input DNA single strands for desired output DNA single strands through their combination with multi-stranded DNA molecules called fuel gates. This is implemented by having a toehold on the input strand combine with its open compliment toehold on another DNA molecule. After the toeholds combine, the rest of the input strand will combine with the DNA molecule while detaching other species attached on the DNA molecule. As explained later, fuel gates are intermediate DNA complexes designed to control how the input strands combine and produce the desired output strand. To use this method, both the input and output DNA strands must be structured the same, as shown in Fig. 2. D1 and D2 are domain strands and T1 and T2 are toehold strands. Both domains and toeholds are sequences of nucleotides A, C, G, and T. The length of the toehold is shorter than the domain's length. The toehold is initiating a DNA strand displacement in the method we use in this paper.

D1 T1 D2 T2

Fig. 2. The basic structure for input and output DNA strands where D1 and D2 are domain strands and T1 and T2 are toehold strands.

The domain D1 in the defined structure is referred to as the history of the DNA strand. The history of a DNA strand does not matter in identifying an input or output DNA strand, as the input and output strands are identified only by T1, D2, and T2 in Fig. 2. For example, DNA strand A_0 structured as <5 1 2 3> has a history of 5 but its identifier is <1 2 3>. If there is another DNA strand structured as <6 1 2 3>, although it has history of 6, it would be considered the same as DNA strand A_0 since its identifier is <1 2 3> and it is identified as A_0.

As explained in Sect. 2, the CRNs for each molecular computing unit has four reactions. In this implementation, the setup for each unit includes four input strands (A_0, A_1, B_0, B_1), two output strands (C_0, C_1), and eight fuel gates $(Gate_n, Produce_n)$, where $n = 1, 2, 3, 4$ defines which fuel gates are used for each chemical reaction. Each chemical reaction uses two appropriately designed fuel gates out of the eight required based on the combination of input strands. For Example, $Gate_1$ and $Produce_1$ are used for the chemical reaction, $A_0 + B_0 \rightarrow C_0$.

We initialize the concentrations of input strands such that the initial sum of $[A_0]$ and $[A_1]$ and the sum of $[B_0]$ and $[B_1]$ are set equal to 4 nM, where M is mole per liter. All output strands are initialized with zero concentration and all fuel gates are initialized with 100 nM; this is much larger than the initial concentration of input strands to allow enough fuel molecule for the input molecules to bond with. The binding rate of the DNA reactions are $3.0 * 10^{-4} \text{n}M^{-1}S^{-1}$ and $0.1126 M^{-1}S^{-1}$ for the unbinding rate, where S represents time unit as seconds. These are the default rates that Visual DSD uses.

3.1 DNA Pathway for M-OR Molecular Computing Unit

We explain the details of the designed DNA strands for the M-OR unit in this section. For a M-OR computing unit, the designed input strands and output strands are illustrated in Fig. 3. Although the output strands are defined in Fig. 3, it does not show all

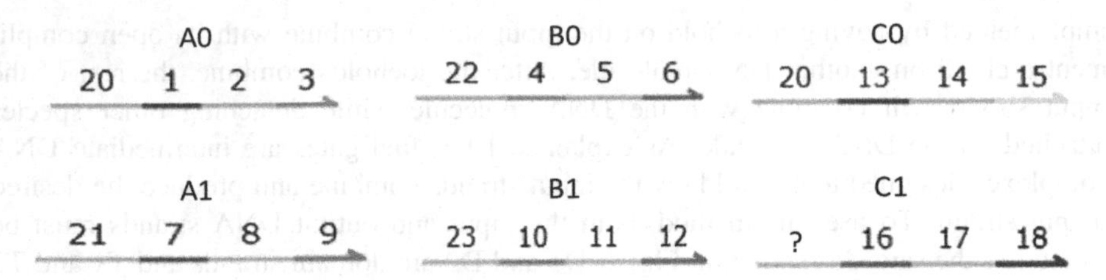

Fig. 3. The input (A_0, A_1, B_0, B_1) and output strands (C_0, C_1) used during the experiment for the M-OR computing unit. For C_1, the history domain is labeled by ? since it could be 20 or 21.

output strands the pathway produces, as the history is different for some strands. As it is shown later, the history for output C_1, indicated by the question mark in Fig. 3, could be 20 or 21.

Figure 4 illustrates each fuel gate needed for the M-OR computing unit. Gate 1 and Produce 1 are used in the first chemical reaction $A_0 + B_0 \rightarrow C_0$ while Gate 2 and Produce 2 are used for the second chemical reaction $A_0 + B_1 \rightarrow C_1$ and so on.

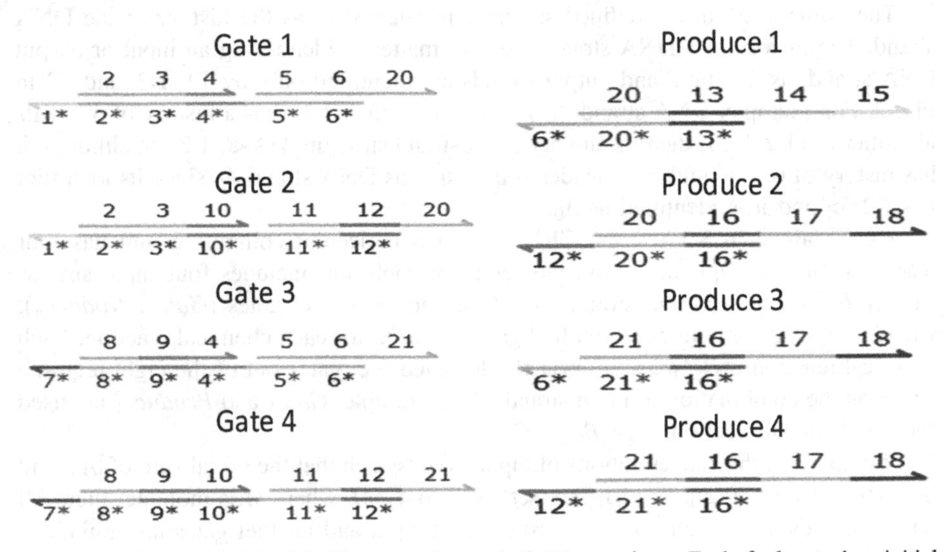

Fig. 4. Fuel gates for the M-OR computing unit DNA reactions. Each fuel gate has initial concentration of 100 n*M* in the simulation.

After designing the domains and toeholds for input, output, fuel gates, and their initial concentrations, Visual DSD is used to obtain the sequence of DNA reactions implemented by these strands. The details of the DNA reaction pathway generated by Visual DSD and used for simulations are illustrated in Figs. 5 and 6. These figures show the whole pathway of all DNA reactions for the M-OR unit.

The top row of DNA molecules are the molecules used in the first reaction. After they combine, they will briefly stay together and then unbind. This process produces an open compliment toehold for B_0 or B_1 to combine with. Following a path from top to bottom, in Figs. 5 and 6, will show what DNA reactions are possible and at what moment they happen. For example, the reaction that involves B_0 or B_1 cannot happen unless the first reaction of the pathway happens. Each black outlined box shows a defined molecule used in the simulation. Input strands and output strands are labeled with a red marking while each line in between two boxes shows the path of the reaction. The boxes with numbers show the unbinding and binding rate of each molecule.

The DNA reaction pathway in Figs. 5 and 6 can be explained by considering DNA reactions corresponding to each of the chemical reactions in the CRN of the M-OR unit. Each chemical reaction is mapped to five DNA reactions. Therefore, there are twenty DNA reactions implementing the M-OR unit. Figures 7, 8, 9 and 10 show the DNA reactions for the mapped chemical reactions $A_0 + B_0 \rightarrow C_0$, $A_0 + B_1 \rightarrow C_1$, $A_1 + B_0 \rightarrow C_1$, and $A_1 + B_1 \rightarrow C_1$ of the M-OR computing unit, respectively. In Fig. 7, the first reaction has A_0 combines with Gate1 by having its toehold 1, combine with Gate1's open toehold 1* which is the compliment of toehold 1. As shown in the second row, this produces a DNA molecule with open toehold 4*. The toehold is

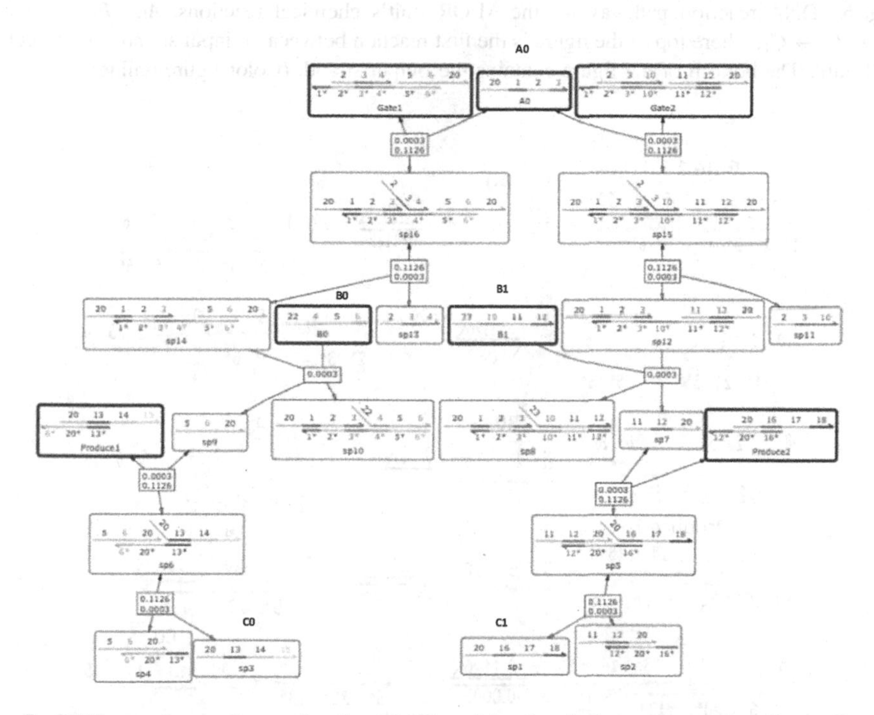

Fig. 5. DNA reaction pathway for the M-OR unit's chemical reactions, $A_0 + B_0 \rightarrow C_0$ and $A_0 + B_1 \rightarrow C_1$, where top of the figure is the first reaction between an input strand and respective fuel gate. The bottom of the figure contains the output strand. (Color figure online)

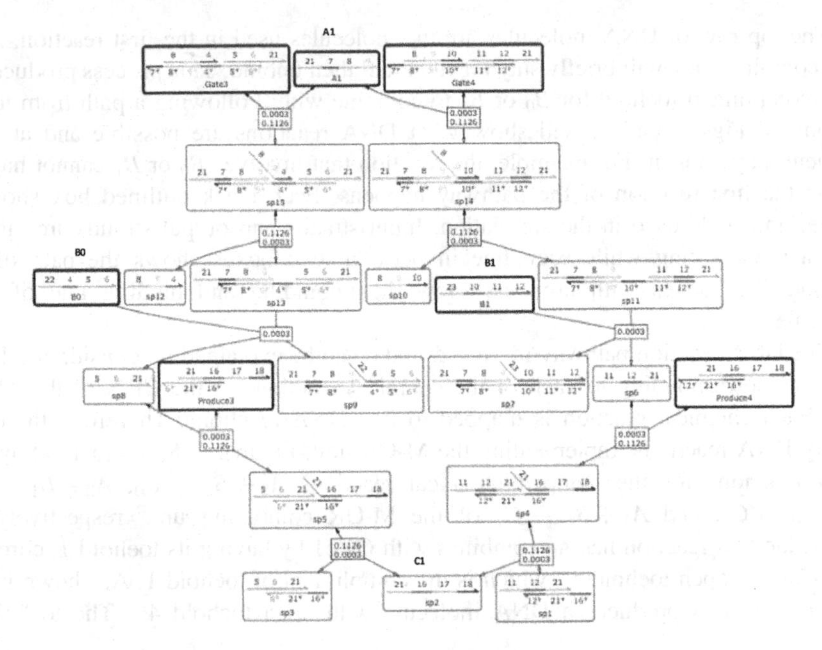

Fig. 6. DNA reaction pathway for the M-OR unit's chemical reactions, $A_0 + B_0 \rightarrow C_0$ and $A_0 + B_1 \rightarrow C_1$, where top of the figure is the first reaction between an input strand and respective fuel gate. The bottom of the figure contains the output strand. (Color figure online)

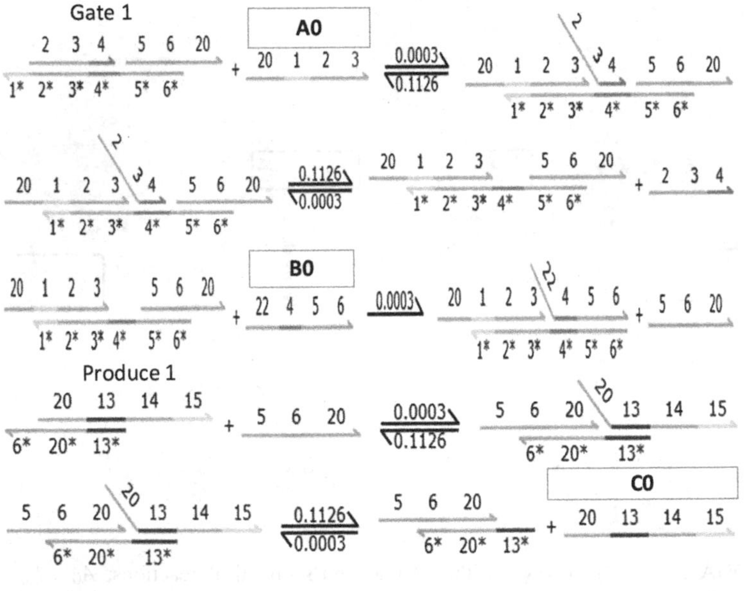

Fig. 7. DNA reactions for $A_0 + B_0 \rightarrow C_0$ of the M-OR computing unit. Input, fuel gates, and output strands are labeled.

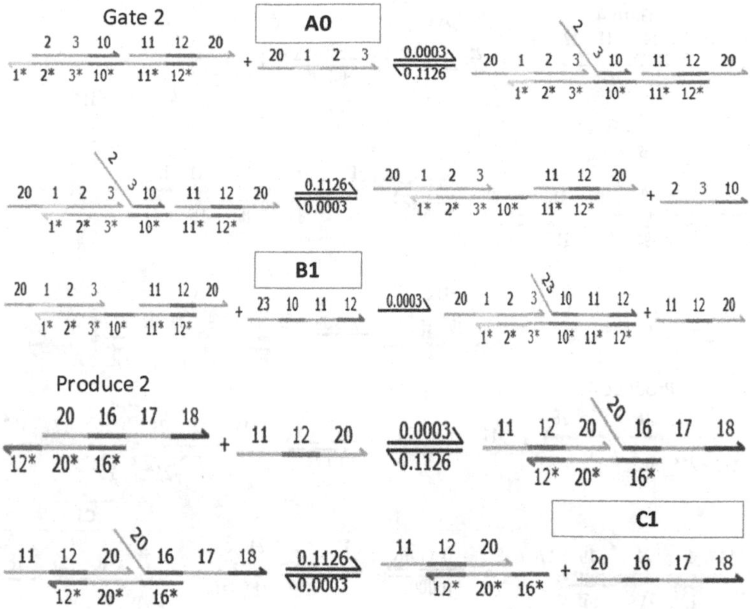

Gate 2

A0

2 3 10 11 12 20

1* 2* 3* 10* 11* 12*

+ 20 1 2 3

0.0003 / 0.1126

20 1 2 3 10 11 12 20

1* 2* 3* 10* 11* 12*

20 1 2 3 10 11 12 20

1* 2* 3* 10* 11* 12*

0.1126 / 0.0003

20 1 2 3 11 12 20

1* 2* 3* 10* 11* 12*

+ 2 3 10

B1

20 1 2 3 11 12 20

1* 2* 3* 10* 11* 12*

+ 23 10 11 12

0.0003

20 1 2 3 23 10 11 12

1* 2* 3* 10* 11* 12*

+ 11 12 20

Produce 2

20 16 17 18

12* 20* 16*

+ 11 12 20

0.0003 / 0.1126

11 12 20 20 16 17 18

12* 20* 16*

C1

11 12 20 20 16 17 18

12* 20* 16*

0.1126 / 0.0003

11 12 20

12* 20* 16*

+ 20 16 17 18

Fig. 8. DNA reactions for $A_0 + B_1 \rightarrow C_1$ of the M-OR computing unit. Input, fuel gates, and output strands are labeled.

Gate 3

A1

8 9 4 5 6 21

7* 8* 9* 4* 5* 6*

+ 21 7 8 9

0.0003 / 0.1126

21 7 8 9 8 9 4 5 6 21

7* 8* 9* 4* 5* 6*

21 7 8 9 8 9 4 5 6 21

7* 8* 9* 4* 5* 6*

0.1126 / 0.0003

21 7 8 9 5 6 21

7* 8* 9* 4* 5* 6*

+ 8 9 4

B0

21 7 8 9 5 6 21

7* 8* 9* 4* 5* 6*

+ 22 4 5 6

0.0003

21 7 8 9 22 4 5 6

7* 8* 9* 4* 5* 6*

+ 5 6 21

Produce 3

21 16 17 18

6* 21* 16*

+ 5 6 21

0.0003 / 0.1126

5 6 21 21 16 17 18

6* 21* 16*

C1

5 6 21 21 16 17 18

6* 21* 16*

0.1126 / 0.0003

5 6 21

6* 21* 16*

+ 21 16 17 18

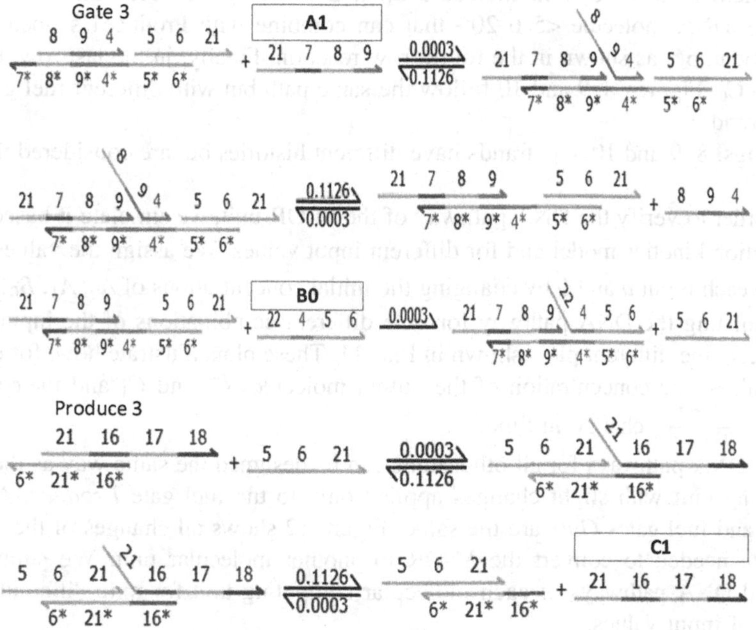

Fig. 9. DNA reactions for $A_1 + B_0 \rightarrow C_1$ of the M-OR computing unit. Input, fuel gates, and output strands are labeled.

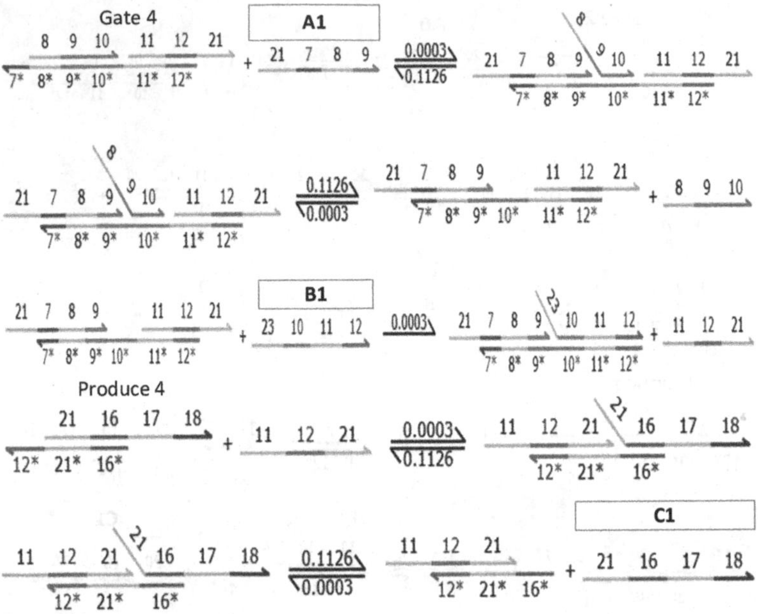

Fig. 10. DNA reactions for $A_1 + B_1 \rightarrow C_1$ of the M-OR computing unit. Input, fuel gates, and output strands are labeled.

compliment to toehold 4 in molecule B_0 and allows B_0 to combine with it. This produces DNA molecule <5 6 20> that can combine with Produce1's open toehold compliment, 6*, as shown in the fourth row reaction. Finally, in the last row, this will produce C_0. Figures 8, 9 and 10 follow the same path but with different fuel gates and input strands.

In Figs. 8, 9 and 10, C_1 strands have different histories but are considered the same strand.

In order to verify the DNA pathway of the M-OR unit, we simulate it based on the mass-action kinetics model and for different input values. We assign the values of $\frac{1}{4}$, $\frac{1}{2}$, and $\frac{3}{4}$ to each input a and b by changing the initial concentrations of A_0, A_1, B_0, and B_1.

Simulating the DNA pathway for nine different combinations of the input values, we collect nine different plots shown in Fig. 11. These plots illustrate how, for different input values, the concentration of the output molecules C_0 and C_1 and their unipolar ratio $c = \frac{[C_1]}{[C_0]+[C_1]}$ change in time.

The DNA pathways for all other units can be designed the same way as the M-OR computing unit with slight changes applied only to the fuel gate *Produce*. All input strands and fuel gates *Gate* are the same. Figure 12 shows all changes of the *Produce* fuel gate needed to convert the M-OR to another molecular unit. We simulate the designed DNA pathways of each molecular computing unit for nine different combinations of input values.

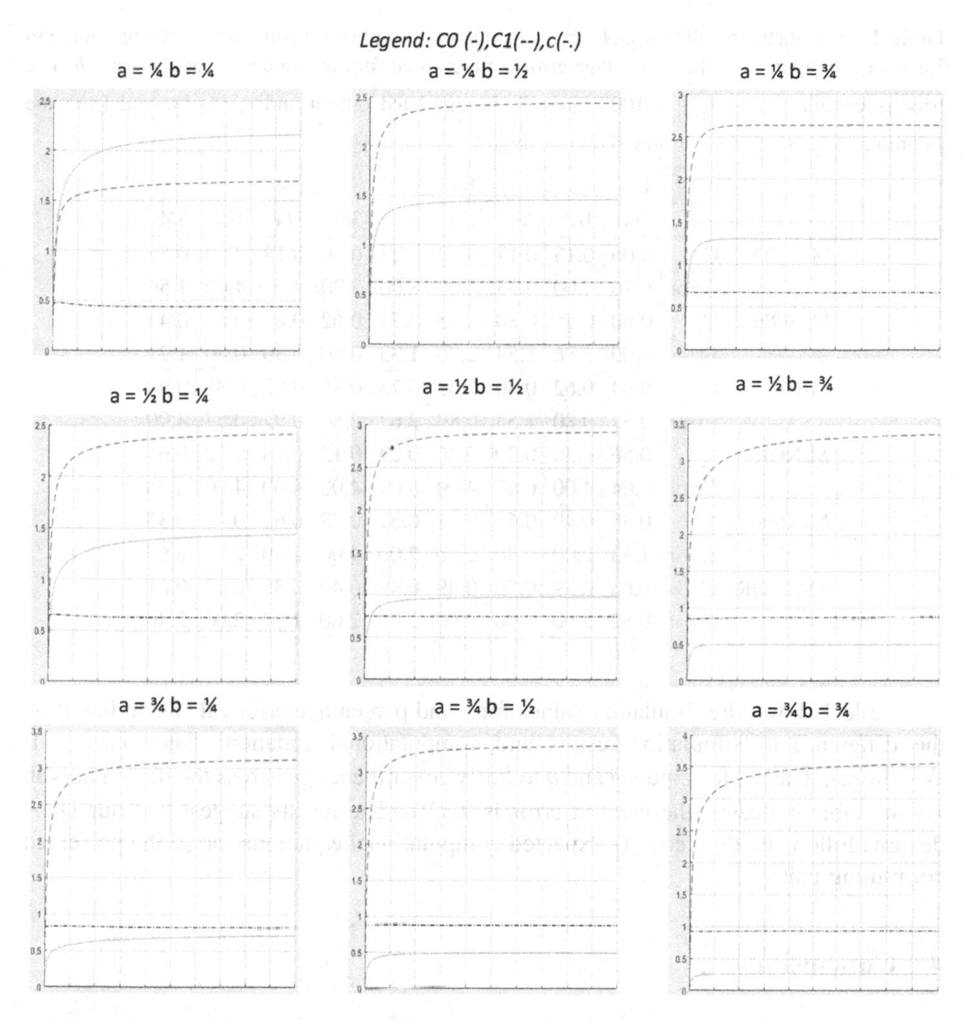

Fig. 11. Simulation plots for the M-OR computing unit for nine different values of inputs *a* and *b*. The x-axis of each plot represents time and ranges from 0 to 200,000 arbitrary units (a.u.) while the y-axis represents the concentration (a.u.).

	Produce 1	Produce 2	Produce 3	Produce 4
M-AND	20 13 14 15 / 6* 20* 13*	20 13 14 15 / 12* 20* 13*	21 13 14 15 / 6* 21* 13*	21 16 17 18 / 12* 21* 16*
M-NAND	20 16 17 18 / 6* 20* 16*	20 16 17 18 / 12* 20* 16*	21 16 17 18 / 6* 21* 16*	21 13 14 15 / 12* 21* 13*
M-NOR	20 16 17 18 / 6* 20* 16*	20 13 14 15 / 12* 20* 13*	21 13 14 15 / 6* 21* 13*	21 13 14 15 / 12* 21* 13*
M-XOR	20 13 14 15 / 6* 20* 13*	20 16 17 18 / 12* 20* 16*	21 16 17 18 / 6* 21* 16*	21 13 14 15 / 12* 21* 13*
M-XNOR	20 16 17 18 / 6* 20* 16*	20 16 17 18 / 12* 20* 16*	21 16 17 18 / 6* 21* 16*	21 13 14 15 / 12* 21* 13*

Fig. 12. Produce Fuel gates designed for each molecular computing unit.

Table 1. Simulation results for DNA implementation of molecular computing units presented in this paper. Output c and the percentage error are measured for nine values of inputs a and b. The error is calculated as $\left|\frac{c_s - c_t}{c_t}\right| \times 100$, where c_s is simulated output and c_t is obtained from the theoretical equation of each unit shown in Fig. 1.

	A	1/4			1/2			3/4		
	B	1/4	1/2	3/4	1/4	1/2	3/4	1/4	1/2	3/4
M-AND	C	0.06	0.13	0.17	0.12	0.23	0.36	0.18	0.60	0.55
	Error	6.70	4.00	9.33	4.00	8.00	3.20	4.00	4.00	1.50
M-NAND	C	0.90	0.85	0.80	0.85	0.74	0.62	0.81	0.62	0.43
	Error	4.00	2.86	1.54	2.86	1.33	0.80	0.30	0.80	1.71
M-OR	C	0.44	0.62	0.80	0.63	0.73	0.85	0.82	0.88	0.90
	Error	0.57	0.80	1.54	0.80	2.67	2.86	0.92	0.57	4.00
M-NOR	C	0.56	0.36	0.20	0.36	0.23	0.13	0.18	0.12	0.07
	Error	0.04	4.00	6.67	4.00	8.00	4.00	4.00	4.00	8.48
M-XOR	C	0.38	0.49	0.63	0.51	0.51	0.48	0.61	0.49	0.35
	Error	1.33	2.00	0.80	2.00	2.00	4.00	2.40	2.00	6.67
M-XNOR	C	0.63	0.49	0.36	0.49	0.49	0.49	0.37	0.49	0.61
	Error	0.80	2.00	4.00	2.00	2.00	2.00	1.33	2.00	2.40

Table 1 shows the simulation values for c and percentage error calculated based on the difference of simulation results and computational equations listed in Fig. 1. We chose $\frac{1}{4}$, $\frac{1}{2}$, and $\frac{3}{4}$ for inputs a and b to fairly cover the range across the [0, 1] interval. As the table indicates, the greatest error is 9.33%. The results suggest that our DNA designs follow the theoretically expected computational equations for all the presented computing units.

4 Conclusion

In this paper, based on the unipolar fractional coding that relates molecular computing to stochastic computing, we have studied the CRNs, computational equations, and DNA realization of new molecular units such as M-OR, M-NOR, M-XOR, and M-XNOR. The DNA pathway of these units are on the basis of DNA strand-displacement mechanism. The simulation results show that these units perform the same computations calculated by their counterpart gates in stochastic computing circuits. We can extend the molecular computing units discussed in this paper for bipolar fractional coding. We can design different molecular computing units with different computational equations based on the different combinations of C_0 and C_1 as their product. In this paper, we have presented only six of them corresponding to common logic gates: AND, NAND, OR, NOR, XOR, and XNOR. We can extend the discussion for other units similarly.

Although we have verified our designs using Visual DSD software tool, future work can be the physical implementation of these units in test tube. For the experimental implementation, we can use a fluorescent reporter strategy [20] to detect output strands. In this strategy, an increase in the amount of output strands leads to proportional increase in fluorescence and let us follow their concentrations.

Acknowledgement. This work was supported by the "UK ECE Undergraduate Research Fellowship".

References

1. Chen, H., Doty, D., Soloveichik, D.: Rate-independent computation in continuous chemical reaction networks. In: Conference on Innovations in Theoretical Computer Science, pp. 313–326 (2014)
2. Cardelli, et al.: Chemical reaction network designs for asynchronous logic circuits. Nat. Comput. **17**, 109–130 (2018)
3. Salehi, S.A., Parhi, K.K., Riedel, M.D.: Chemical reaction networks for computing polynomials. ACS Synth. Biol. **6**(1), 76–83 (2017)
4. Chen, H.L., Doty, D., Soloveichik, D.: Deterministic function computation with chemical reaction networks. Nat. Comput. **13**, 517–534 (2013)
5. Chalk, C., Kornerup, N., Reeves, W., Soloveichik, D.: Composable rate-independent computation in continuous chemical reaction networks. In: 16th International Conference on Computational Methods in Systems Biology (CMSB) (2018)
6. Doty, D., Hajiaghayi, M.: Leaderless deterministic chemical reaction networks. Natural Comput. **14**(2), 213–223 24 (2015)
7. Cummings, R., Doty, D., Soloveichik, D.: Probability 1 computation with chemical reaction networks. Nat. Comput. **15**(2), 245–261 (2016)
8. Chen, H.L., Cummings, R., Doty, D., Soloveichik, D.: Speed faults in computation by chemical reaction networks. Distrib. Comput. **30**(5), 373–390 (2017)
9. Chou, C.T.: Chemical reaction networks for computing logarithm. Synthetic Biol. **2**(1), ysx002 (2017)
10. Qian, L., Soloveichik, D., Winfree, E.: Efficient turing-universal computation with dna polymers. In: Sakakibara, Y., Mi, Y. (eds.) DNA 2010. LNCS, vol. 6518, pp. 123–140. Springer, Heidelberg (2011). https://doi.org/10.1007/978-3-642-18305-8_12
11. Salehi, S.A., Liu, X., Riedel, M.D., Parhi, K.K.: Computing mathematical functions using DNA via fractional coding, Scientific Reports, vol. 8, Article 8312, May 2018
12. Gaines, B.R.: Stochastic Computing. In: Proceedings of AFIPS Spring Joint Computer Conference, pp. 149–156. ACM (1967)
13. Poppelbaum, W.J., Afuso, C., Esch. J.W.: Stochastic computing elements and systems. In: Proceedings of the Joint Computer Conference, AFIPS 1967 (Fall), pp. 635–644. ACM, New York (1967)
14. Gaines, B.R.: Stochastic computing systems. In: Tou, J.T. (ed.) Advances in Information Systems Science, pp. 37–172. Springer, Boston (1969). https://doi.org/10.1007/978-1-4899-5841-9_2
15. Salehi, S.A., Liu, Y., Riedel, M., Parhi, K.K.: Computing polynomials with positive coefficients using stochastic logic by DoubleNAND expansion. In: Proceedings of the 2017 ACM Great Lakes Symposium on VLSI (GLSVLSI), pp. 471–474 (2017)

16. Yurke, B., Turberfeld, A.J., Mills, A.P., Simmel, F.C., Neumann, J.L.: A DNA-fuelled molecular machine made of DNA. Nature **406**, 605–608 (2000)
17. Soloveichik, D., Seelig, G., Winfree, E.: DNA as a universal substrate for chemical kinetics. PNAS **107**(12), 5393–5398 (2010)
18. Lakin, M.R., Youssef, S., Polo, F., Emmott, S., Phillips, A.: Visual DSD: a design and analysis tool for DNA strand displacement systems. Bioinformatics **27**, 3211–3213 (2011)
19. Visual DSD Homepage. https://www.microsoft.com/en-us/research/project/programming-dna-circuits/. Accessed 28 Jan 2019
20. Chen, Y.J., et al.: Programmable chemical controllers made from DNA. Nat. Nanotechnol. **8**, 755–762 (2013)

The Role of the Representational Entity in Physical Computing

Susan Stepney[1(✉)] and Viv Kendon[2]

[1] Department of Computer Science, University of York, York, UK
susan.stepney@york.ac.uk

[2] Department of Physics, Durham University, Durham, UK

Abstract. We have developed abstraction/representation (AR) theory to answer the question "When does a physical system compute?" AR theory requires the existence of a *representational entity* (RE), but the vanilla theory does not explicitly include the RE in its definition of physical computing. Here we extend the theory by showing how the RE forms a linked complementary model to the physical computing model, and demonstrate its use in the case of intrinsic computing in a non-human RE: a bacterium.

1 Introduction

Many and diverse physical substrates are proposed for unconventional computing, from relativistic and quantum systems to chemical reactions and slime moulds, from carbon nanotubes to non-linear optical reservoir systems, from amorphous substrates to highly engineered devices, from general purpose analogue computers to one-shot devices. In another domain, biological systems are often said to perform information processing. In all these cases is crucial to be able to determine when such substrates and systems are specifically computing, as opposed to merely undergoing the physical processes of that substrate.

In order to address this question, we have been developing *abstraction/representation* theory (AR theory). This is a framework in which science, engineering/technology, computing, and communication/signalling are all defined as a form of *representational activity*, requiring the fundamental use of the representation relation linking physical system and abstract model in order to define their operation [3,7]. Within this framework, it is possible to distinguish scientific experimentation on a novel substrate, from the performance of computation by that substrate.

In work following on from the original definitions, [6] provides a high level overview, [4] delves into more philosophical aspects, and [5] presents an example of intrinsic computation: signalling in bacteria. Also see [3,4] for references to the wider unconventional computing and philosophical literature.

AR theory requires the existence of a representational entity (RE) to support the representation relation. One issue glossed over in our previous descriptions of AR theory that becomes crucial when analysing computation in systems where

I. McQuillan and S. Seki (Eds.): UCNC 2019, LNCS 11493, pp. 219–231, 2019.
https://doi.org/10.1007/978-3-030-19311-9_18

the RE is not a human or conscious user, is the relationship between the physical RE and the physical computer. Here we enrich AR theory by incorporating the RE explicitly, and showing how it relates to the physical computing process.

The structure of the paper is as follows. In Sect. 2 we summarise the current formulation of AR theory. In Sect. 3 we extend the theory to include the RE explicitly. In Sect. 4 we demonstrate how the extended theory allows us to capture and model intrinsic computing in a bacterium.

2 AR Theory in a Nutshell

2.1 Our View of Physical Computing

AR theory has been developed to answer the specific question of when a physical system is computing [3]. Its answer hinges on the relationship between an abstract object (a computation) and a physical object (a computer). It employs a language of relations, not from mathematical objects to mathematical objects (as is usual in mathematics and theoretical computer science), but between physical objects and those in the abstract domain. The core of AR theory is the *representation relation*, mapping from physical objects to abstract objects. Experimental science, engineering, and computing all require the interplay of abstract and physical objects via representation in such a way that their descriptive diagrams *commute* such that the same result can be gained through either physical or abstract evolutions (see Sect. 2.3). From this, Horsman et al [3] define computing as *the use of a physical system to predict the outcome of an abstract evolution*.

2.2 Representation

AR theory has physical objects in the domain of material systems, abstract objects (including mathematical and logical entities), and the representation relation that mediates between the two. The distinction between the two spaces, abstract and physical, is fundamental in the theory, as is their connection *only* by the (directed) representation relation. An intuitive example is given in Fig. 1: a physical switch is *represented* by an abstract bit, which in this case takes the value 0 for switch state up, and 1 for switch state down. Note, however, that AR theory is not a dualist theory in the sense of Descartes. Everything in the theory is physical in some form. The symbols in the *Abstract* domain in Fig. 1 are instantiated as ink on paper or pixels on the screen as you read this. What makes them *abstract* in AR theory is that this physical form is to some degree arbitrary, and can change, while still corresponding to the same abstract object.

An example of a physical object in the domain of material entities is a *computer*. It has, usually, internal degrees of freedom, and a physical time evolution that transforms initial input to final output states. An example of an abstract object is a *computation*, which is a set of objects and relations as described in one of the logical formalisms of theoretical computer science. Likewise, an object

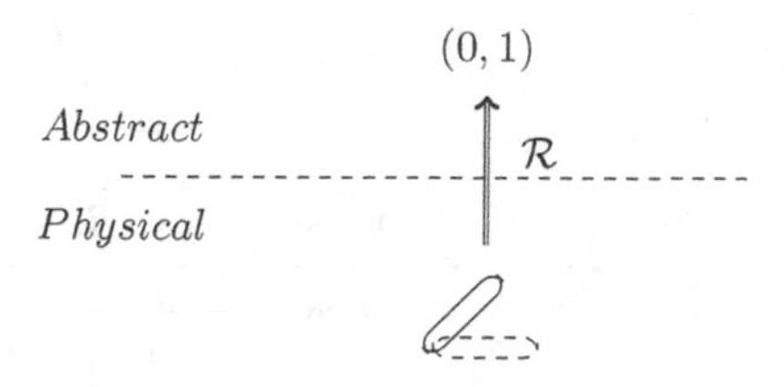

Fig. 1. Basic representation has three components: (i) the space of physical objects (here, a switch with two settings); (ii) the space of abstract objects (here, a binary digit); (iii) the directed representation relation $\mathcal{R}$ mediating between the spaces.

such as a bacterium is a physical entity, and its theoretical representation within biology is an object in the domain of abstract entities.

The central role of representation leads to the requirement for a *representational entity* (RE). The RE supports the representation relation between physical and abstract. AR theory does not require the RE to be human, or conscious; see [5] for an example of a bacterial RE, which is expanded on in Sect. 4 here.

The elementary *representation relation* is the directed map from physical objects to abstract objects, $\mathcal{R}_\mathcal{T} : \mathbf{P} \to M$, where $\mathbf{P}$ is the set of physical objects, and M is the set of abstract objects. We subscript the relation $\mathcal{R}$ with a theory $\mathcal{T}$ to indicate that the relation is theory-dependent. When two objects are connected by $\mathcal{R}_\mathcal{T}$ we write them as $\mathcal{R}_\mathcal{T} : \mathbf{p} \to m_\mathbf{p}$. The abstract object $m_\mathbf{p}$ is then said to be the *abstract representation* (under the given theory) of the physical object $\mathbf{p}$, and together they form one of the basic composites of AR theory, the *representational triple* $\langle \mathbf{p}, \mathcal{R}_\mathcal{T}, m_\mathbf{p} \rangle$. The basic representational triple is shown in Fig. 2(a).

Abstract evolution takes abstract objects to abstract objects, which we write as $C_\mathcal{T} : M \to M$. Again, we subscript with theory $\mathcal{T}$ to indicate that C is theory-dependent. An individual example is shown in Fig. 2(b), for the mapping $C_\mathcal{T}(m_\mathbf{p})$ taking $m_\mathbf{p} \to m'_\mathbf{p}$. The corresponding physical evolution map is given by $\mathbf{H} : \mathbf{P} \to \mathbf{P}$. For individual elements in Fig. 2(c) this is $\mathbf{H}(\mathbf{p})$ which takes $\mathbf{p} \to \mathbf{p}'$.

2.3 ε-Commuting Diagrams

In order to link the final abstract and physical objects, we apply the representation relation to the outcome state of the physical evolution, to give its abstract representation $m_{\mathbf{p}'}$, Fig. 2(d). We now have two abstract objects: $m'_\mathbf{p}$, the result of the abstract evolution, and $m_{\mathbf{p}'}$, the representation of the result of the physical evolution. For some (problem-dependent) error quantity ε and norm $|.|$, if $|m_{\mathbf{p}'} - m'_\mathbf{p}| \leq \varepsilon$ (or, more briefly, $m_{\mathbf{p}'} =_\varepsilon m'_\mathbf{p}$), then we say that the diagram in Fig. 2(d) *ε-commutes*.

Commuting diagrams are fundamental to the use of AR theory. If the relevant abstract and physical objects form an ε-commuting diagram under representation, then $m_\mathbf{p}$ is a *faithful abstract representation* (up to ε) of physical system $\mathbf{p}$ for the evolutions $C_\mathcal{T}(m_\mathbf{p})$ and $\mathbf{H}(\mathbf{p})$.

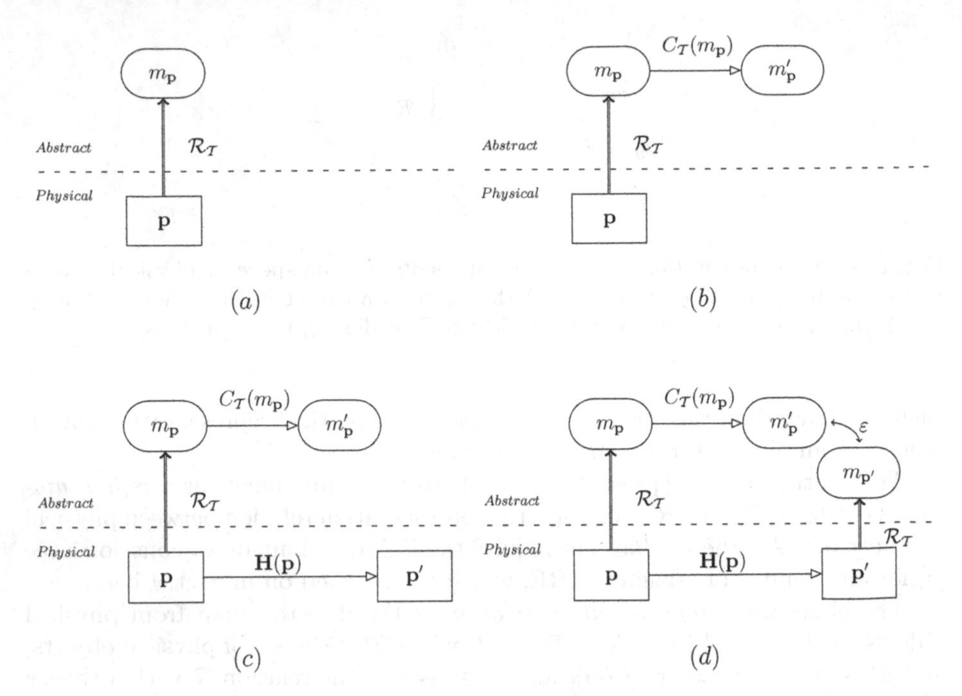

Fig. 2. Parallel evolution of an abstract object and the physical system it represents. (a) The basic representational triple, $\langle \mathbf{p}, \mathcal{R}, m_{\mathbf{p}} \rangle$: physical system $\mathbf{p}$ is represented abstractly by $m_{\mathbf{p}}$ using the modelling representation relation $\mathcal{R}_{\mathcal{T}}$ of theory $\mathcal{T}$. (b) Abstract dynamics $C_{\mathcal{T}}(m_{\mathbf{p}})$ give the evolved abstract state $m'_{\mathbf{p}}$. (c) Physical dynamics $\mathbf{H}(\mathbf{p})$ give the final physical state $\mathbf{p}'$. (d) $\mathcal{R}_{\mathcal{T}}$ is used again to represent $\mathbf{p}'$ as the abstract output $m_{\mathbf{p}'}$. If $m_{\mathbf{p}} =_{\varepsilon} m_{\mathbf{p}'}$, the diagram ε-commutes. (Adapted from [3].)

The existence of such ε-commuting diagrams defines what is meant by a faithful abstract representation of a physical system. The final state of a physical object undergoing time evolution can be known *either* by tracking the physical evolution and then representing the output abstractly, *or* by evolving the abstract representation of the system; and the two results differ by less than the problem-dependent ε. In the first case, the 'lower path' of the diagram is followed; in the latter, the 'upper path'.

Finding out which diagrams ε-commute is the business of basic experimental science; once commuting diagrams have been established they can be exploited through engineering and technology.

2.4 Compute Cycle

Figure 2(d) shows the basic 'science cycle', of representing a physical system, and determining whether $C_{\mathcal{T}}$ is a sufficiently good abstract model of its behaviour, by requiring that $m_{\mathbf{p}} =_{\varepsilon} m_{\mathbf{p}'}$ for sufficiently many different initial states $\mathbf{p}$ to have confidence in $C_{\mathcal{T}}$ and $\mathcal{R}_{\mathcal{T}}$. There are derived variants of this diagram that capture the 'engineering cycle', and the related 'compute' cycle. See the original references for details; here we focus on the compute cycle.

An ε-commuting diagram in the context of *computation* also connects the physical computing device, $\mathbf{p}$, and its abstract representation $m_{\mathbf{p}}$. But to do so it makes use of the *instantiation* relation $\widetilde{\mathcal{R}}_{\mathcal{T}} : M \to \mathbf{P}$. Here, instead of saying abstract object $m_{\mathbf{p}}$ represents physical system $\mathbf{p}$, we say that physical system $\mathbf{p}$ instantiates abstract object $m_{\mathbf{p}}$. Whereas the representation relation is primitive, the instantiation relation is a derived relation, based on multiple science cycles, abbreviated as $\widetilde{\mathcal{R}}_{\mathcal{T}}$; see original references for full details.

The use of $\widetilde{\mathcal{R}}_{\mathcal{T}}$ acknowledges that a computer is physical system engineered (or possibly evolved) to have a particular behaviour, rather than a natural physical system being scientifically modelled. The full compute cycle is shown in Fig. 3, starting from initial abstract problem, through instantiation into a physical computer, physical evolution of the device, followed by representation of the final physical state as the abstract answer the problem.

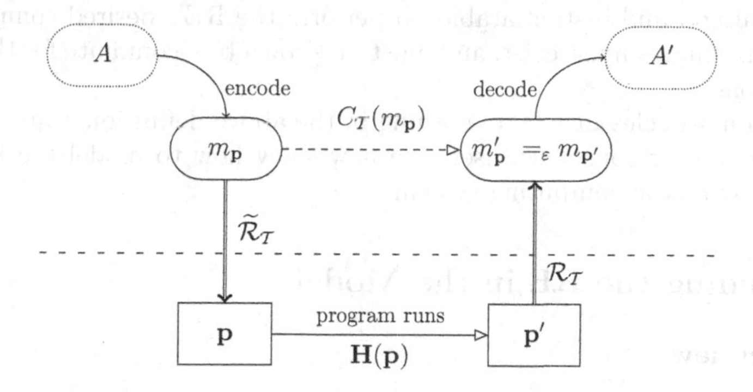

Fig. 3. Physical computing in AR theory. An abstract problem A is encoded into the model $m_{\mathbf{p}}$; the model is instantiated into the physical computer state $\mathbf{p}$; the computer calculates via $\mathbf{H}(\mathbf{p})$, evolving into physical state $\mathbf{p}'$; the final state is represented as the final abstract model $m_{\mathbf{p}'} =_{\varepsilon} m'_{\mathbf{p}}$; this is decoded as the solution to the problem, A'. The instantiation, physical evolution, and representation together implement the desired abstract computation $C_{\mathcal{T}}(m_{\mathbf{p}})$. (From now on we omit the dashed line separating the physical and abstract world, and rely on the different shaped boxes to indicate what components lie in which domain.)

Ensuring that the diagram ε-commutes is a process of debugging the physical system, including how it is instantiated (engineered, programmed and provided with input data), and how its output is represented. This shows another key difference from the science cycle: there the diagram is made to ε-commute by instead debugging the abstract model.

The most important use of a computing system is when the abstract outcome $m'_{\mathbf{p}}$ is unknown: when computers are used to solve problems. Consider as an example the use of a computer to perform the abstract arithmetical calculation $2 + 3$. If the outcome were unknown, and the computing device were being used to compute it, the final abstract state, $m'_{\mathrm{p}} = 5$, would not be calculated abstractly. Instead, confidence in the technological capabilities of the computer and the correctness of the instantiation would enable the user to reach the final, abstract, output state $m_{\mathrm{p}'} =_0 m'_{\mathrm{p}}$ using the physical evolution of the computing device alone. (One advantage of digital computers is that we can achieve $\varepsilon = 0$.)

This use of a physical computer is the compute cycle, Fig. 3: *the use of a physical system* (the computer) *to predict the outcome of an abstract evolution* (the computation).

2.5 Generality of AR Theory

Nothing in the above definition requires the physical computer to be digital, or electronic, or universal, or pre-existing. The computer could be a continuous analogue device; it could be a mechanical or organic device; it could be a hard-wired device with limited capabilities; it could be a 'one-shot' device constructed for a particular computation. It simply needs to be sufficiently powerful, sufficiently accurate, and instantiatable, to perform the RE's desired computations: the relevant squares must exist, and must be known to ε-commute for the desired computations.

And, of most relevance here, nothing in the above definition requires the RE to be a human, or conscious, user. We now show how to model the RE in the same context as the computing system.

3 Including the RE in the Model

3.1 Overview

As mentioned above, the representational entity (RE) supports the representation relation $\mathcal{R}$. Although it does not appear explicitly in the compute cycle of Fig. 3, it is the *physical entity* that 'owns' the abstract problem A and desires the abstract solution A'.

To help clarify the issues, consider a (human) RE who has the problem "I have two apples in my left hand, and three in my right hand; how many apples do I have in total?" We model this physical RE's problem, how they encode it as a computational problem, how this is instantiated in a physical computer, how the computer finds the answer, how the answer is represented back as an abstract computational result, and how that result is decoded as an answer to the RE's problem.

In this section, we add the RE to the overall model of physical computing as defined in AR theory. As before, we have objects in two domains: the physical RE, and (our) abstract model of the RE. First we show how we model the RE in a manner analogous to how we model a physical computation (Sect. 3.2). Then we show how to integrate the RE and full physical compute cycle models, and how to interpret various parts of the resulting model (Sect. 3.3).

3.2 The Physical RE

The RE is a physical system $\mathbf{p}_{RE}$. The relevant part of the RE here is the physical states that it uses to represent its abstract problem A. (There are several levels of indirection at play here, that will become clearer with later examples.) Our abstract model of these relevant parts in AR theory diagrams is $m_{\mathbf{p}_{RE}}$; the RE does not in general itself construct AR diagrams. See Fig. 4. Our model $m_{\mathbf{p}_{RE}}$ may incorrectly capture the RE's physical state, in which case the model needs to be modified; $m_{\mathbf{p}_{RE}}$ is our *scientific* model of $\mathbf{p}_{RE}$.

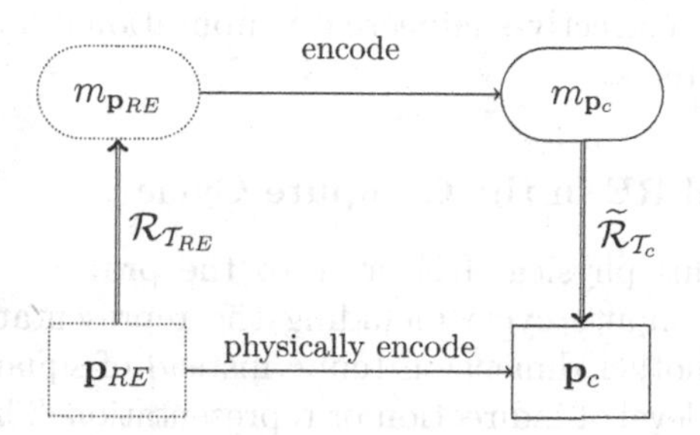

Fig. 4. The relationship between the physical representational entity $\mathbf{p}_{RE}$ and the physical computer $\mathbf{p}_c$ via abstract models of each. There is an *encoding* of the abstract model $m_{\mathbf{p}_{RE}}$ into $m_{\mathbf{p}_c}$. In a correctly working system, this encoding is appropriately implemented by the respective physical systems: the square should ε-commute. Note that the models of the RE and the computer are potentially with respect to different theories.

We model the computational system as before. There is the abstract model $m_{\mathbf{p}_c}$ that forms the 'specification' of the RE's problem $m_{\mathbf{p}_{RE}}$ encoded as a computational problem. (This is the model $m_{\mathbf{p}}$ in Fig. 3.) This is our model of the RE's model of the computer and encoding: the RE's model is also part of its physical state $\mathbf{p}_{RE}$.

The computer's physical state may incorrectly implement the RE's model, in which case the physical state needs to be modified; the RE is using an engineering model of $\mathbf{p}_c$. However, our model $m_{\mathbf{p}_c}$ of the RE's model may incorrectly represent the RE's model: we are using a scientific model of the RE and its computer. We model the RE's problem being encoded into the computer by $m_{\mathbf{p}_{RE}}$ being *encoded* into the computational model $m_{\mathbf{p}_c}$. There is no guarantee that such an encoding is possible: not all problems are computable.

The two representation/instantiation relation arrows in Fig. 4 are with respect to two different theories. The representation $\mathcal{R}_{T_{RE}} : \mathbf{p}_{RE} \to m_{\mathbf{p}_{RE}}$ is based on the theory of how the physical RE forms abstract problem specifications; the instantiation $\widetilde{\mathcal{R}}_{T_c} : m_{\mathbf{p}_c} \to \mathbf{p}_c$ is based on the theory of how the physical computer implements abstract computations.

In a correctly implemented computer, the diagram in Fig. 4 should ε-commute: the instantiated state of the physical computer should correctly mirror the desired state of the physical RE: it should *physically encode* the desired state. The establishment of this physical encoding link is part of the engineering process.

During the execution, this physical encoding link is not necessarily established immediately. There may be some delay, for example in updating a record to reflect reality, or in opening or closing a valve to reflect changed demand. In, for example, a mechanical control system, with feedback, there can be an immediate coupling: the behaviour of the physical controlled system changes its state, which is directly communicated to the physical controller though their physical mechanical coupling. We do not consider this aspect further here, although it is a key feature of correctly-engineered computational 'mirror worlds' and of feedback control systems.

3.3 The Physical RE in the Compute Cycle

We can now add this physical RE layer to the previous compute cycle. See Fig. 5 for the full compute cycle including the representational entity. Notice how the RE adds another dimension (cube instead of square) to the diagrams. Each dimension is a level of indirection or representation. The full compute cycle involves traversal of many faces and edges of the displayed cube. Each face has its own place in the model.

Consider again a (human) RE who has the problem "I have two apples in my left hand, and three in my right hand; how many apples do I have in total?"

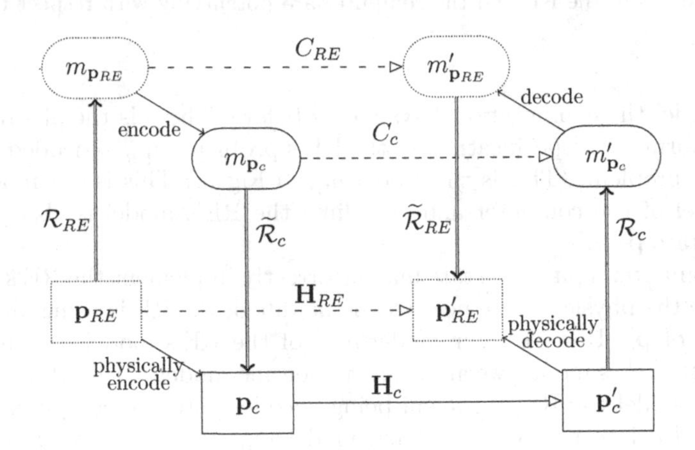

Fig. 5. The full compute cycle including the representational entity and the physical computer. The desired change in the RE's state, from posed problem to perceived solution, is $\mathbf{p}_{RE} \rightarrow \mathbf{p}'_{RE}$. The physical computer performs $\mathbf{p}_c \rightarrow \mathbf{p}'_c$. The full compute cycle from AR theory is: represent RE's physical state $\mathbf{p}_{RE}$ (desired computation) as abstract model $m_{\mathbf{p}_{RE}}$; encode to computational model $m_{\mathbf{p}_c}$; instantiate into physical computer state $\mathbf{p}_c$; physical computer evolves to final state $\mathbf{p}'_c$; represent physical solution as abstract computational solution $m'_{\mathbf{p}_c}$; decode to final abstract problem solution $m'_{\mathbf{p}_{RE}}$; which models the instantiation of the final state of the RE. Each set of squares (between representational entity and physical computer, and across the compute cycle) should ε-commute.

Back face; RE's view of the computation (Fig. 6): the RE's desired states, starting from a problem state (abstract initial state, $m_{p_{RE}}$, "how many apples?"; physical state, p_{RE}, a brain state that is represented by that abstract question) and resulting in a solution state (abstract final state, $m'_{p_{RE}}$,"five apples!"; physical state, p'_{RE}, a brain state that captures that abstract solution). There is no direct path from initial to final state, either abstractly or physically, as a separate computer is used to achieve the desired state changes.

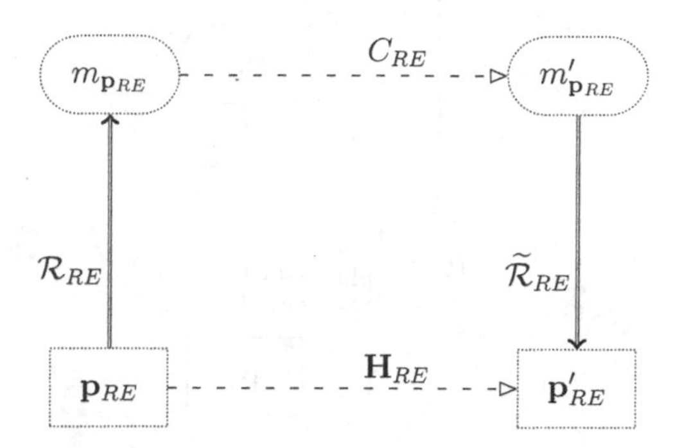

Fig. 6. The RE's view of the problem solution (back face of Fig. 5). The RE has an initial physical state $\mathbf{p}_{RE}$, modelled as $m_{\mathbf{p}_{RE}}$. It has a desired final state $\mathbf{p}'_{RE}$, modelled as $m'_{\mathbf{p}_{RE}}$. Both the horizontal arrows are dashed, as they are implemented in a different medium: the computer.

Left face; encoding the problem (Fig. 4): the RE's initial physical and abstract state encoded into the computer's initial physical and abstract states. The RE's abstract problem $m_{\mathbf{p}_{RE}}$ of "how many apples?" can be encoded as the computer's initial abstract state $m_{\mathbf{p}_c}$ "2+3". This is instantiated as the computer's initial physical state $\mathbf{p}_c$, [2+3]. The RE $\mathbf{p}_{RE}$ physically encodes the problem in the computer's initial state $\mathbf{p}_c$ by, for example, pressing the keys labelled [2] then [+] then [3]. (How this human RE manages to press the keys, given the apples it is currently holding, is an exercise left to the reader.)

Front face; compute cycle (Fig. 3, which also includes the back face RE abstract models as its 'abstract problem' components): the original simple AR theory compute cycle, ignoring the role of the RE. The abstract computational problem $m_{\mathbf{p}_c}$ is instantiated in the computer's initial physical state $\mathbf{p}_c$, [2+3]. The physical computer evolves as given by its physical structure, $\mathbf{H}_c$, which results in the final physical state $\mathbf{p}'_c$ of [5]. This is represented as the final abstract state $m'_{\mathbf{p}_c}$ of "5". These three steps (instantiation, physical evolution, representation) implement the desired abstract computation C_c: "2+3 = 5".

Right face; decoding the solution (Fig. 7): the RE's final physical and abstract state decoded from the computer's final physical and abstract state. The computer's final physical state $\mathbf{p}'_c$ (some kind of pattern of lights in the shape of a figure 5) is represented as the final abstract state $m'_{\mathbf{p}_c}$ of "5". This is decoded to the RE's final abstract state $m'_{\mathbf{p}_{RE}}$ of "five apples!". The RE's final physical brain state $\mathbf{p}'_{RE}$ is an instantiation of this, physically achieved by the RE looking at and physically decoding the output from the computer.

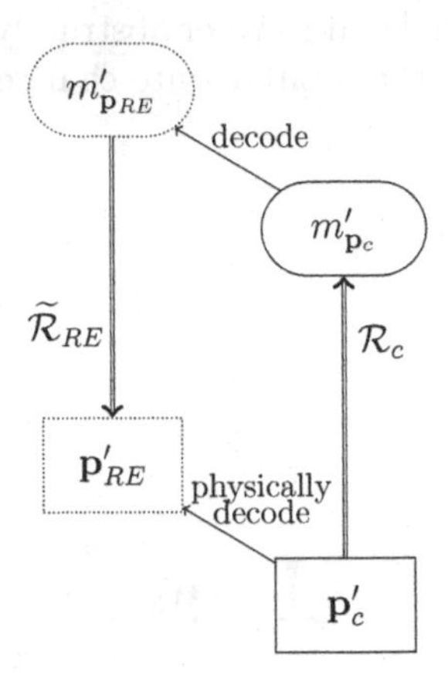

Fig. 7. Decoding the solution from the computer to the RE (right face of Fig. 5). The final state of the computer, $\mathbf{p}'_c$, is represented as the final abstract state $m'_{\mathbf{p}_c}$; this is decoded to the final abstract state of the RE, $m'_{\mathbf{p}_{RE}}$; and instantiated as the RE's final physical state. This is the model of the physical decoding lower arrow, achieved by the RE physically interrogating the computer.

Top face; abstract use of a computer (Fig. 8): the purely abstract view of the (modelled) RE encoding its problem into a (modelled) computation, and decoding the desired solution. There is no direct path from initial to final abstract states as the physical computer is used to achieve the desired abstract state changes. In terms of classical refinement theory [1,2], C_{RE} can be thought of as the 'global-to-global' requirement (although here this need not be captured in a formal manner), with "encode, computation C_c, decode" corresponding to the "initialisation, operation, finalisation" steps.

Bottom face; physical use of a computer (Fig. 9): the purely physical view of the RE encoding its problem in a physical computer, and decoding the desired solution. That this is a *computation*, rather than some other activity, is established by the abstract models and the various ε-commuting squares.

All of these relationships must be correctly implemented and modelled (the relevant squares containing encoding, decoding, instantiation, and representation must ε-commute) for the actual physical RE final state $\mathbf{p}'_{RE}$ to be the desired physical RE final state, that is, for the physical computer to have been used correctly, and for it to have performed correctly, to solve the RE's problem.

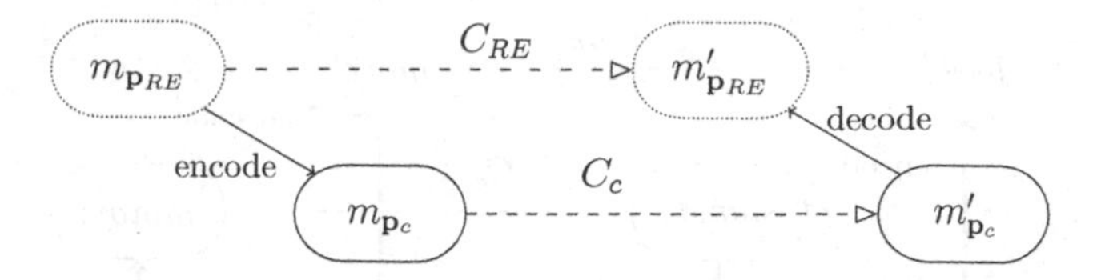

Fig. 8. The abstract model of the RE's use of the computer to solve its problem (top face of Fig. 5). The RE has an initial abstract state $m_{\mathbf{p}_{RE}}$; this is encoded into the initial abstract state of the computer $m_{\mathbf{p}_c}$. The computer performs its calculations to produce its final state $m'_{\mathbf{p}_c}$, which is decoded to produce the desired final state of the RE, $m'_{\mathbf{p}_{RE}}$. Both the horizontal arrows are dashed, as they are implemented in a different medium: the physical computer.

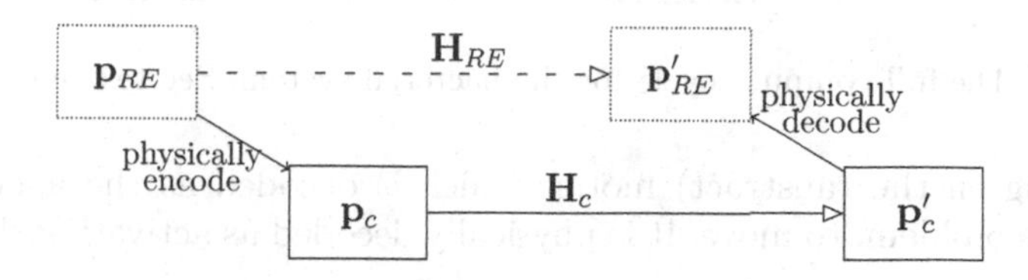

Fig. 9. The physical system of the RE's use of the computer to solve its problem (bottom face of Fig. 5). The physical RE has an initial physical state $\mathbf{p}_{RE}$; this is physically encoded into the initial physical state of the computer $\mathbf{p}_c$. The computer evolves over time to produce its final state $\mathbf{p}'_c$, which is decoded to produce the desired final state of the physical RE, $\mathbf{p}'_{RE}$.

4 Example: Intrinsic Computing in Bacteria

Figure 10 shows the RE and the compute cycle in the case of the problem of bacterial computing. This example was originally studied in [5] to demonstrate that it is possible to have computation with a non-human RE. However, without the explicit modelling of the bacterial RE, it resulted in a somewhat circuitous description. With the RE here explicitly present, the model is much clearer.

The physical RE, $\mathbf{p}_{RE}$, is a bacterium, with a receptor at the front, and a flagellar motor at the back. In the absence of input at the receptor, the motor is off; input causes the motor to switch on, propelling the bacterium towards food. (As ever, the biology is more complicated than this. The original reference should be consulted for the biological details.) The abstract problem, C_{RE}, that the RE wants to solve is "if there is food, move towards it". This is encoded as the abstract computational problem C_c: "if there is a signal, switch the motor on". The abstract signal is instantiated as a particular chemical; the physical *RE* physically encodes the reception of exterior food as the presence of an internal chemical, chem X.

This chemical physically propagates through the bacterium, undergoing transformation via a biochemical pathway, such that another chemical becomes present at the rear. The presence of this other chemical, chem Y, is represented

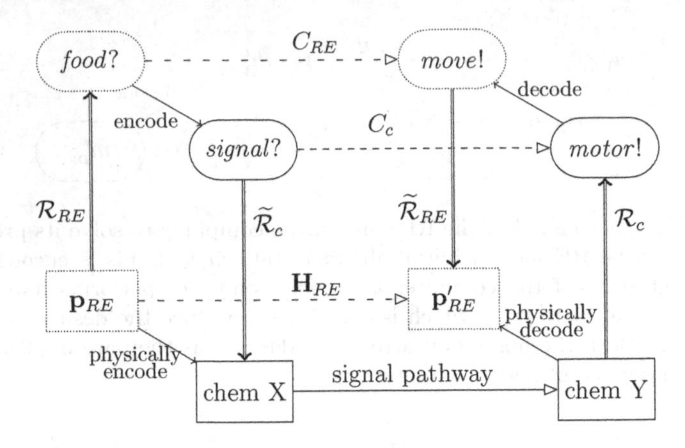

Fig. 10. The full compute cycle for the bacterial system. See text for details.

as switching on the (abstract) motor, which is decoded as the answer to the bacterium's problem: to move. It is physically decoded as activating the flagellar motor.

The bottom face of this bacterial-compute cube shows the purely physical computing: The bacterial RE physically encodes the detection of food by its receptor as a chemical chem X; the biochemical pathway moves and transforms this chemical signal to the rear where it appears as chem Y. The resulting chemical is physically decoded: it attaches to and activates the flagellar motor. The physical problem, of detecting food and moving towards it, has been solved.

That this is indeed a *computation*, rather than a purely physical process, is argued in [5]: chem X, chem Y, and the pathway are in some sense 'arbitrary' (they comprise different molecules in different bacterial species), and so it is not their specific *physical* properties, but their *representational*, informational properties, that are being exploited. We are able to model the part of the bacterium that represents the problem as $m_{\mathbf{p}_{RE}}$, and the part that encodes into the computer $m_{\mathbf{p}_c}$ in a way that convincingly contains the right sorts of representation. With representation (and hence a representational entity) identified, we can conclude that there is *data* being processed, not mere *material* being exploited. With ε-commuting diagrams present, we can conclude that computation is present.

5 Conclusion

We have shown how the RE in AR theory can be incorporated into the compute cycle, and how this can illuminate how the physical RE can use a physical device as a physical computer. The RE does not need to be a human brain: the example here shows intrinsic computing by a bacterial RE. This demonstrates how computing, whether conventional or unconventional, can be broader than human use of computers (external or brain-based), but is narrower than pan-computationalism, in requiring the existence of an RE in addition to the computer itself.

It may be that the RE does not even need to be organic, or 'alive'; it might potentially appear in the loop as an engineered 'proxy' for the ultimate RE, for example, the plant in a control system using the controller as a computer to maintain itself in a particular behaviour. In future work we will investigate how far the concept of the RE can be removed from a living user.

Acknowledgements. We thank our colleagues Dom Horsman, Tim Clarke, and Peter Young, for illuminating discussions. VK is funded by UK Engineering and Physical Sciences Research Council (EPSRC) grant EP/L022303/1.

References

1. Clark, J.A., Stepney, S., Chivers, H.: Breaking the model: finalisation and a taxonomy of security attacks. In: REFINE 2005 Workshop, Guildford, UK, April 2005. ENTCS, vol. 137(2), pp. 225–242. Elsevier (2005)
2. He, J., Hoare, C.A.R., Sanders, J.W.: Data refinement refined resume. In: Robinet, B., Wilhelm, R. (eds.) ESOP 1986. LNCS, vol. 213, pp. 187–196. Springer, Heidelberg (1986). https://doi.org/10.1007/3-540-16442-1_14
3. Horsman, C., Stepney, S., Wagner, R.C., Kendon, V.: When does a physical system compute? Proc. R. Soc. A **470**(2169), 20140182 (2014)
4. Horsman, D., Kendon, V., Stepney, S.: Abstraction/representation theory and the natural science of computation. In: Cuffaro, M.E., Fletcher, S.C. (eds.) Physical Perspectives on Computation, Computational Perspectives on Physics, pp. 127–149. Cambridge University Press, Cambridge (2018)
5. Horsman, D., Kendon, V., Stepney, S., Young, J.P.W.: Abstraction and representation in living organisms: when does a biological system compute? In: Dodig-Crnkovic, G., Giovagnoli, R. (eds.) Representation and Reality in Humans, Other Living Organisms and Intelligent Machines. SAPERE, vol. 28, pp. 91–116. Springer, Cham (2017). https://doi.org/10.1007/978-3-319-43784-2_6
6. Horsman, D., Stepney, S., Kendon, V.: The natural science of computation. Commun. ACM **60**(8), 31–34 (2017)
7. Horsman, D.C.: Abstraction/representation theory for heterotic physical computing. Philos. Trans. R. Soc. **373**, 20140224 (2015)

OIM: Oscillator-Based Ising Machines for Solving Combinatorial Optimisation Problems

Tianshi Wang(✉) and Jaijeet Roychowdhury

Department of Electrical Engineering and Computer Sciences,
University of California, Berkeley, USA
{tianshi,jr}@berkeley.edu

Abstract. We present a new way to make Ising machines, *i.e.*, using networks of coupled self-sustaining nonlinear oscillators. Our scheme is theoretically rooted in a novel result that establishes that the phase dynamics of coupled oscillator systems, under the influence of subharmonic injection locking, are governed by a Lyapunov function that is closely related to the Ising Hamiltonian of the coupling graph. As a result, the dynamics of such oscillator networks evolve naturally to local minima of the Lyapunov function. Two simple additional steps (*i.e.*, adding noise, and turning subharmonic locking on and off smoothly) enable the network to find excellent solutions of Ising problems. We demonstrate our method on Ising versions of the MAX-CUT and graph colouring problems, showing that it improves on previously published results on several problems in the G benchmark set. Our scheme, which is amenable to realisation using many kinds of oscillators from different physical domains, is particularly well suited for CMOS IC implementation, offering significant practical advantages over previous techniques for making Ising machines. We present working hardware prototypes using CMOS electronic oscillators.

1 Introduction

The Ising model [1,2] takes any weighted graph and uses it to define a scalar function called the Ising Hamiltonian. Each vertex in the graph is associated with a *spin*, *i.e.*, a binary variable taking values ± 1. The Ising problem is to find an assignment of spins that minimises the Ising Hamiltonian (which depends on the spins and on the graph's weights). Solving the Ising problem in general has been shown to be very difficult [3], but devices that can solve it quickly using specialised hardware have been proposed in recent years [4–9]. Such Ising machines have attracted much interest because many classically difficult combinatorial optimisation problems (including all 21 of Karp's well-known list of NP-complete problems [10]) can be mapped to Ising problems [11]. Hence, as Moore's Law nears its limits, Ising machines offer promise as a novel alternative paradigm for solving difficult computational problems efficiently.

We present a new and attractive means for realising Ising machines, *i.e.*, using networks of coupled, self-sustaining nonlinear oscillators. We first establish a key

I. McQuillan and S. Seki (Eds.): UCNC 2019, LNCS 11493, pp. 232–256, 2019.
https://doi.org/10.1007/978-3-030-19311-9_19

theoretical result that relates the (continuous) phase dynamics of an oscillator network with the (discrete/combinatorial) Ising Hamiltonian of the graph representing the oscillator couplings [12]. We then build on this result to develop practical oscillator-based Ising machines and demonstrate their effectiveness by solving MAX-CUT and graph colouring combinatorial optimisation problems [13,14]. We also present working hardware prototypes of our oscillator-based Ising machines.

We first show that the phase dynamics of any network of coupled, self-sustaining, amplitude-stable oscillators can be abstracted using the Generalised Adler model [15,16], a generalisation of the well-known Kuramoto model [17–19]. The model's phase dynamics are governed by an associated Lyapunov function, *i.e.*, a scalar function of the oscillators' phases that is always non-increasing and settles to stable local minima as phase dynamics evolve. If each oscillator's phase settles to either 0 or π (radians) and these values are associated with spins of ± 1, we show that this Lyapunov function is essentially identical to the Ising Hamiltonian of the oscillator network's connectivity graph. In general, however, oscillator phases do not settle to the discrete values $0/\pi$, but span a continuum of values instead. In order to binarise oscillator phases (*i.e.*, get them to settle to values near $0/\pi$), we inject each oscillator with a second harmonic signal (dubbed SYNC) that induces subharmonic injection locking (SHIL), which makes the phase of each oscillator settle to a value near either 0 or π [15,20–22]. We devise a new Lyapunov function that governs the network's dynamics with SHIL, then show its equivalence to the Ising Hamiltonian at phase values of $0/\pi$.

Thus we show that when SHIL binarisation is applied, coupled oscillator network dynamics settle naturally to *local* minima of a continuised version of the associated Ising Hamiltonian. To evolve the system out of local minima towards the global minimum, we show that a simple scheme, in which the binarising second-harmonic SYNC signal's amplitude is ramped up and down together with judicious amounts of noise added, works well. We present simulation results on a standard MAX-CUT benchmark set of 54 large problems, demonstrating not only that it finds the best-known results in many cases, but finds **better results than seem to have been previously published for 17 of the 54 problems**. We also demonstrate our method on the graph-colouring problem and present small (up to 32 CMOS oscillators) prototypes built on breadboard that function perfectly, testifying to the ease with which practical hardware implementations can be built.

Our scheme is different from previous Ising machine approaches, which are of 3 types (see Sect. 2): (1) a fibre-optic laser-based scheme known as the Coherent Ising Machine [4–6], (2) the D-WAVE quantum Ising machine [7,8] and (3) CMOS hardware accelerated simulated annealing chips for solving Ising problems [9,23–25]. Unlike CIM and D-WAVE, which are large, expensive and ill-suited to low-cost mass production, our approach is a purely classical scheme that does not rely on quantum phenomena or novel nano-devices. Indeed, it can be implemented using conventional CMOS electronics, which has many advantages: scalability/miniaturisability (*i.e.*, very large numbers of spins in a physically small system), well-established design processes and tools that essentially guarantee first-time working hardware, very low power operation, seamless integration with control and I/O logic, easy programmability via standard interfaces like USB, and low cost mass production. CMOS implementations of our scheme also allow complete flexibility in introducing controlled noise and programming SYNC ramping schedules. Furthermore, implementing oscillator coupling by physical connectivity makes our scheme inherently parallel, unlike CIM, where coupling is implemented via FPGA-based digital computation and is inherently serial. The advantages of CMOS also apply, of course, to hardware simulated annealing engines [9,23–25], but our scheme has additional attractive features. One key advantage relates to variability, a significant problem in nanoscale CMOS. For oscillator networks, device- and circuit-level variability impacts the system by causing a spread in the natural frequencies of the oscillators. Unlike other schemes, where performance deteriorates due to variability [9], we can essentially eliminate variability by means of simple VCO-based calibration to bring all the oscillators to the same frequency.[1] Another key potential advantage stems from the continuous/analog nature of our scheme (as opposed to purely digital simulated annealing schemes). Computational experiments indicate that the time our scheme takes to find good solutions of the Ising problem grows only very slowly with respect to the number of spins. This is a significant potential advantage over simulated annealing schemes [23] as hardware sizes scale up to large numbers of spins. Note that we can use virtually any type of nonlinear oscillator (not just CMOS) to implement our scheme, including optical, MEMS, biochemical, spin-based, *etc.*, oscillators; however, CMOS seems the easiest and most advantageous implementation route given the current state of technology.

In the remainder of this paper, we first provide a brief summary of the Ising problem and existing Ising machine schemes in Sect. 2. We then present our oscillator-based Ising machine scheme (dubbed OIM, for Oscillator Ising Machine) in Sect. 3, explaining the theory that enables it to work. Then in Sect. 4, we present both computational and hardware examples showing the effectiveness of our scheme for solving several combinatorial optimisation problems.

2 The Ising Problem and Existing Ising Machine Approaches

The Ising model is named after the German physicist Ernest Ising. It was first studied in the 1920s as a mathematical model for explaining domain formation in ferromagnets [1]. It comprises a group of discrete variables $\{s_i\}$, *aka* spins, each taking a binary value ± 1, such that an associated "energy function", known as the Ising Hamiltonian, is minimised:

$$\min \quad H \triangleq -\sum_{1 \le i < j \le n} J_{ij} s_i s_j - \sum_{i=1}^{n} h_i s_i, \quad \text{such that} \quad s_i \in \{-1, \ +1\}, \tag{1}$$

where n is the number of spins; $\{J_{ij}\}$ and $\{h_i\}$ are real coefficients.

[1] Moreover, as we show in Sect. 3.4, our scheme is inherently resistant to variability even without such calibration.

The Ising model is often simplified by dropping the $\{h_i\}$ terms. Under this simplification, the Ising Hamiltonian becomes

$$H = -\sum_{i,j,\ i<j} J_{ij} s_i s_j. \tag{2}$$

What makes the Ising model particularly interesting is that many hard optimisation problems can be shown to be equivalent to it [26]. In fact, all of Karp's 21 NP-complete problems can be mapped to it by assigning appropriate values to the coefficients [11]. Physical systems that can directly minimise the Ising Hamiltonian, namely Ising machines, thus become very attractive for potentially outperforming conventional algorithms run on CPUs for these problems.

Several schemes have been proposed recently for realising Ising machines in hardware. One well-known example is from D-Wave Systems [7,8]. Their quantum Ising machines use superconducting loops as spins and connect them using Josephson junction devices [27]. As the machines require a temperature below 80 mK (−273.07 °C) to operate [7], they all have a large footprint to accommodate the necessary cooling system. While many question their advantages over simulated annealing run on classical computers [28], proponents believe that through a mechanism known as quantum tunnelling, they can offer large speedups on problems with certain energy landscapes [29].

Other proposals use novel non-quantum devices as Ising spins instead, so that the machines can function at room temperature. Most notable among them is a scheme based on lasers and kilometre long optical fibres [4–6]. The Ising spins are represented using time-multiplexed optical parametric oscillators (OPOs), which are laser pulses travelling on the same fibre. The coupling between these pulses is implemented digitally by measurement and feedback using an FPGA. While these machines can potentially be more compact than D-Wave's machines, it is unclear how they can be miniaturised and integrated due to the use of long fibres. Recent studies have also proposed the use of several novel nanodevices as Ising spins, including MEMS (Micro-Electro-Mechanical Systems) resonators [30] and nanomagnets from Spintronics [31]. Physical realisation of these machines still awaits future development of these emerging device technologies.

Another broad direction is to build Ising model emulators using digital circuits. A recent implementation [9] uses CMOS SRAM cells as spins, and couples them using digital logic gates. The authors point out, however, that "the efficacy in achieving a global energy minimum is limited" [9] due to variability. The speed-up and accuracy reported by [9] are instead based on deterministic on-chip computation paired with an external random number generator—a digital hardware implementation of the simulated annealing algorithm. More recently, similar digital accelerators have also been tried on FPGAs [32]. These implementations are not directly comparable to the other Ising machine implementations discussed above, which attempt to use interesting intrinsic physics to minimise the Ising Hamiltonian for achieving large speedups.

3 Oscillator-Based Ising Machines

In this section, we show that a network of coupled self-sustaining oscillators can function as an Ising machine. To do so, we first study the response of a single oscillator under injection locking in Sect. 3.1. Specifically, we examine the way the oscillator's phase locks to that of the external input. While regular injection locking typically aligns the oscillator's phase with the input, as illustrated in Fig. 1(a) and (b), its variant—subharmonic injection locking (SHIL)—can make the oscillator develop multiple stable phase-locked states (Fig. 1(c) and (d)). As we show in Sect. 3.1, these phenomena can be predicted accurately using the Gen-Adler model [16].

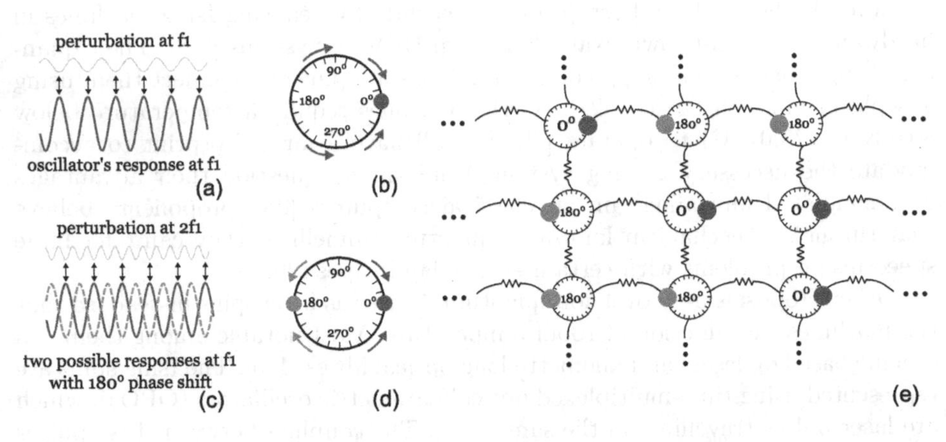

Fig. 1. Illustration of the basic mechanism of oscillator-based Ising machines: (a) oscillator shifts its natural frequency from f_0 to f_1 under external perturbation; (b) oscillator's phase becomes stably locked to the perturbation; (c) when the perturbation is at $2f_1$, the oscillator locks to its subharmonic at f_1; (d) bistable phase locks under subharmonic injection locking; (e) coupled subharmonically injection-locked oscillators settle with binary phases representing an optimal spin configuration for an Ising problem.

The Gen-Adler equation of a single oscillator, when extended to the phase dynamics of coupled oscillator networks, becomes equivalent to a variant of the Kuramoto model. In Sect. 3.2, we show that the model's dynamics are governed by a global Lyapunov function, a scalar "energy like" quantity that is naturally minimised by the coupled oscillator network. Then in Sect. 3.3, we introduce SHIL into the system to binarise the phases of oscillators. As illustrated in Fig. 1(e), SHIL induces each oscillator to settle to one of two stable phase-locked states. Due to the coupling between them, a network of such binarised oscillators will prefer certain phase configurations over others. We confirm this intuition in Sect. 3.3 by deriving a new Lyapunov function that such a system (*i.e.*, with SHIL) minimises. By examining this function's equivalence to the Ising Hamiltonian, we show that such a coupled oscillator network under SHIL indeed physically implements an Ising machine. Finally, in Sect. 3.4, we consider the effect of variability on the system's operation. We show that a spread in the natural frequencies of the oscillators contributes a linear term in the global Lyapunov function, and does affect Ising machine performance by much if the variability is not extreme.

3.1 Injection Locking in Oscillators

When an oscillator with a natural frequency ω_0 is perturbed by a small periodic input at a similar frequency ω_1, its phase response can be predicted well using the Generalised Adler's model (Gen-Adler) [16]. Gen-Adler has the following form:

$$\frac{d}{dt}\phi(t) = \omega_0 - \omega_1 + \omega_0 \cdot c(\phi(t) - \phi_{in}), \tag{3}$$

where $\phi(t)$ and ϕ_{in} are the phases of the oscillator and the perturbation. $c(.)$ is a 2π-periodic function derived based on an intrinsic quantity of the oscillator known as the Phase Response Curve (PRC) [33] or the Perturbation Projection Vector (PPV) [34]. A detailed derivation of Gen-Adler from the low-level differential equations of an oscillator is provided in a preprint of this paper [35].

The Gen-Adler equation governs the dynamics of the oscillator's phase under periodic inputs; its equilibrium states can be used to accurately predict the injection-locked states of the oscillator. The equilibrium Gen-Adler equation can be derived by rearranging (3):

$$\frac{\omega_1 - \omega_0}{\omega_0} = c(\phi^* - \phi_{in}). \tag{4}$$

where ϕ^* is the solution of phase in equilibrium.

The Left Hand Side (LHS) of (4) is a constant representing the frequency detuning of the oscillator from the input; the Right Hand Side (RHS) is a periodic function of ϕ^*; its magnitude depends on both the PPV of the oscillator and the strength of the input [16]. By plotting both terms and looking for intersections, one can easily predict whether injection locking will occur, and if it does, what the locked phase of the oscillator ϕ^* will be. Figure 2(a) plots a few examples of LHS/RHS, showing their shapes and magnitudes under different conditions.

As mentioned in Sect. 1, SHIL can occur when the external input is about twice as fast as the oscillator. When the input is at frequency $2\omega_1$, it can be shown that the corresponding $c(.)$ becomes a π-periodic function [15,36]; a typical example is given in Fig. 2(b), where $c(.)$ takes the shape of $-\sin(2\phi)$. In this case, two of the four LHS-RHS intersections represent stable phase-locked states; it can be shown that they are separated by a phase difference of 180° [15]. Gen-Adler is a powerful technique for predicting and understanding injection locking in oscillators and constitutes an important foundation for the analyses that follow.

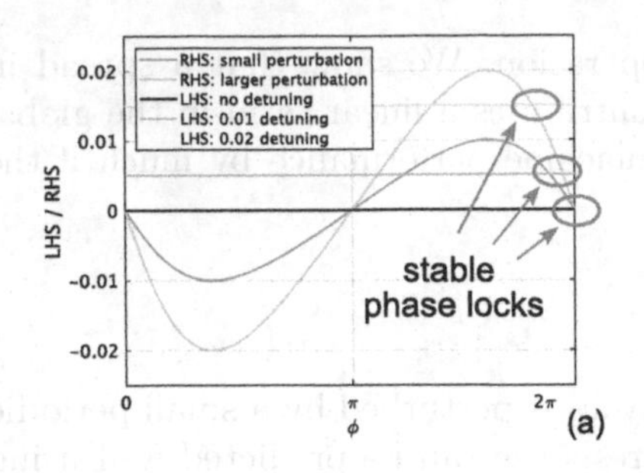

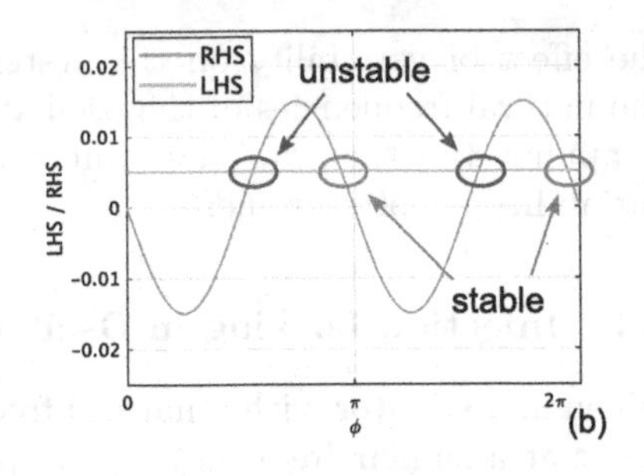

Fig. 2. Illustration of the LHS and RHS of the equilibrium Gen-Adler equation. (a) Under normal injection locking, the intersection of LHS and RHS predicts the only solution of ϕ under different scenarios. (b) Perturbation at $2\omega_1$ changes the shape of $c(.)$ in Gen-Adler; the intersections now predict the locations of two stable phase-locked states.

3.2 Global Lyapunov Function

For an oscillator in a coupled oscillator network, its external perturbations come from the other oscillators connected to it. Its Gen-Adler equation can be written as

$$\frac{d}{dt}\phi_i(t) = \omega_i - \omega^* + \omega_i \cdot \sum_{j=1,\ j\neq i}^{n} c_{ij}(\phi_i(t) - \phi_j(t)), \tag{5}$$

where $\{\phi_i\}$ represents the phases of n oscillators; ω_i is the frequency of the oscillator whereas ω^* is the central frequency of the network. $c_{ij}(.)$ is a 2π-periodic function for the coupling between oscillator i and oscillator j.

To simplify exposition, we now assume that the c_{ij} functions are sinusoidal, although in [35], we show that this does not have to be the case for the analysis to hold true.[2] We further assume zero spread in the natural frequencies of oscillators, *i.e.*, $\omega_i \equiv \omega^*$, and discuss the effect of frequency variability later in Sect. 3.4. With these simplifications, (5) can be written as

$$\frac{d}{dt}\phi_i(t) = -K \cdot \sum_{j=1,\ j\neq i}^{n} J_{ij} \cdot \sin(\phi_i(t) - \phi_j(t)). \tag{6}$$

Here, we are using the coefficients $\{J_{ij}\}$[3] from the Ising model (1) to set the connectivity of the network, *i.e.*, the coupling strength between oscillators i and j is proportional to J_{ij}. The parameter K modulates the overall coupling strength of the network.

There is a global Lyapunov function associated with (6) [37]:

$$E(\vec{\phi}(t)) = -K \cdot \sum_{i,j,\ i\neq j} J_{ij} \cdot \cos(\phi_i(t) - \phi_j(t)), \tag{7}$$

[2] More generally, c_{ij}s can be any 2π-periodic odd functions, which are better suited to practical oscillators.

[3] In the Ising Hamiltonian (1), J_{ij} is only defined when $i < j$; here we assume that $J_{ij} = J_{ji}$ for all $i,\ j$.

where $\vec{\phi}(t) = [\phi_1(t), \cdots, \phi_n(t)]^T$. Being a global Lyapunov function, it is an objective function the coupled oscillator system always tends to minimise as it evolves over time [38].

If we look at the values of this continuous function $E(\vec{\phi}(t))$ at some discrete points, we notice that it shares some similarities with the Ising Hamiltonian. At points where every ϕ_i is equal to either 0 or π,[4] if we map $\phi_i = 0$ to $s_i = +1$ and $\phi_i = \pi$ to $s_i = -1$, we have

$$E(\vec{\phi}(t)) = -K \cdot \sum_{i,j,\ i \neq j} J_{ij} \cdot \cos(\phi_i(t) - \phi_j(t)) = -K \cdot \sum_{i,j,\ i \neq j} J_{ij} s_i s_j = -2K \cdot \sum_{i,j,\ i<j} J_{ij} s_i s_j. \quad (8)$$

If we choose $K = 1/2$, the global Lyapunov function in (7) exactly matches the Ising Hamiltonian in (2) at these discrete points. But this does not mean that coupled oscillators are naturally minimising the Ising Hamiltonian, as there is no guarantee at all that the phases $\{\phi_i(t)\}$ are settling to these discrete points. In fact, networks with more than two oscillators almost always synchronise with analog phases, *i.e.*, $\{\phi_i(t)\}$ commonly settle to continuous values spread out in the phase domain as opposed to converging towards 0 and π. As an example, Fig. 3(a) shows the phase responses of 20 oscillators connected in a random graph. As phases do not settle to the discrete points discussed above, the Lyapunov function they minimise becomes irrelevant to the Ising Hamiltonian, rendering the system ineffective for solving Ising problems. While one may think that the analog phases can still serve as solutions when rounded to the nearest discrete points, experiments in Sect. 4.2 show that the quality of these solutions is very poor compared with our scheme of Ising machines proposed in this paper.

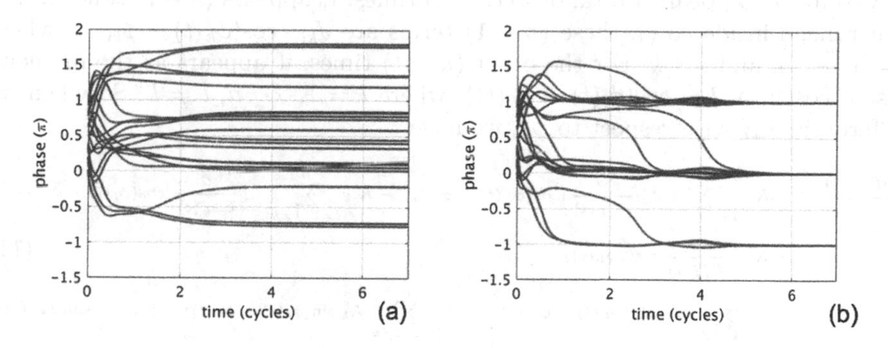

Fig. 3. Phases of 20 oscillators with random $\{J_{ij}\}$ generated by `rudy -rnd_graph 20 50 10001`: (a) without SYNC; (b) with $K_s = 1$.

[4] More generally, we can use $\{2k\pi \mid k \in \mathbf{Z}\}$ and $\{2k\pi + \pi \mid k \in \mathbf{Z}\}$ to represent the two states for each oscillator's phase.

3.3 Network of Coupled Oscillators Under SHIL and Its Global Lyapunov Function

In our scheme, a common SYNC signal at $2\omega^*$ is injected to every oscillator in the network. Through the mechanism of SHIL, the oscillator phases are binarised. The example shown in Fig. 3(b) confirms that this is indeed the case: under SHIL, the phases of 20 oscillators connected in the same random graph now settle very close to discrete points. To write the model for such a system, we recall from Sect. 3.1 that a $2\omega^*$ perturbation introduces a π-periodic coupling term (*e.g.*, $\sin(2\phi)$) in the phase dynamics. Therefore, we directly write the model as follows and show its derivation in [35].

$$\frac{d}{dt}\phi_i(t) = -K \cdot \sum_{j=1,\ j\neq i}^{n} J_{ij} \cdot \sin(\phi_i(t) - \phi_j(t)) - K_s \cdot \sin(2\phi_i(t)), \tag{9}$$

where K_s represents the strength of coupling from SYNC.

Remarkably, there is a global Lyapunov function for this new type of coupled oscillator system. It can be written as

$$E(\vec{\phi}(t)) = -K \cdot \sum_{i,j,\ i\neq j} J_{ij} \cdot \cos(\phi_i(t) - \phi_j(t)) - K_s \cdot \sum_{i=1}^{n} \cos\left(2\phi_i(t)\right). \tag{10}$$

Now, we show that E in (10) is indeed a global Lyapunov function. To do so, we first differentiate E with respect to $\vec{\phi}$. We observe that the first component of E is the sum of $(n^2 - n)$ number of cos() terms. Among them, for any given index k, variable ϕ_k appears a total of $2 \cdot (n-1)$ times. It appears $(n-1)$ times as the subtrahend inside cos(): these $(n-1)$ terms are $J_{kl} \cdot \cos(\phi_k(t) - \phi_l(t))$, where $l = 1, \cdots, n$ and $l \neq k$. For the other $(n-1)$ times, it appears as the minuend inside cos(): in $J_{lk} \cdot \cos(\phi_l(t) - \phi_k(t))$, where $l = 1, \cdots, n$, $l \neq k$. So when we differentiate E with respect to ϕ_k, we have

$$\frac{\partial E(\vec{\phi}(t))}{\partial \phi_k(t)} = -K \cdot \sum_{l=1,\ l\neq k}^{n} J_{kl} \frac{\partial}{\partial \phi_k(t)} [\cos(\phi_k(t) - \phi_l(t))] - K \cdot \sum_{l=1,\ l\neq k}^{n} J_{lk} \frac{\partial}{\partial \phi_k(t)} [\cos(\phi_l(t) - \phi_k(t))]$$
$$- K_s \cdot \frac{\partial}{\partial \phi_k(t)} \cos(2\phi_k(t)) \tag{11}$$

$$= K \cdot \sum_{l=1,\ l\neq k}^{n} J_{kl} \sin(\phi_k(t) - \phi_l(t)) - K \cdot \sum_{l=1,\ l\neq k}^{n} J_{lk} \sin(\phi_l(t) - \phi_k(t)) + K_s \cdot 2 \cdot \sin(2\phi_k(t)) \tag{12}$$

$$= K \cdot \sum_{l=1,\ l\neq k}^{n} J_{kl} \cdot 2 \cdot \sin(\phi_k(t) - \phi_l(t)) + K_s \cdot 2 \cdot \sin(2\phi_k(t)) \tag{13}$$

(using $\sin(x) = -\sin(-x)$ and $J_{lk} = J_{kl}$)

$$= -2 \cdot \frac{d\phi_k(t)}{dt}. \tag{14}$$

Therefore,

$$\frac{\partial E(\vec{\phi}(t))}{\partial t} = \sum_{k=1}^{n} \left[\frac{\partial E(\vec{\phi}(t))}{\partial \phi_k(t)} \cdot \frac{d\phi_k(t)}{dt} \right] \tag{15}$$

$$= -2 \cdot \sum_{k=1}^{n} \left(\frac{d\phi_k(t)}{dt} \right)^2 \leq 0. \tag{16}$$

Thus, we have proved that (10) is indeed a global Lyapunov function the coupled oscillators under SHIL naturally minimise over time. A similar but more detailed proof for the general case where we do not assume sinusoidal coupling functions is given in [35].

At the discrete points (phase values of $0/\pi$), because $\cos(2\phi_i) \equiv 1$, (10) reduces to

$$E(\vec{\phi}(t)) \approx -K \cdot \sum_{i,j,\ i \neq j} J_{ij} \cdot \cos(\phi_i(t) - \phi_j(t)) - n \cdot K_s, \tag{17}$$

where $n \cdot K_s$ is a constant. By choosing $K = 1/2$, we can then make (17) equivalent to the Ising Hamiltonian in (2) with a constant offset.

Note that the introduction of SYNC does not change the relative E levels between the discrete points, but modifies them by the *same* amount. However, with SYNC, all phases can be forced to eventually take values near either 0 or π—the system now tries to reach a binary state that minimises the Ising Hamiltonian, thus functioning as an Ising machine. We emphasise that this is *not* equivalent to running the system without SHIL and then rounding the analog phase solutions to discrete values as a post-processing step. Instead, the introduction of SHIL modifies the energy landscape of E, changes the dynamics of the coupled oscillator system, and as we show in Sect. 4, results in greatly improved minimisation of the Ising Hamiltonian.

Our approach here is analogous to several existing schemes to map a system's Lyapunov function to a minimisation objective, including the Hopfield-Tank neural network proposed for the travelling salesman problem [39], and the more recent work on designing differential equations for satisfiability (SAT) problems [40,41]. Based on the equivalence established above of the Ising Hamiltonian and the Lyapunov function (10) of coupled oscillators, our OIM can be used on a wide class of problems with an Ising formulation [11], including the problems addressed in these other schemes.

It is worth noting, also, that the Lyapunov function in (10) will, in general, have many local minima and there is no guarantee the oscillator-based Ising machine will settle at or near any global optimal state. However, as we show in [35], when judicious amounts of noise are introduced via a noise level parameter K_n, it becomes more likely to settle to lower minima. Indeed, the several parameters in the Ising machine—K, K_s and K_n—all play an important role in its operation and should be given suitable values. Furthermore, K, K_s, K_n can also be time varying, creating various "annealing schedules". As we show in

Sect. 4, this feature gives us considerable flexibility in operating oscillator-based Ising machines for good performance.

3.4 Coupled Oscillator Networks with Frequency Variations

A major obstacle to the practical implementation of large-scale Ising machines is variability. While few analyses exist for assessing the effects of variability for previous Ising machine technologies (Sect. 2), the effect of variability on our oscillator-based Ising machine scheme is easy to analyse, predicting that performance degrades gracefully.

One very attractive feature of oscillators is that variability, regardless of the nature and number of elemental physical sources, eventually manifests itself essentially in only one parameter, namely the oscillator's natural frequency. As a result, the effect of variability in an oscillator network is that there is a spread in the natural frequencies of the oscillators. Taking this into consideration, our model can be revised as

$$\frac{d}{dt}\phi_i(t) = \omega_i - \omega^* - \omega_i \cdot K \cdot \sum_{j=1,\ j\neq i}^{n} J_{ij} \cdot \sin(\phi_i(t) - \phi_j(t)) - \omega_i \cdot K_s \cdot \sin(2\phi_i(t)). \quad (18)$$

As it turns out, there is also a global Lyapunov function associated with this system.

$$E(\vec{\phi}(t)) = -K \cdot \sum_{i,j,\ i\neq j} J_{ij} \cdot \cos(\phi_i(t) - \phi_j(t)) - K_s \cdot \sum_{i=1}^{n} \cos\left(2\phi_i(t)\right) - 2\sum_{i=1}^{n} \frac{\omega_i - \omega^*}{\omega_i}\phi_i. \quad (19)$$

This can be proven as follows.

$$\frac{\partial E(\vec{\phi}(t))}{\partial \phi_k(t)} = K \cdot \sum_{l=1,\ l\neq k}^{n} J_{kl} \cdot 2 \cdot \sin(\phi_k(t) - \phi_l(t)) + K_s \cdot 2 \cdot \sin(2\phi_k(t)) - 2\frac{\omega_k - \omega^*}{\omega_k} \quad (20)$$

$$= -\frac{2}{\omega_k} \cdot \frac{d\phi_k(t)}{dt}. \quad (21)$$

Therefore,

$$\frac{dE(\vec{\phi}(t))}{dt} = -\sum_{k=1}^{n} \frac{2}{\omega_k}\left(\frac{d\phi_k(t)}{dt}\right)^2 \leq 0. \quad (22)$$

Note that (19) differs from (10) only by a weighted sum of ϕ_i—it represents essentially the same energy landscape but tilted linearly with the optimisation variables. While it can still change the locations and values of the solutions, its effects are easy to analyse given a specific combinatorial optimisation problem. Also, as the coupling coefficient K gets larger, the effect of variability can be reduced. Small amounts of variability merely perturb the locations of minima a little, *i.e.*, the overall performance of the Ising machine remains essentially unaffected. Very large amounts of variability can, of course, eliminate minima that would exist if there were no variability. However, another great advantage of using oscillators is that even in the presence of large variability, the oscillator frequencies can be calibrated (*e.g.*, using a voltage-controlled oscillator (VCO) scheme) prior to each run of the machine. As a result, the spread in frequencies can be essentially eliminated in a practical and easy-to-implement way.

4 Examples

In this section, we demonstrate the feasibility and efficacy of our oscillator-based Ising machine scheme by applying it to several MAX-CUT examples and a graph-colouring problem.

4.1 Small MAX-CUT Problems

Given an undirected graph, the MAX-CUT problem [13,42] asks us to find a subset of vertices such that the total weights of the cut set between this subset and the remaining vertices are maximised. As an example, Fig. 4 shows a size-8 cubic graph, where each vertex is connected to three others—neighbours on both sides and the opposing vertex. As shown in Fig. 4, dividing the 8 vertices randomly yields a cut size of 5; grouping even and odd vertices, which one may think is the best strategy, results in a cut size of 8; the maximum cut is actually 10, with one of the solutions shown in the illustration. Changing the edge weights to non-unit values can change the maximum cut and also make the solution look less regular, often making the problem more difficult to solve. While the problem may not seem challenging at size 8, it quickly becomes intractable as the size of the graph grows. In fact, MAX-CUT is one of Karp's 21 NP-complete problems [10].

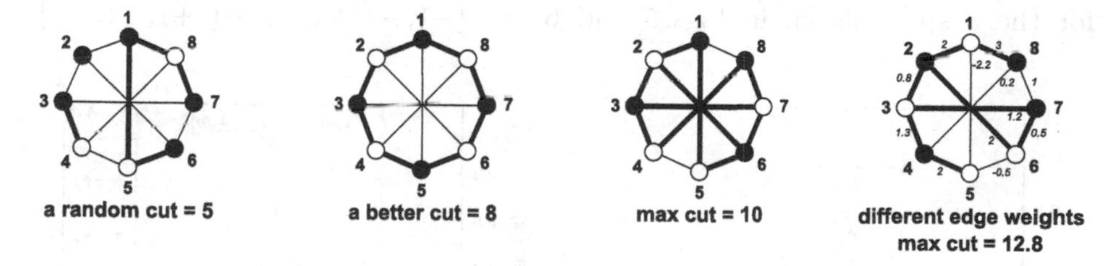

Fig. 4. Illustration of different cut sizes in a 8-vertex cubic graph with unit edge weights, and another one with random weights (rightmost).

The MAX-CUT problem has a direct mapping to the Ising model [10], by choosing J_{ij} to be the opposite of the weight of the edge between vertices i and j, *i.e.*, $J_{ij} = -w_{ij}$. To explain this mapping scheme, we can divide the vertices into two sets—V_1 and V_2. Accordingly, all the edges in the graph are separated into three groups—those that connect vertices within V_1, those within V_2, and the cut set containing edges across V_1 and V_2. The sums of the weights in these three sets are denoted by S_1, S_2 and S_{cut}. Together, they constitute the total edge weights of the graph, which is also the negation of the sum of all the J_{ij}s:

$$S_1 + S_2 + S_{cut} = \sum_{i,j,\ i<j} w_{ij} = - \sum_{i,j,\ i<j} J_{ij}. \tag{23}$$

We then map this division of vertices to the values of Ising spins, assigning +1 to a spin i if vertex $v_i \in V_1$, and −1 if the vertex is in V_2. The Ising Hamiltonian in (2) can then be calculated as

$$\begin{aligned} H &= \sum_{i,j,\ i<j} J_{ij} s_i s_j \\ &= \sum_{i<j,\ v_i,v_j \in V_1} J_{ij}(+1)(+1) + \sum_{i<j,\ v_i,v_j \in V_2} J_{ij}(-1)(-1) + \sum_{i<j,\ v_i \in V_1, v_j \in V_2} J_{ij}(+1)(-1) \\ &= \sum_{i<j,\ v_i,v_j \in V_1} J_{ij} + \sum_{i<j,\ v_i,v_j \in V_2} J_{ij} - \sum_{i<j,\ v_i \in V_1, j \in V_2} J_{ij} \\ &= -(S_1 + S_2 - S_{cut}) = \sum_{i,j,\ i<j} J_{ij} - 2 \cdot S_{cut}. \end{aligned} \tag{24}$$

Therefore, when the Ising Hamiltonian is minimised, the cut size is maximised.

To show that an oscillator-based Ising machine can indeed be used to solve MAX-CUT problems, we simulated the Kuramoto model in (9) while making the J_{ij}s represent the unit-weight cubic graph in Fig. 4. The magnitude of SYNC is fixed at $K_s = 3$, while we ramp up the coupling strength K from 0 to 5. Results from the deterministic model ($K_n = 0$) and the stochastic model ($K_n = 0.1$) are shown in Figs. 5 and 6 respectively. In the simulations, oscillators started with random phases between 0 and π; after a while, they all settled to one of the two phase-locked states separated by π. These two groups of oscillators represent the two subsets of vertices in the solution. The results for the 8 spins shown in Figs. 5 and 6 are $\{+1, -1, +1, -1, -1, +1, -1, +1\}$

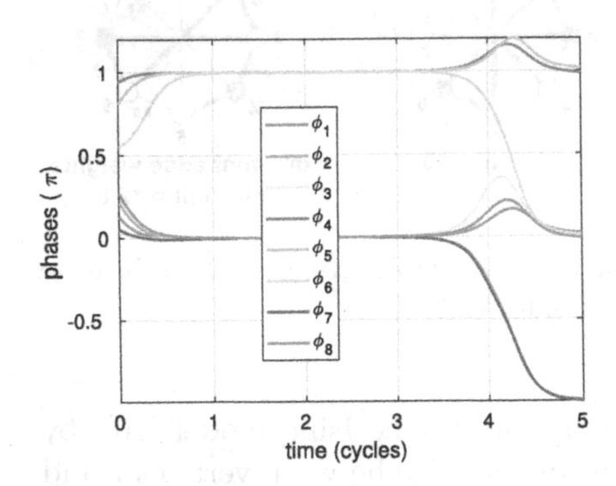

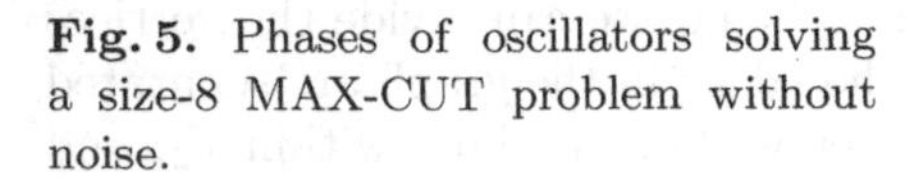

Fig. 5. Phases of oscillators solving a size-8 MAX-CUT problem without noise.

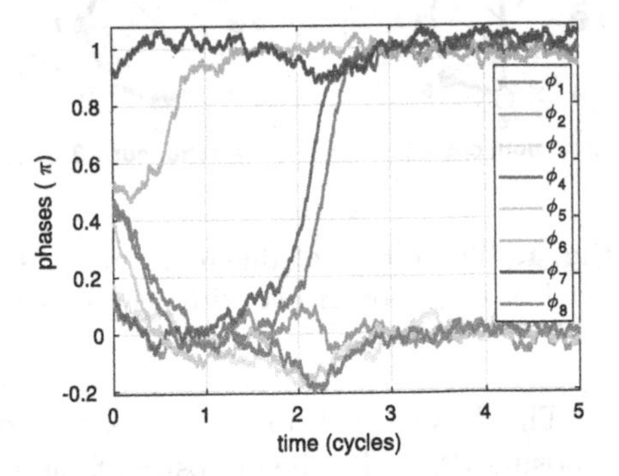

Fig. 6. Phases of oscillators solving a size-8 MAX-CUT problem with noise.

and $\{-1, +1, +1, -1, +1, -1, -1, +1\}$ respectively; both are globally optimal solutions.

A minimal code for reproducing these results is show in [35]. Note that these are simulations on stochastic differential equations with random initial conditions. Every run will return different waveforms; there is no guarantee that the global optimum will be reached on every run.

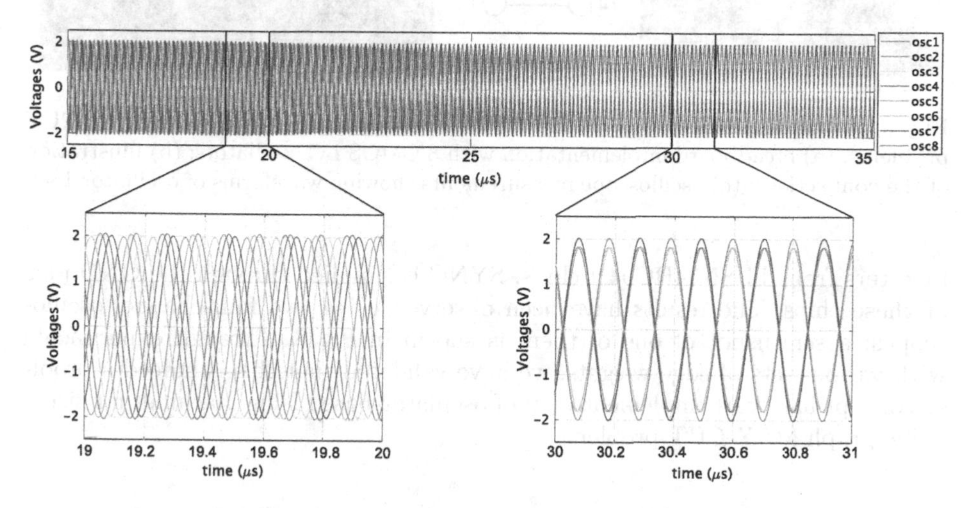

Fig. 7. Simulation results from ngspice on 8 coupled oscillators.

We have also directly simulated coupled oscillators at the SPICE level to confirm the results obtained on phase macromodels. Such simulations are at a lower level than phase macromodels and are less efficient. But they are closer to physical reality and are useful for circuit design. In the simulations, 8 cross-coupled LC oscillators are tuned to a frequency of 5 MHz. They are coupled through resistors, with conductances proportional to the coupling coefficients; in this case, we use $J_{ij} \cdot 1/100\,\text{k}\Omega$. Results from transient simulation using ngspice-28 are shown in Fig. 7. The 8 oscillators' phases settle into two groups {1, 4, 6, 7} and {2, 3, 5, 8}, representing one of the optimal solutions for the MAX-CUT problem. They synchronise within 20 μs after oscillation starts, which is about 100 cycles. We have tried this computational experiment with different random initial conditions; like phase-macromodels, the SPICE-level simulations of these coupled oscillators reliably return optimal solutions for this size-8 MAX-CUT problem.

We have also implemented these 8 coupled LC oscillators on a breadboard; a photo of it is shown in Fig. 8. The inductance of the LC oscillators is provided by fixed inductors of size 33 μH. The capacitance is provided by trimmer capacitors with a maximum value of 50 pF; we have tuned them to around 30 pF such that the natural frequencies of all oscillators are about 5 MHz. The nonlinearity for sustaining the LC oscillation is implemented by cross-coupled CMOS

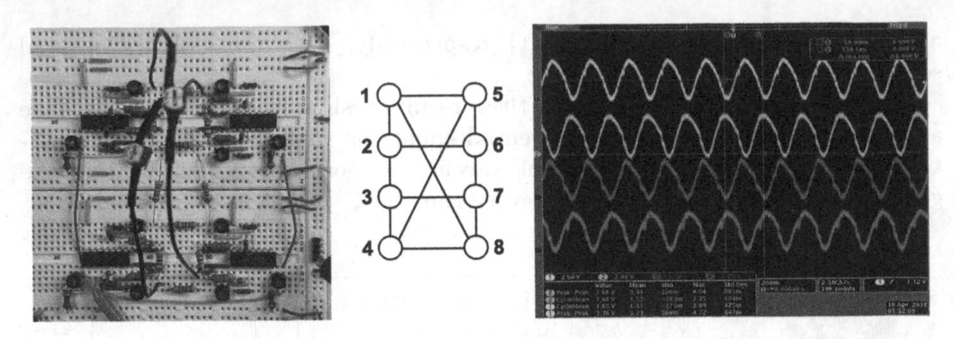

Fig. 8. A simple oscillator-based Ising machine solving size-8 cubic graph MAX-CUT problems: (a) breadboard implementation with 8 CMOS LC oscillators; (b) illustration of the connections; (c) oscilloscope measurements showing waveforms of oscillator 1∼4.

inverters from TI SN74HC04N chips. SYNC is supplied through the GND pins of these chips. The results have been observed using two four-channel oscilloscopes; a screenshot of one of them is shown in Fig. 8. Through experiments with various sets of edge weights, we have validated that this is indeed a proof-of-concept hardware implementation of oscillator-based Ising machines for size-8 cubic-graph MAX-CUT problems.

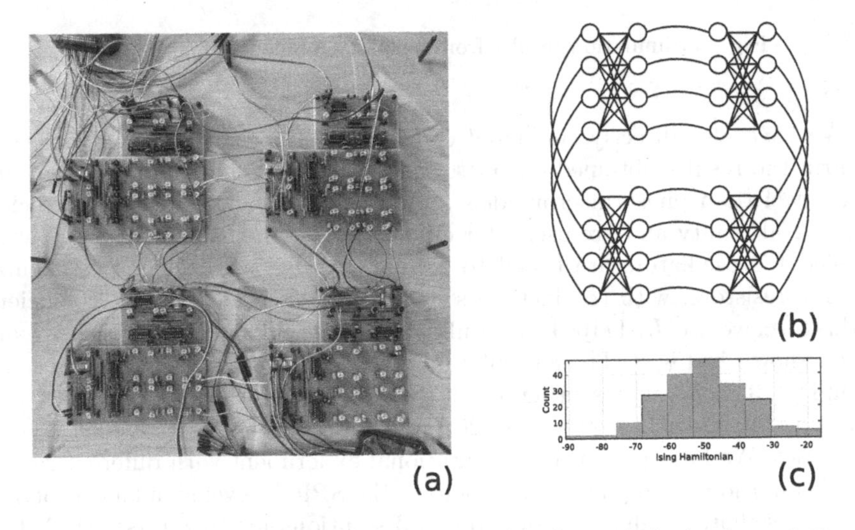

Fig. 9. A size-32 oscillator-based Ising machine: (a) photo of the implementation on perfboards; (b) illustration of the connectivity; (c) a typical histogram of the energy values achieved in 200 runs on a random size-32 Ising problem; the lowest energy level is -88 and is achieved once in this case.

Using the same type of oscillators, we have built hardware Ising machines of larger sizes. Figure 9 shows a size-32 example implementing a type of connectivity known as the Chimera graph, much like the quantum Ising machines manufactured by D-Wave Systems. In this graph, oscillators are organised into groups of 8, with denser connections within the groups and sparse ones in between. The hardware is on perfboards, with components soldered on the boards so that the setup is more permanent than those on breadboards. Connections are implemented using rotary potentiometers. Next to each potentiometer we have designed male pin connectors soldered on the board such that the polarity of each connection can be controlled by shorting different pins using female jumper caps. When encoding Ising problems, we have also colour-coded the jumper caps to make debugging easier, as can be seen in the photo as red and green dots next to the four arrays of white round potentiometers. To read the phases of the oscillators, instead of using multichannel oscilloscopes, we have soldered TI SN74HC86N Exclusive-OR (XOR) gate chips on board. The XOR operation of an oscillator's response and a reference signal converts the oscillating waveform into a high or low voltage level, indicating if the oscillator's phase is aligned with or opposite to the reference phase. The voltage level can then be picked up by a multichannel logic analyzer. The entire setup is powered by two Digilent Analog Discovery 2 devices, which are portable USB devices that integrate power supplies, logic analyzers and function generators. We have tried random Ising problems by programming each connection with a random polarity using the jumper caps. A typical histogram of the Ising Hamiltonians achieved is shown in Fig. 9(c). Note that because J_{ij}s have random polarities, a random solution would have an average energy level of zero. In comparison, the results measured from the hardware are always below 0, and sometimes achieve the global minimum. While such a hand-soldered system is nontrivial to assemble and operate, and its size of 32 cannot be characterised as large scale, it is a useful proof of concept for implementing oscillator-based Ising machines using standard CMOS technologies, and serves as a very solid basis for our future plans to scale the implementations with custom PCBs and custom ICs.

4.2 MAX-CUT Benchmark Problems

In this section, we demonstrate the efficacy of oscillator-based Ising machines for solving larger-scale MAX-CUT problems. Specifically, we have run simulations on all the problems in a widely used set of MAX-CUT benchmarks known as the G-set [43].[5] Problem sizes range from 800 to 3000.[6] In the experiments, we operated the Ising machine for all the problems with a single annealing schedule, *i.e.*, we did not tune our Ising machine parameters individually for different problems. Each problem was simulated with 200 random instances. In Table 1, we list the results and runtime alongside those from several heuristic algorithms developed for MAX-CUT—Scatter Search (SS) [44], CirCut [45], and Variable

[5] The G-set problems are available for download as set1 at http://www.optsicom.es/maxcut.

[6] G1∼21 are of size 800; G22∼42 are of size 2000; G43∼47, G51∼54 are of size 1000; G48∼50 are of size 3000.

Neighbourhood Search with Path Relinking (VNSPR) [13].[7] We also list the performances of simulated annealing from a recent study [42], the only one we were able to find that contains results for all the G-set problems.

From Table 1, we observe that our oscillator-based Ising machine is indeed effective—it finds best-known cut values for 38 out of the 54 problems, 17 of which are even better than those reported in the above literature. Moreover, in the 200 random instances, the best cut is often reached more than once—the average n_{max} for all benchmarks is 20 out of 200. If we relax the objective and look at the number of instances where 99.9% of the cut value is reached, represented by $n_{0.999}$, the average is 56, more than a quarter of the total trials. The results can in fact be improved further if we tailor the annealing schedule for each problem. But to show the effectiveness and generality of our scheme, we have chosen to use the same annealing schedule for all the problems.

In the annealing schedule we used, the coupling strength K increases linearly, the noise level K_n steps up from 0 to 1, while SYNC's amplitude K_s ramps up and down multiple times. Such a schedule was chosen empirically and appears to work well for most G-set problems. Figure 10 shows the behaviour of oscillator phases and the instantaneous cut values under this schedule for solving benchmark problem G1 to its best-known cut size. Some MATLAB® code to illustrate the annealing schedule is shown in [35]. The code uses MATLAB®'s SDE solver and is thus much slower than an implementation in C++ we used to generate the results in Table 1. We plan to release all our code as open-source software so that others can verify and build on our work.

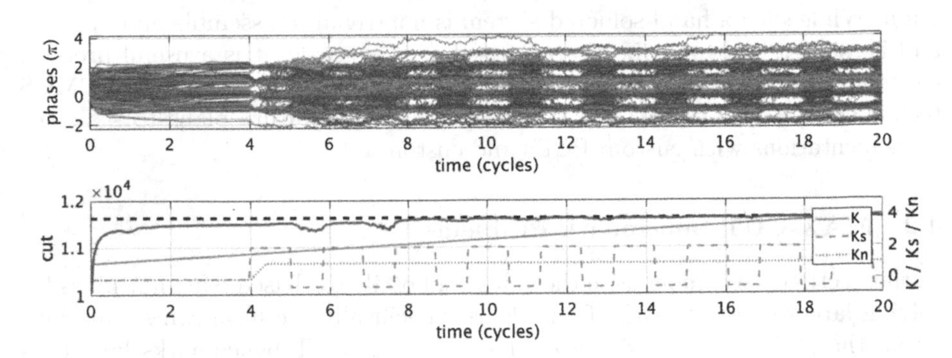

Fig. 10. Coupled oscillators solving MAX-CUT benchmark problem G1 [43] to its best-known cut size 11624.

The fact that we were using a fixed schedule also indicates that the actual hardware time for the Ising machine to solve all these benchmarks is the same, regardless of problem size and connectivity. Note that in Fig. 10, the end time 20 means 20 oscillation cycles, but this end time is predicated on a coupling

[7] Their results and runtime are available for download at http://www.optsicom.es/maxcut in the "Computational Experiences" section.

Table 1. Results of oscillator-based Ising machines run on MAX-CUT benchmarks in the G-set, compared with several heuristic algorithms. Time reported in this table is for a single run. n_{max} is the number of runs out of the 200 trials where the cut reaches the maximum of 200; $n_{0.999}$ is the number for it to reach 99.9% of the maximum.

Benchmark	SS	Time	CirCut	Time	VNSPR	Time	SA	Time	OIM	Time	n_{max}	$n_{0.999}$
G1	**11624**	139	**11624**	352	11621	22732	11621	295	**11624**	52.6	43	123
G2	**11620**	167	11617	283	11615	22719	11612	327	**11620**	52.7	1	87
G3	**11622**	180	**11622**	330	**11622**	23890	11618	295	**11622**	52.4	10	117
G4	**11646**	194	11641	524	11600	24050	11644	294	**11646**	52.7	20	133
G5	**11631**	205	11627	1128	11598	23134	11628	300	**11631**	52.6	3	121
G6	2165	176	**2178**	947	2102	18215	**2178**	247	**2178**	52.8	4	46
G7	1982	176	2003	867	1906	17716	**2006**	205	2000	52.9	17	21
G8	1986	195	2003	931	1908	19334	**2005**	206	2004	52.8	2	26
G9	2040	158	2048	943	1998	15225	**2054**	206	**2054**	52.6	2	2
G10	1993	210	1994	881	1910	16269	1999	205	**2000**	52.9	21	58
G11	562	172	560	74	**564**	10084	**564**	189	**564**	6.7	6	6
G12	552	242	552	58	**556**	10852	554	189	**556**	6.3	25	25
G13	578	228	574	62	580	10749	580	195	**582**	6.4	3	3
G14	3060	187	3058	128	3055	16734	**3063**	252	3061	14.6	27	91
G15	**3049**	143	**3049**	155	3043	17184	**3049**	220	**3049**	16.1	41	145
G16	3045	162	3045	142	3043	16562	3050	219	**3052**	14.5	8	53
G17	3043	313	3037	366	3030	18555	3045	219	**3046**	14.6	5	52
G18	988	174	978	497	916	12578	**990**	235	**990**	14.7	3	3
G19	903	128	888	507	836	14546	904	196	**906**	14.5	13	13
G20	**941**	191	**941**	503	900	13326	**941**	195	**941**	14.7	160	160
G21	930	233	**931**	524	902	12885	927	195	**931**	14.6	10	10
G22	13346	1336	13346	493	13295	197654	13158	295	**13356**	58.7	3	93
G23	13317	1022	13317	457	13290	193707	13116	288	**13333**	58.6	8	54
G24	13303	1191	13314	521	13276	195749	13125	289	**13329**	59.0	6	23
G25	13320	1299	**13326**	1600	12298	212563	13119	316	**13326**	58.7	6	45
G26	13294	1415	**13314**	1569	12290	228969	13098	289	13313	58.9	4	81
G27	3318	1438	3306	1456	3296	35652	**3341**	214	3323	59.0	18	24
G28	3285	1314	3260	1543	3220	38655	**3298**	252	3285	61.2	1	5
G29	3389	1266	3376	1512	3303	33695	3394	214	**3396**	58.9	2	8
G30	3403	1196	3385	1463	3320	34458	**3412**	215	3402	59.0	12	16
G31	3288	1336	3285	1448	3202	36658	**3309**	214	3296	59.1	5	15
G32	1398	901	1390	221	1396	82345	**1410**	194	1402	17.5	5	5
G33	1362	926	1360	198	**1376**	76282	**1376**	194	1374	15.9	1	1
G34	1364	950	1368	237	1372	79406	**1382**	194	1374	15.9	24	24
G35	7668	1258	7670	440	7635	167221	7485	263	**7675**	37.1	5	29
G36	7660	1392	7660	400	7632	167203	7473	265	**7663**	37.6	3	58
G37	7664	1387	7666	382	7643	170786	7484	288	**7679**	37.8	1	15
G38	**7681**	1012	7646	1189	7602	178570	7479	264	7679	37.7	7	18

(*continued*)

Table 1. (*continued*)

Benchmark	SS	Time	CirCut	Time	VNSPR	Time	SA	Time	OIM	Time	n_{max}	$n_{0.999}$
G39	2393	1311	2395	852	2303	42584	**2405**	209	2404	37.2	1	1
G40	2374	1166	2387	901	2302	39549	2378	208	**2389**	38.1	7	7
G41	2386	1017	2398	942	2298	40025	**2405**	208	2401	37.8	20	71
G42	2457	1458	**2469**	875	2390	41255	2465	210	**2469**	37.3	4	4
G43	6656	406	6656	213	6659	35324	6658	245	**6660**	29.1	17	129
G44	**6648**	356	6643	192	6642	34519	6646	241	**6648**	29.2	21	129
G45	6642	354	6652	210	6646	34179	6652	241	**6653**	29.1	10	53
G46	6634	498	6645	639	6630	38854	6647	245	**6649**	29.1	9	13
G47	6649	359	**6656**	633	6640	36587	6652	242	**6656**	29.1	16	91
G48	**6000**	20	**6000**	119	**6000**	64713	**6000**	210	**6000**	23.2	194	194
G49	**6000**	35	**6000**	134	**6000**	64749	**6000**	210	**6000**	23.2	180	180
G50	**5880**	27	**5880**	231	**5880**	147132	5858	211	5874	25.6	10	94
G51	**3846**	513	3837	497	3808	89966	3841	234	**3846**	18.4	23	68
G52	**3849**	551	3833	507	3816	95985	3845	228	3848	18.4	10	49
G53	**3846**	424	3842	503	3802	92459	3845	230	**3846**	18.4	9	102
G54	3846	429	3842	524	3820	98458	3845	228	**3850**	18.5	3	40

strength of $K \sim 1$. The actual value of K for each oscillator depends on the PPV function as well as the amplitude of perturbation from other oscillators, as we show in the derivation of Gen-Adler in [35]. As an example, for the LC oscillators we use in Sect. 4.1 with 100k resistive coupling, $K \approx 0.02$. This indicates that it takes less than 100 cycles for the oscillators to synchronise in phase, which is consistent with measurements. For such a coupled LC oscillator network, a hardware time of 20 in Fig. 10 represents approximately 2000 cycles of oscillation; for 5 MHz oscillators that takes 0.4 ms. If we use GHz nano-oscillators, the computation time can be well within a microsecond. In comparison, the runtime of the several heuristic algorithms listed in Table 1, even with faster CPUs and parallel implementations in the future, is unlikely to ever drop to this range.

As the hardware time is fixed, the runtime we report in Table 1 for our Ising machines is the time for simulating the SDEs of coupled oscillators on CPUs for one run. While we list runtime results for each algorithm in Table 1, note that they come from different sources and are measured on different platforms. Results for SS, CirCut and VNSPR were obtained from Dual Intel Xeon at 3.06 GHz with 3.2 GB of RAM; SA was run on Intel Xeon E3-1245v2 at 3.4 GHz with 32GB of RAM [42]. To make the results generally comparable, we ran our simulations on a modest personal desktop with Intel Xeon E5-1603v3 at 2.8 GHz with 16 GB of RAM. Even so, it came as a nice surprise to us that even by simulating SDEs we were able to solve the benchmarks efficiently. Another notable feature of our method is that unlike other algorithms, SDE simulation does not know about the Ising Hamiltonian or cut value—it never needs to evaluate the energy function or relative energy changes, which are implicit in the dynamics of differential equations, yet it proves effective and fast.

We also ran more computational experiments on the G-set benchmarks in order to study the mechanism of oscillator-based Ising machines. We created several variants of the Ising machine used above by removing different components in its operation. For each variant, we re-ran 200 random instances for each of the 54 benchmarks, generating 10800 cut values. In Fig. 11, we compare the quality of these cut values with results from the unaltered Ising machine by plotting histograms of the distances of the cut values to their respective maxima. In the first variant, we removed noise from the model by setting $K_n \equiv 0$. The solutions become considerably worse, confirming that noise helps the coupled oscillator system settle to lower energy states.

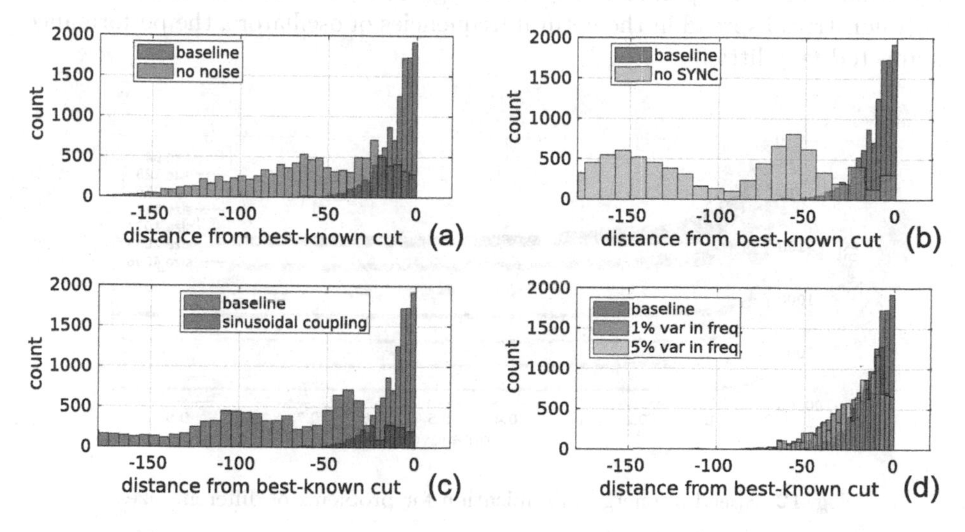

Fig. 11. Histograms of the cut values achieved by several variants of the Ising machine, compared with the baseline results used in Table 1.

In the next variant, we removed SYNC by setting $K_s \equiv 0$. Without SYNC, the system becomes a simple coupled oscillator system with phases that take a continuum of values, as discussed in Sect. 3.2. The settled analog values of the phases that were then thresholded to 0 or π to correspond to Ising spins. As shown in Fig. 11, the results become significantly worse; indeed, none of the best-known results were reached. This indicates that the SYNC signal and the mechanism of SHIL we introduce to the coupled oscillator networks are indeed essential for them to operate as Ising machines.

Our baseline Ising machine actually uses a smoothed square function tanh(sin(.)) for the coupling, as opposed to the sin(.) used in the original Kuramoto model, as shown in the code in [35]. This changes the cos(.) term in the energy function (10) to a triangle function. Such a change appears to give better results than the original, as shown in Fig. 11(c). The change requires designing oscillators with special PPV shapes and waveforms such that their cross-correlation is a square wave, which is not difficult in practice based on our derivation in [35]. As an example, rotary traveling wave oscillators naturally have square PPVs. Ring oscillators can also be designed with various PPVs and waveforms by sizing each stage individually. We cannot say definitively that the square function we have used is optimal for Ising solution performance, but the significant improvement over sinusoidal coupling functions indicates that a fruitful direction for further exploration may be to look beyond the original Kuramoto model for oscillator-based computing.

The last variant we report here added variability to the natural frequencies of the oscillators, as in (18). We assigned Gaussian random variables to ω_is with ω^* as the mean, and 0.01 (1%) and 0.05 (5%), respectively, as the standard deviations for two separate runs. From Fig. 11(d), we observe that even with such non-trivial spread in the natural frequencies of oscillators, the performance is affected very little.

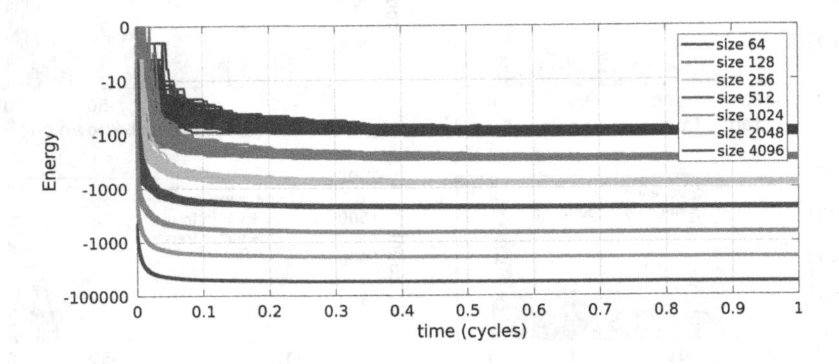

Fig. 12. Speed of energy minimisation for problems of different sizes.

Finally, we conducted a preliminary study of the scaling of the time taken by the Ising machine to reach good solutions as problem sizes increase. As the G-set benchmarks have only a few sizes (800, 1000, 2000 and 3000), we used the program (named `rudy`) that generated them to create more problems of various sizes. All generated problems used random graphs with 10% connectivity and ± 1 coupling coefficients. We simulated all of them, each for 200 instances, with fixed parameters $K = 1$, $K_s = 0.1$, $K_n = 0.01$, and show all their Ising Hamiltonians over time in Fig. 12. Much to our surprise, the speed in which the values settle appears almost constant, regardless of the problem size. While this does not necessarily mean they all converge to the global optima within the same time, this preliminary study is encouraging as it confirms the massively parallel nature of the system. For larger Ising problems, our Ising machine only needs to scale linearly in hardware size with the number of spins, but does not necessarily require much more time to reach a solution.

4.3 A Graph Colouring Example

As mentioned in Sect. 2, many problems other than MAX-CUT can be mapped to the Ising model [11] and solved by an oscillator-based Ising machine. Here we show an example of a graph colouring problem—assigning four colours to the 51 states (including a federal district) of America such that no two adjacent states have the same colour.[8]

Each state is represented as a vertex in the graph. When two states are adjacent, there is an edge in the graph that connects the corresponding vertices. For every vertex i, we assign four spins—s_{iR}, s_{iG}, s_{iB} and s_{iY} to represent its colouring scheme; when only one of them is +1, the vertex is successfully coloured as either red, green, blue or yellow. Then we write an energy function H associated with these $4 \times 51 = 204$ spins as follows:

$$H = \sum_{i}^{n}(2 + s_{iR} + s_{iG} + s_{iB} + s_{iY})^2 \\ + \sum_{(i,j)\in\mathbb{E}}^{n_{\mathbb{E}}} [(1 + s_{iR})(1 + s_{jR}) + (1 + s_{iG})(1 + s_{jG}) + (1 + s_{iB})(1 + s_{jB}) + (1 + s_{iY})(1 + s_{jY})], \tag{25}$$

where $n = 51$ is the number of vertices, $\mathbb{E}$ represents the edge set, $n_{\mathbb{E}}$ is the number of edges and in this case equal to 220.[9]

The first term of H is a sum of squares never less than zero; it reaches zero only when $\{s_{iR}, s_{iG}, s_{iB}, s_{iY}\}$ contains three -1s and one $+1$ for every i, *i.e.*, each state has a unique colour. The latter term is also a sum that is always greater than or equal to zero, as each spin can only take a value in $\{-1, +1\}$; it is zero when $s_{iX} = s_{jX} = +1$ never occurs for any edge connecting i and j, and for any colour $X \in \{R, G, B, Y\}$, *i.e.*, adjacent states do not share the same colour. Therefore, when H reaches its minimum value 0, the spin configuration represents a valid colouring scheme—following the indices of the +1 spins $\{i, X \mid s_{iX} = +1\}$, we can then assign colour X to state i.

Note that when expanding the sum of squares in (25), we can use the fact $s_{iX}^2 \equiv 1$ to eliminate the square terms. This means H contains only products of two spins—modelled by J_{ij}s, and self terms—modelled by h_i. These Ising coefficients can then be used to determine the couplings in an oscillator-based Ising machine.

We simulated these 204 coupled oscillators and show the results in Fig. 13. In the simulation, we kept K and K_n constant while ramping K_s up and down 5 times. We found the Ising machine to be effective with this schedule as it could colour the map successfully in more than 50% of the random trials and returned many different valid colouring schemes.

[8] Ising machines can be used on general graph colouring problems, and this four-colouring problem is chosen here for illustrative purposes. Four-colouring a planar graph is actually not NP-hard and there exist polynomial-time algorithms for it [46].

[9] Hawaii and Alaska are considered adjacent such that their colours will be different in the map.

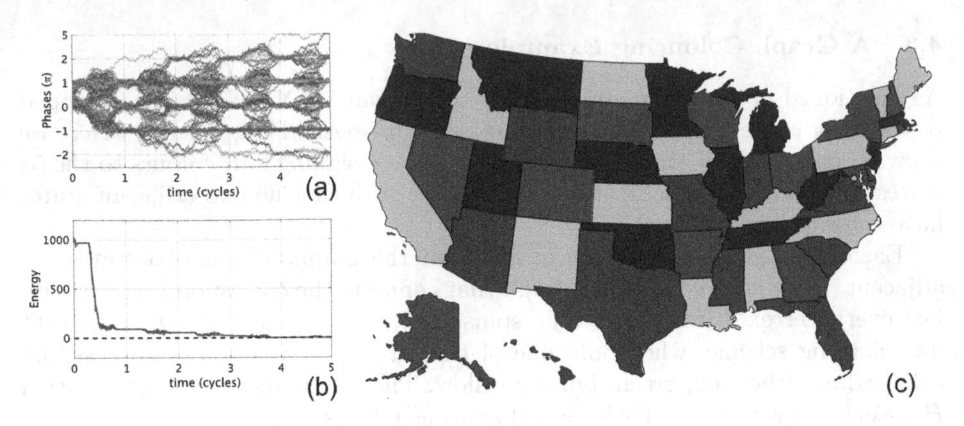

Fig. 13. Coupled oscillators colouring the states in the US map: (a) phases of oscillators evolve over time; (b) energy function (25) decreases during the process; (c) the resulting US map colouring scheme.

5 Conclusion

In this paper, we have proposed a novel scheme for implementing Ising machines using self-sustaining nonlinear oscillators. We have shown how coupled oscillators naturally minimise an "energy" represented by their global Lyapunov function, and how introducing the mechanism of subharmonic injection locking modifies this function to encode the Ising Hamiltonian for minimisation. The validity and feasibility of the scheme have been examined via multiple levels of simulation and proof-of-concept hardware implementations. Simulations run on larger-scale benchmark problems have also shown promising results in both speed and the quality of solutions. We believe that our scheme constitutes an important and practical means for the implementation of scalable Ising machines.

Acknowledgements. The authors would like to thank the reviewers for the useful comments and in particular anonymous reviewer No. 2 for pointing us to Ercsey-Ravasz/Toroczkai and Yin's work on designing dynamical systems to solve NP-complete problems.

References

1. Ising, E.: Beitrag zur theorie des ferromagnetismus. Zeitschrift für Physik A Hadrons and Nuclei **31**(1), 253–258 (1925)
2. Brush, S.G.: History of the Lenz-Ising Model. Rev. Mod. Phys. **39**, 883–893 (1967)
3. Barahona, F.: On the computational complexity of Ising spin glass models. J. Phys. A: Math. Gen. **15**(10), 3241 (1982)
4. Marandi, A., Wang, Z., Takata, K., Byer, R.L., Yamamoto, Y.: Network of time-multiplexed optical parametric oscillators as a coherent Ising machine. Nat. Photonics **8**(12), 937–942 (2014)

5. McMahon, P.L., et al.: A fully-programmable 100-spin coherent Ising machine with all-to-all connections. Science **354**, 5178 (2016)
6. Inagaki, T., et al.: A coherent Ising machine for 2000-node optimization problems. Science **354**(6312), 603–606 (2016)
7. Johnson, M.W., et al.: Quantum annealing with manufactured Spins. Nature **473**(7346), 194 (2011)
8. Bian, Z., Chudak, F., Israel, R., Lackey, B., Macready, W.G., Roy, A.: Discrete optimization using quantum annealing on sparse Ising models. Front. Phys. **2**, 56 (2014)
9. Yamaoka, M., Yoshimura, C., Hayashi, M., Okuyama, T., Aoki, H., Mizuno, H.: A 20k-spin Ising chip to solve combinatorial optimization problems with CMOS annealing. IEEE J. Solid-State Circuits **51**(1), 303–309 (2016)
10. Karp, R.M.: Reducibility among combinatorial problems. In: Miller, R.E., Thatcher, J.W., Bohlinger, J.D. (eds.) Complexity of Computer Computations, pp. 85–103. Springer, Boston (1972)
11. Lucas, A.: Ising formulations of many NP problems. arXiv preprint arXiv:1302.5843 (2013)
12. Wang, T., Roychowdhury, J.: Oscillator-based Ising Machine. arXiv preprint arXiv:1709.08102 (2017)
13. Festa, P., Pardalos, P.M., Resende, M.G.C., Ribeiro, C.C.: Randomized heuristics for the MAX-CUT problem. Optim. Methods Softw. **17**(6), 1033–1058 (2002)
14. Jensen, T.R., Toft, B.: Graph Coloring Problems. Wiley, New York (2011)
15. Neogy, A., Roychowdhury, J.: Analysis and Design of Sub-harmonically Injection Locked Oscillators. In: Proceedings of the IEEE DATE, March 2012. http://dx.doi.org/10.1109/DATE.2012.6176677
16. Bhansali, P., Roychowdhury, J.: Gen-Adler: The generalized Adler's equation for injection locking analysis in oscillators. In: Proceedings of the IEEE ASP-DAC, pp. 522–227, January 2009. http://dx.doi.org/10.1109/ASPDAC.2009.4796533
17. Kuramoto, Y.: Self-entrainment of a population of coupled non-linear oscillators. In: Araki, H. (ed.) International Symposium on Mathematical Problems in Theoretical Physics, pp. 420–422. Springer, Heidelberg (1975)
18. Kuramoto, Y.: Chemical Oscillations, Waves and Turbulence. Dover, New York (2003)
19. Acebrón, J.A., Bonilla, L.L., Vicente, C.J.P., Ritort, F., Spigler, R.: The kuramoto model: a simple paradigm for synchronization phenomena. Rev. Mod. Phys. **77**(1), 137 (2005)
20. Wang, T., Roychowdhury, J.: PHLOGON: PHase-based LOGic using oscillatory nano-systems. In: Ibarra, O.H., Kari, L., Kopecki, S. (eds.) UCNC 2014. LNCS, vol. 8553, pp. 353–366. Springer, Cham (2014). https://doi.org/10.1007/978-3-319-08123-6_29
21. Wang, T.: Sub-harmonic Injection Locking in Metronomes. arXiv preprint arXiv:1709.03886 (2017)
22. Wang, T.: Achieving Phase-based Logic Bit Storage in Mechanical Metronomes. arXiv preprint arXiv:1710.01056 (2017)
23. Aramon, M., Rosenberg, G., Valiante, E., Miyazawa, T., Tamura, H., Katzgraber, H.G.: Physics-inspired optimization for quadratic unconstrained problems using a digital annealer. arXiv:1806.08815 [physics.comp-ph] August 2018
24. Gyoten, H., Hiromoto, M., Sato, T.: Area efficient annealing processor for ising model without random number generator. IEICE Trans. Inf. Syst. **E101.D**(2), 314–323 (2018)

25. Gyoten, H., Hiromoto, M., Sato, T.: Enhancing the solution quality of hardware Ising-model solver via parallel tempering. In: Proceedings of the ICCAD, ICCAD 2018, pp. 70:1–70:8. ACM, New York (2018)
26. Bian, Z., Chudak, F., Macready, W.G., Rose, G.: The Ising model: teaching an old problem new tricks. D-Wave Syst. **2**, 1–32 (2010)
27. Harris, R., et al.: Experimental demonstration of a robust and scalable flux qubit. Phys. Rev. B **81**(13), 134–510 (2010)
28. Rønnow, T.F., et al.: Defining and detecting quantum speedup. Science **345**(6195), 420–424 (2014)
29. Denchev, V.S., et al.: What is the computational value of finite-range tunneling? Phys. Rev. X **6**(3), 031015 (2016)
30. Mahboob, I., Okamoto, H., Yamaguchi, H.: An electromechanical Ising Hamiltonian. Sci. Adv. **2**(6), e1600236 (2016)
31. Camsari, K.Y., Faria, R., Sutton, B.M., Datta, S.: Stochastic p-bits for invertible logic. Phys. Rev. X **7**(3), 031014 (2017)
32. Yamamoto, K., Huang, W., Takamaeda-Yamazaki, S., Ikebe, M., Asai, T., Motomura, M.: A time-division multiplexing Ising machine on FPGAs. In: Proceedings of the 8th International Symposium on Highly Efficient Accelerators and Reconfigurable Technologies, p. 3. ACM (2017)
33. Winfree, A.: Biological rhythms and the behavior of populations of coupled oscillators. Theor. Biol. **16**, 15–42 (1967)
34. Demir, A., Mehrotra, A., Roychowdhury, J.: Phase noise in oscillators: a unifying theory and numerical methods for characterization. IEEE Trans. Circuits Syst.- I: Fund. Th. Appl. **47**, 655–674 (2000). http://dx.doi.org/10.1109/81.847872
35. Wang, T., Roychowdhury, J.: OIM: Oscillator-based Ising Machines for Solving Combinatorial Optimisation Problems. arXiv preprint arXiv:1903.07163 (2019)
36. Wang, T., Roychowdhury, J.: Design tools for oscillator-based computing systems. In: Proceedings IEEE DAC, pp. 188:1–188:6 (2015). http://dx.doi.org/10.1145/2744769.2744818
37. Shinomoto, S., Kuramoto, Y.: Phase transitions in active rotator systems. Progress Theoret. Phys. **75**(5), 1105–1110 (1986)
38. Lyapunov, A.M.: The general problem of the stability of motion. Int. J. Control **55**(3), 531–534 (1992)
39. Hopfield, J.J., Tank, D.W.: "Neural" computation of decisions in optimization problems. Biol. Cybern. **52**(3), 141–152 (1985)
40. Ercsey-Ravasz, M., Toroczkai, Z.: Optimization hardness as transient chaos in an analog approach to constraint satisfaction. Nat. Phys. **7**(12), 966 (2011)
41. Yin, X., Sedighi, B., Varga, M., Ercsey-Ravasz, M., Toroczkai, Z., Hu, X.S.: Efficient analog circuits for boolean satisfiability. IEEE Trans. Very Large Scale Integr. (VLSI) Syst. **26**(1), 155–167 (2018)
42. Myklebust, T.: Solving maximum cut problems by simulated annealing. arXiv preprint arXiv:1505.03068 (2015)
43. Helmberg, C., Rendl, F.: A spectral bundle method for semidefinite programming. SIAM J. Optim. **10**(3), 673–696 (2000)
44. Martí, R., Duarte, A., Laguna, M.: Advanced scatter search for the max-cut problem. INFORMS J. Comput. **21**(1), 26–38 (2009)
45. Burer, S., Monteiro, R., Zhang, Y.: Rank-two relaxation heuristics for max-cut and other binary quadratic programs. SIAM J. Optim. **12**(2), 503–521 (2002)
46. Robertson, N., Sanders, D.P., Seymour, P., Thomas, R.: Efficiently four-coloring planar graphs. In: Proceedings of the Twenty-Eighth Annual ACM Symposium on Theory of Computing, pp. 571–575. ACM (1996)

Relativizations of Nonuniform Quantum Finite Automata Families

Tomoyuki Yamakami(✉)

Faculty of Engineering, University of Fukui, 3-9-1 Bunkyo, Fukui 910-8507, Japan
TomoyukiYamakami@gmail.com

Abstract. Theory of relativization provides profound insights into the structural properties of various collections of mathematical problems by way of constructing desirable oracles that meet numerous requirements of the problems. This is a meaningful way to tackle unsolved questions on relationships among computational complexity classes induced by machine-based computations that can relativize. Slightly different from an early study on relativizations of uniform models of finite automata in [Tadaki, Yamakami, and Li (2010); Yamakami (2014)], we intend to discuss relativizations of state complexity classes (particularly, 1BQ and 2BQ) defined in terms of nonuniform families of time-unbounded quantum finite automata with polynomially many inner states. We create various relativized worlds where certain nonuniform state complexity classes become equal or different. By taking a nonuniform family of promise decision problems as an oracle, we can define a Turing reduction witnessed by a certain nonuniform finite automata family. We demonstrate closure properties of certain nonuniform state complexity classes under such reductions. Turing reducibility further enables us to define a hierarchy of nonuniform nondeterministic state complexity classes.

Keywords: Quantum finite automata · Nonuniform state complexity · Oracle finite automata · Promise problems · Turing reducibility

1 Prelude: Background and Perspectives

1.1 Nonuniform State Complexity Classes

Finite automata can be generally seen as a mathematical model of a programmable device producing memoryless computation to solve mathematical problems. In the past literature, numerous structural properties have been discovered for various types of finite automata. Although finite automata are viewed simply as Turing machines with no work tape, the behaviors of the finite automata are quite different from those of the Turing machines. There have been numerous approaches taken in automata theory toward the full understanding of the behavioral characteristics of various finite automata.

This paper aims at promoting our full understanding of the state complexity of various nonuniform families of finite automata having polynomially many

I. McQuillan and S. Seki (Eds.): UCNC 2019, LNCS 11493, pp. 257–271, 2019.
https://doi.org/10.1007/978-3-030-19311-9_20

inner states. The study of such complexity measure was initiated by Berman and Lingas [4] as well as Sakoda and Sipser [13]. For readability, we refer to the state complexity of nonuniform family of finite automata as *nonuniform state complexity* (as in [18]). A family of finite automata behaves quite differently from a family of Boolean circuits. For instance, each circuit of the family is limited to inputs of fixed length whereas each automaton can work on inputs of any length. A collection of families of such nonuniform state complexity naturally induces a complexity class of "promise" decision problems, in which only certain inputs (not necessarily all inputs) are "promised" to have clear outcomes. A typical such class is 1D, which is composed of all families of promise decision problems solvable by nonuniform families of *one-way deterministic finite automata* (or 1dfa's, for short) with polynomially many inner states. If we take *one-way nondeterministic finite automata* (or 1nfa's) instead of 1dfa's, we immediately obtain 1N. Since a certain 1nfa cannot be converted into an equivalent 1dfa of polynomial size, the class separation 1D $\neq$ 1N follows immediately. The use of two-way finite automata analogously introduces 2D and 2N in place of 1D and 1N, respectively. When the length of inputs to promise problems is upper-bounded by a certain polynomial in n, we use the notation, e.g., 1N/poly in place of 1N. A still unsolved question of Sakoda and Sipser [13] concerns with the inclusion of 1N to 2D.

Lately, there has been a surge of research activities regarding nonuniform state complexity classes by Kapoutsis [8–10], Kapoutsis and Pighizzini [11], and Yamakami [18,19]. The aforementioned 1D and 1N were further extended into various types of finite automata, including probabilistic and quantum finite automata and their nonuniform state complexity classes, 1BP, 1P, 1BQ, and 1Q, were newly invented in [18] (whereas certain restricted variants had been already discussed in [8]). The classes 1BQ and 1Q are the most relevant complexity classes, which are founded on one-way quantum finite automata equipped with garbage tapes (or 2qfa's, for short) with bounded-error probability and unbounded-error probability, respectively. In a similar way, two-way finite automata of various types form 2D, 2N, 2BP, 2BQ, 2P, and 2Q [8,9,13,18]. We note that an early systematic study on nonuniform families of quantum finite automata was already reported in [15]. Most one-way classes mentioned above were shown to be different from each other [8,9,18]. For two-way classes, when restricted to polynomial-size inputs, few separations have been shown since the corresponding two-way nonuniform state complexity classes are related to logarithmic-space advised complexity classes [10,11,18] and these close relationships therefore make it difficult to prove any collapse or separation. Moreover, an extension of 2qfa's, called *two-way super quantum finite automata* (abbreviated as 2sqfa's), was introduced in [18]. We write 2sBQ and 2sQ for the nonuniform state complexity classes obtained respectively from 2BQ and 2Q using such 2sqfa's.

While we are waiting for a new tool to be discovered for proving the collapses and separations among two-way nonuniform state complexity classes, we boldly take a new step to tackle those collapse and separation issues by resorting to *relativizations*, which will be discussed in the next subsection.

1.2 Relativizations of Finite Automata

A practical notion of *oracle machines* relative to various oracles was introduced into the field of polynomial-time computational complexity by Baker, Gill, and Solovay [3] in 1975, in order to present a circumstantial evidence that the P = ?NP problem cannot be solved by any argument that can relativize. Since many known proofs in computational complexity theory actually relativize, such proofs cannot be used to solve this open problem.

As for finite automata models, an early notion of their relativizations was discussed in [14,17] with the use of a simple query mechanism incorporated with *oracles*, which are external information sources. Finite automata with such a mechanism are customarily called *oracle finite automata*. A relativization of finite automata requires a special attention because the machines do not have any memory tape to remember any pre-query configuration. Here, we relativize nonuniform state complexity classes by supplementing an oracle as an external device as well as a mechanism of accessing such a device. By relativizing to any given oracle A, we naturally expand 1N and 2D to $1\mathrm{N}^A$ and $2\mathrm{D}^A$, respectively.

From a slightly different viewpoint, relativizations can be seen as *Turing reducibility*, opposed to a merely functional mechanism of *many-one reducibility*. To realize this viewpoint for families of promise decision problems, we first modify each underlying oracle finite automata so that it can take another family of promise decision problems as an oracle. This modification can tremendously expand the scope of our study on families of promise decision problems. Notationally, for two given families $\mathcal{L}$ and $\mathcal{K}$ of promise decision problems, we write $\mathcal{L} \leq_T^{\mathrm{2BQ}} \mathcal{K}$ to indicate that $\mathcal{L}$ is Turing reducible to $\mathcal{K}$ by families of bounded-error oracle 2qfa's having polynomially many inner states. In a similar fashion, we can define $\mathcal{L} \leq_T^{\mathrm{2D}} \mathcal{K}$, $\mathcal{L} \leq_T^{\mathrm{1BP}} \mathcal{K}$, etc. Collectively, we further denote by $2\mathrm{BQ}^{\mathcal{K}}$ the collection of all nonuniform families $\mathcal{L}$ of promise decision problems satisfying $\mathcal{L} \leq_T^{\mathrm{2BQ}} \mathcal{K}$. This notation further leads to a way of introducing, e.g., $2\mathrm{BQ}^{2\mathrm{BQ}}$ as the union of $2\mathrm{BQ}^{\mathcal{K}}$ for all $\mathcal{K} \in 2\mathrm{BQ}$.

1.3 Overview of Main Contributions

We quickly give an overview of our main results of this exposition. We first note that, in the unrelativized world, we do not know whether $1\mathrm{N} \nsubseteq 2\mathrm{D}$ or $1\mathrm{BQ} \nsubseteq 2\mathrm{D}$ although we know that $1\mathrm{N} \subseteq 2^{1\mathrm{D}}$ [8] and $1\mathrm{BQ} \subseteq 2^{1\mathrm{D}}$ [18]. Refer to [18] for comprehensive relationships among nonuniform state complexity classes. Our approach to these unsolved questions is to resort to relativizations. We want to show the existence of oracles for which these separations are indeed the case.

First, we discuss the inclusion and exclusion relationships among 1N, 1BQ, and 2D when transition amplitudes are limited to approximately computable numbers.

Theorem 1. *There are recursive oracles A and B that meet the following conditions when 1qfa's are assumed to take approximately computable amplitudes.*
(a) $1\mathrm{N}^A \not\subseteq 2\mathrm{D}^A$ *and* $1\mathrm{BQ}^A \subseteq 2\mathrm{D}^A$.
(b) $1\mathrm{N}^B \subseteq 2\mathrm{D}^B$ *and* $1\mathrm{BQ}^B \not\subseteq 2\mathrm{D}^B$ *(more strongly,* $1\mathrm{BP}^B \not\subseteq 2\mathrm{D}^B$*).*

Theorem 1 demonstrates a clear difference between 1N and 1BQ in computational power. Our oracle construction in the proof of the theorem, presented in Sect. 3.1, resembles a diagonalization argument used for the associated polynomial-time setting where a single machine is used at each stage of the construction. However, since we need to deal with nonuniform families of finite automata, our oracle construction requires a diagonalization of "multiple" machines at each stage of the construction.

Next, let us compare among 2D, 2BP, and 2BQ. Since it is not yet known whether they are equal or different, we want to present oracles that make them collapse as well as separated.

Theorem 2. *There exist recursive oracles A and B satisfying the following conditions when transitions of 2pfa's and 2qfa's take approximately computable probabilities and amplitudes.*

(a) $2\mathrm{D}^A = 2\mathrm{BP}^A \neq 2\mathrm{BQ}^A$.
(b) $2\mathrm{D}^B \neq 2\mathrm{BP}^B = 2\mathrm{BQ}^B$.

To prove Theorem 2 in Sect. 3.1 requires a new insight of a time-unbounded nature of probabilistic finite automata. Different from time-bounded computations, 2pfa's can run for more than exponentially many steps [5] and so do quantum finite automata. This extremely long runtime allows those finite automata to make more than exponentially many queries. Therefore, it seems difficult for us to construct oracles that demonstrate the desired separations among nonuniform state complexity classes because, for example, a standard query complexity argument does not seem to work. To prove Theorem 2(a), therefore, we cannot apply the same oracle-construction strategy used in the polynomial-time setting.

The nonuniform state complexity class 2sBQ in [18] is closely related to quantum advice for BQL or the class BQL/Qpoly. Next, we look into a relationship between 2qfa's and 2sqfa's. Unfortunately, we do not seem to be able to construct a deterministic oracle that separates 2sBQ from 2BQ. Instead, we try to construct a *quantum oracle* that makes the desired separation, where a quantum oracle is simply a family of unitary matrices.

Theorem 3. *There exists a quantum oracle U such that* $2\mathrm{sBQ}^U \neq 2\mathrm{BQ}^U$ *(more strongly,* $1\mathrm{sBQ}^U \not\subseteq 2\mathrm{BQ}^U$*).*

We will verify the above theorem in Sect. 3.2.

As noted in Sect. 1.2, relativizations can be seen as Turing reductions between two families $\mathcal{L}$ and $\mathcal{K}$ of promise decision problems. The nonuniform state complexity class $1\mathrm{BP}^{\mathcal{K}[k]}$ is obtained from $1\mathrm{BP}^{\mathcal{K}}$ by limiting the number of queries to $\mathcal{K}$ along each computation path within k. Moreover, we set $1\mathrm{BP}^{1\mathrm{BP}[k]}$ to be the union $\bigcup_{\mathcal{K}\in 1\mathrm{BP}} 1\mathrm{BP}^{\mathcal{K}[k]}$. Similarly, we obtain $1\mathrm{P}^{1\mathrm{P}[k]}$.

Theorem 4. *(a)* $2\mathrm{D}^{2\mathrm{D}} = 2\mathrm{D}$.
(b) $1\mathrm{BP}^{1\mathrm{BP}[1]} = 1\mathrm{BP}$ *and* $1\mathrm{P}^{1\mathrm{P}[1]} = 1\mathrm{P}$.

Unfortunately, it is not known but is expected to hold that $2\mathrm{N}^{2\mathrm{N}} \neq 2\mathrm{N}$, $2\mathrm{BP}^{2\mathrm{BP}} \neq 2\mathrm{BP}$, and $2\mathrm{BQ}^{2\mathrm{BQ}} \neq 2\mathrm{BQ}$. The proof of Theorem 4 will appear in Sect. 4.1.

Although Turing reducibility embodies an oracle mechanism, we further relativize Turing reductions by adding an extra oracle. Given an oracle A, we inductively define $2\Sigma_1^A = 2\mathrm{N}^A$ and $2\Sigma_{k+1}^A = \{\mathcal{L} \in \mathrm{FPDP} \mid \exists \mathcal{K} \in 2\Sigma_k^A\ [\mathcal{L} \leq_T^{2\mathrm{N}} \mathcal{K}]\}$ for each index $k \geq 1$, where FPDP is the set of all families of promise decision problems. Clearly, $2\Sigma_k^A \subseteq 2\Sigma_{k+1}^A$ holds for any $k \geq 1$. We show in Sect. 4.2 the following oracle separation of 2BQ from $2\Sigma_2$.

Theorem 5. *There exists a recursive oracle A such that $2\mathrm{BQ}^A \nsubseteq 2\Sigma_2^A$, when transition amplitudes are limited to approximately computable numbers.*

In the subsequent sections, we will prove the aforementioned theorems after explaining the notions and notation necessary to read through the sections.

Due to the page limit, in what follows, we will provide only proof sketches to the theorems and leave the detailed proofs to a complete version of this exposition.

2 Fundamental Notions and Notation

2.1 Numbers, Strings, and Promise Decision Problems

For numbers, $\mathbb{N}$, $\mathbb{Z}$, $\mathbb{R}$, and $\mathbb{C}$ respectively denote the sets of all nonnegative integers, of all integers, of all real numbers, and of all complex numbers. For two numbers $m, n \in \mathbb{Z}$ with $m \leq n$, we write $[m,n]_{\mathbb{Z}}$ for the integer interval $\{m, m+1, m+2, \ldots, n\}$. When $n \in \mathbb{N}$ and $n \geq 1$, in particular, we abbreviate $[1,n]_{\mathbb{Z}}$ as $[n]$. A complex number is *approximately computable* if its real and imaginary parts are approximated algorithmically to within 2^{-n} for any given input 1^n. All *polynomials* are assumed to take coefficients of nonnegative integers. Moreover, for any finite set A, $\#A$ denotes the *cardinality* of A. An *alphabet* is a nonempty finite set and a *string* over an alphabet Σ is a finite sequence of symbols in Σ. The *length* of a string x, which is the total number of symbols in x, is denoted $|x|$. A *language* over an alphabet Σ is a set of strings over Σ. We freely identify any decision problem with its associated language. Given a string x and an index $j \in [|x|]$, the notation $x_{(j)}$ denotes the jth symbol of x. Given two binary strings x and y of length n, $x \oplus y$ denotes the bitwise XOR of x and y, i.e., $(x \oplus y)_{(i)} = x_{(i)} + y_{(i)} \pmod 2$ for any $i \in [n]$. Given a string x and a language L over the same alphabet, xL and Lx respectively denote the sets $\{xz \mid z \in L\}$ and $\{zx \mid z \in L\}$. Moreover, a language L is thought as a $\{0,1\}$-valued function $L(x)$ defined by, for a string x, 1 if $x \in L$, and 0 otherwise.

Our finite automata are generally designed to solve a "promise" version of decision problems. Formally, a *promise decision problem* is a pair (A, R) of disjoint subsets of Σ^* for a certain alphabet Σ, where A and R consist of YES instances (positive instances or accepted instances) and of NO instances (negative instances or rejected instances), respectively. For simplicity, if a string x belongs to $A \cup B$, then we say that x is *promised* to (A, R) (or x is a promised word of (A, B)).

2.2 Basics of Quantum Finite Automata

We assume the reader's familiarity with classical and quantum computing (see, e.g., [7,12]). In particular, we use the following model of probabilistic finite automata. A *two-way probabilistic finite automaton* (or a 2pfa, for short) M is a tuple $(Q, \Sigma, \{¢, \$\}, \delta, q_0, Q_{acc}, Q_{rej})$, where Q is a finite set of inner states, Σ is an input alphabet, δ is a probabilistic transition function from $(Q - Q_{halt}) \times \check{\Sigma} \times Q \times D$ to the unit interval $[0, 1]$, q_0 is the initial state in Q, and both Q_{acc} and Q_{rej} are subsets of Q and consist of accepting states and rejecting states, respectively, $Q_{halt} = Q_{acc} \cup Q_{rej}$, $\check{\Sigma} = \Sigma \cup \{¢, \$\}$, and $D = \{-1, 0, +1\}$. An input string, say, x is initially written on an input tape, surrounded by two endmarkers ¢ (left endmarker) and \$ (right endmarker). The 2pfa starts in state q_0 and applies δ at every step to make a probabilistic move. When M enters halting states, it halts and *accepts* (resp., *rejects*) x if the total acceptance (resp., rejection) probability $p_{acc}(x)$ is more than 1/2 (resp., at least 1/2). We say that M makes *bounded error probability* if there exists a constant $\varepsilon \in [0, 1/2)$ such that either $p_{acc}(x) \geq 1 - \varepsilon$ or $p_{rej}(x) \geq 1 - \varepsilon$ holds. Notice that we do not require 2pfa's to terminate within, say, polynomial time (different from [8]).

While probabilistic finite automata use probabilistic choices, quantum finite automata use quantum interference and quantum superposition. Following [18], a *two-way quantum finite automaton*[1] with a garbage tape (or a 2qfa, for short) M is a tuple $(Q, \Sigma, \{¢, \$\}, \Xi, \delta, q_0, Q_{acc}, Q_{rej})$, where Ξ is a garbage alphabet, δ is a quantum transition function from $Q \times \check{\Sigma} \times Q \times D \times \Xi_\lambda$ to $\mathbb{C}$ with $\Xi_\lambda = \Xi \cup \{\lambda\}$, and the rest is the same as those of 2pfa's, where the input tape is always assumed to be *circular*[2] for simplicity of our description. We express the value of δ as $\delta(q, \sigma \mid p, d, \xi)$.

With a use of a garbage tape, writing symbols onto the garbage tape is viewed as discarding unwanted information from the quantum system of a 2qfa into an external environment that surrounds the automaton. This garbage mechanism

[1] In some literature, a quantum finite automaton is allowed to use "classical states" besides "quantum states." Such an automaton is often abbreviated as a qcfa. If we use a garbage tape and a measurement, then we essentially do not need to introduce classical states. To simplify our description of 2qfa's, we use no classical states in this exposition.

[2] A tape is *circular* if both ends of the tape are glued together so that a tape head moves out of \$ to the right, it reaches ¢, and the vice versa.

causes essentially the same effect on the 2qfa's computation as a two-way quantum finite automaton with mixed states and quantum operations, studied in [2,6,16].

A *(surface) configuration* of a 2qfa M on a fixed input x is a tuple $(q, i, w) \in Q \times [0, |x| + 1]_{\mathbb{Z}} \times \Xi^*$, indicating that M is in inner state q, scanning the ith cell of the input tape, and having w in the garbage tape. Since tape heads of a query tape and a garbage tape move only in one direction whenever they write nonempty tape symbols, there is no need to include the positions of those tape heads. The quantum transition function δ naturally induces a *time-evolution operator* $U_M^{(x)}$ on an input x, defined by

$$U_M^{(x)}|q, i, w\rangle = \sum_{p,d,\xi} \delta(q, x_{(i)} \mid p, d, \xi)|p, i + d \ (\mathrm{mod}\ |x| + 2), w\xi\rangle.$$

Technically speaking, this operator $U_M^{(x)}$ is in general an infinite-dimensional matrix because the 2qfa M may run forever. To guarantee the reversibility of each move of M, we need to demand that $U_M^{(x)}$ should be a unitary matrix for any given input x. Hereafter, we implicitly assume the unitarity of M's time-evolution operator.

At every step, we observe the new inner state of M to check whether M enters a halting state. The observation shows us that M enters accepting states as well as rejecting states with the probability that is equal to the square of the ℓ_2-norm of the corresponding quantum state of M. The final criteria of acceptance and rejection of the 2qfa is the same as a 2pfa.

A *transition table* (or a transition matrix) is another tool in describing δ. A transition table T has $|Q \times \check{\Sigma}|$-rows and $|Q \times D \times \Xi_\lambda|$-columns and each $(q, \sigma) \times (p, d, \xi)$-entry is a transition amplitude approximated to within 2^{-n} by a certain quantum circuit made up of a fixed universal set of quantum gates working on inputs of the form 1^n. In a model of *super quantum finite automata*, discussed in [18], the automaton starts its computation with a *superposition* of the encodings of transition tables on an extra tape and, just before its termination, the automaton enters the final process of erasing the transition tables by overwriting any non-blank symbol with a blank symbol. Although this final process is always possible with the use of garbage tape, in certain ideal cases, the process may cause different computation paths following different transition tables to interfere with each other.

Two machines M and N working on the same input alphabet are said to be *equivalent* if, for any input string x, M accepts (resp., rejects) x iff N accepts (resp., rejects) x.

2.3 Oracle Mechanism and Turing Reducibility

One way to empower a finite automaton is to make it access an external information source, known as an *oracle*. An early notion of relativizations of finite automata (including pushdown automata) was sought and studied in [14,17] in a "uniform" setting. We wish to expand this notion to a nonuniform setting.

Given a finite automaton, we append a *query tape* as an extra tape to the automaton. This tape is *write only*[3] and the automaton uses it to access an oracle in the form of query and to draw single-bit information (which corresponds to either "YES" or "NO") from the oracle. The notion of *oracle finite automata* introduced in [14,17] is obtained from finite automata by supplementing such a query tape with its corresponding alphabet (called a *query alphabet*) and a mechanism of making a series of queries to a given oracle during a computation.

Through this exposition, we deal with two types of oracles, one of which is a *deterministic oracle* and the other is a *quantum oracle*. The former is a standard notion of oracle, which is basically a set of strings and is succinctly called an *oracle*. The latter is a family of unitary transformations.

We start with describing (deterministic) oracles and how oracle finite automata access them by means of queries. An *oracle 2qfa* is a tuple $(Q, \Sigma, \{¢, \$\}, \Theta, \Xi, \delta, q_0, Q_{acc}, Q_{rej})$, where Θ is a query alphabet, δ is a map from $Q \times \check{\Sigma} \times Q \times D \times \Theta \times \Xi_\lambda$ to $\mathbb{C}$, and the rest is the same as before.

Before giving the definition of oracle 2qfa's, we make a short discussion on a query-and-answer process of an oracle finite automaton. In the model of Turing machines, relative to an oracle A, an oracle machine writes down a *query word*, say, w on its query tape and then enters one of designated *query states* to make a query of the form "is w in A?" at any time during its computation. The oracle instantly answers by changing the chosen query state to either one of YES states or one of NO states and the query tape is automatically initialized. Unfortunately, an oracle finite automaton has no memory tape, it cannot remember the inner state at the time of making a query.

Unlike oracle finite automata of [14,17] in the classical setting, oracle quantum finite automata require actions permitted by quantum physics. There are two important issues on (1) how to answer each query by an oracle in a reversible manner and (2) how to erase the content of the query tape after each oracle answer is returned.

(1) To make a process of query and answer reversible, we implement the following mechanism to the oracle 2qfa. With an oracle A, the oracle 2qfa produces a string $|y\rangle|b\rangle$ on its query tape, where $y \in \Theta^*$ and $b \in \{0,1\}$, and then enters one of the query states. In return, the oracle transforms $|y\rangle|b\rangle$ to $|y\rangle|A(y) \oplus b\rangle$ on the query tape.
(2) In the models used in [14,17], after each query word is transmitted to an oracle, the query tape is automatically erased and its tape head returns to the initial cell in a single step. Since this mechanism is not physically implemented for 2qfa's, we instead force the 2qfa's to physically erase both y and $A(y) \oplus b$ without reading any input symbol. This special process of clearing the query tape must take place before any symbol of the second query word is written. There are numerous ways to erase them unitarily. For example, a 2qfa may remove them directly to its garbage tape (although this method prevents its computation paths from interfering each other).

[3] A tape is said to be *write only* if a tape head always moves to the next blank cell just after writing any non-blank symbol.

A *quantum oracle* is actually a nonuniform family $\{U_n\}_{n\in\mathbb{N}}$ of unitary transformations, each U_n of which acts on the Hilbert space spanned by $|\Theta|^{\ell(n,|x|)}$ basis vectors $|x\rangle$ for all $x \in \Theta^*$ with $|x| = \ell(n,|x|)$, where ℓ is an appropriately chosen polynomial. When an oracle 2qfa produces a query word $y \in \Theta^*$ on a query tape and enters a query state, a given oracle applies U_n to y and generates $U_n|y\rangle$.

The deterministic oracles are so far thought as sets; however, we can take a family of promise decision problems as an oracle by modifying the above query mechanism. This modification introduces the notion of *Turing reducibility* between two nonuniform families of promise decision problems. Since an oracle is now a promise decision problem, say, (A, R), we conventionally assume that, whenever an oracle machine queries a non-promised word, the oracle returns an "arbitrary" answer.

Given two families $\mathcal{L} = \{(L_n, \overline{L}_n)\}_{n\in\mathbb{N}}$ and $\mathcal{K} = \{(K_n, \overline{K}_n)\}_{n\in\mathbb{N}}$ of promise decision problems, we say that $\mathcal{L}$ is *2BQ-Turing reducible* to $\mathcal{K}$, written as $\mathcal{L} \leq_T^{2BQ} \mathcal{K}$, if there exist a constant $\varepsilon \in [0, 1/2)$ and a suitable nonuniform family $\{M_n\}_{n\in\mathbb{N}}$ of oracle 2qfa's satisfying the following requirement: for any $n \in \mathbb{N}$ and for any $x \in L_n$ (resp., $x \in \overline{L}_n$), M_n accepts (resp., rejects) x with probability at least $1 - \varepsilon$ with any access to $\mathcal{K}$ as an oracle. Unfortunately, the union $\bigcup_{n\in\mathbb{N}}(K_n \cup \overline{K}_n)$ in general does not equal to Σ^*. Hence, when the oracle 2qfa makes a query with a word w, the oracle need to know which promise problem $(K_n, \overline{K}_n)$ to answer.

To specify a pair (n, w), we split a query tape into two tracks. Each tape cell holds two symbols in its upper and lower tracks, and a tape head simultaneously reads the two symbols written in a single tape cell. To express the content of such a tape cell, we use a track notation $[\,{}^{a}_{b}]$ of [14], which indicates that the upper track holds a symbol a and the lower track contains a symbol b. We write $[\,{}^{x}_{y}]$ as a series of such track symbols. In this exposition, we demand that each 1pfa should write down $[\,{}^{1^n}_{w}]$ on a query tape to ask whether either $w \in K_n$ or $w \in \overline{K}_n$. In a similar fashion, we can define $\leq_T^{2D}$, $\leq_T^{2BQ}$, etc. As noted in Sect. 1.2, we define $2BQ^{\mathcal{K}}$ to be the collection of all nonuniform families $\mathcal{L}$ of promise decision problems, each $\mathcal{L}$ of which is 2BQ-Turing reducible to $\mathcal{K}$ by a nonuniform family of oracle 2qfa's.

3 Oracle Separations: Proofs of Theorems 1–3

3.1 Separations by Deterministic Oracles

We begin with the proof of Theorem 1, which asserts the existence of two recursive oracles to which (a) 1N $\nsubseteq$ 2D and 1BQ $\subseteq$ 2D and (b) 1N $\subseteq$ 2D and 1BQ $\nsubseteq$ 2D. Since finite automata may use different input alphabets, we assume the existence of an efficient encoding scheme of each string into its corresponding binary representation.

Our construction of the desired oracles is based on an effective enumeration (with repetitions) of associated finite automata of polynomial state complexity.

Let us recall the track notation $[\,^{a}_{b}]$. Let $\Sigma_{(2)} = \{[\,^{a}_{b}] \mid a, b \in \Sigma\}$ and let $\Sigma_{(4)} = \Sigma \cup \Sigma_{(2)}$.

Proof Sketch of Theorem 1. In what follows, we will verify only the separation result of (a). Our recursive oracle A must satisfy that (i) $1\mathrm{N}^A \nsubseteq 2\mathrm{D}^A$ and (ii) $1\mathrm{BQ}^A \subseteq 2\mathrm{D}^A$. We set $\Sigma = \{0,1\}$ and we split the desired oracle A into two parts, $0\Sigma^*_{(4)} \cap A$ and $1\Sigma^*_{(4)} \cap A$. As an example promise problem, we choose $L^A_n = \{x \in \Sigma^n \mid \exists y[|x| = |y| \wedge 0[\,^{x}_{y}] \in A]\}$ and $\bar{L}^A_n = \Sigma^n - L^A_n$. Finally, we set $\mathcal{L}^A = \{(L^A_n, \bar{L}^A_n)\}_{n\in\mathbb{N}}$.

Let us claim that $\mathcal{L}^A \in 1\mathrm{N}^A$ for any oracle A. To prove this claim, for each index $n \in \mathbb{N}$, we consider the oracle 1nfa N_n that behaves as follows. For any input $x \in \Sigma^n$, we nondeterministically generate $0[\,^{x}_{y}]$ for strings y with $|y| = |x|$ on a query tape, and then transmit this query word to the given oracle A. If $0[\,^{x}_{y}] \in A$, then we accept x; otherwise, we reject x. It is not difficult to verify that, for any oracle A and for any $n \in \mathbb{N}$, N_n solves $(L^A_n, \bar{L}^A_n)$.

A basic idea to achieve (i) is that we pick every oracle 2dfa (according to a fixed effective enumeration of all oracle 2dfa's) and encode its outcome into the desired oracle A. This is possible because the 2dfa makes only polynomially many queries and there is still enough room in $0\Sigma^n_{(4)}$ to insert such an encoding outside of the set of all query words produced by the 2dfa so that the oracle 1nfa N_n can find the encoding nondeterministically and produce the negation of the outcome of the 2dfa.

For (ii), assuming an effective enumeration of oracle 1qfa's whose amplitudes are approximated to within 2^{-n} by deterministic Turing machines on inputs of the form 1^n, we take such oracle 1qfa's N_i one by one and search for an enough room in A where we can insert the encoding of the outcome of N_i so that a certain oracle 2dfa, say, K_i easily finds the encoding. We then claim that, for any N_i and for any $x \in \Sigma^n$, K_i relative to A_n agrees with N_i on the input x.□

Next, we want to show Theorem 2. In its proof sketch that follows shortly, we will meet two different requirements, $2\mathrm{D} = 2\mathrm{BP}$ and $2\mathrm{BP} \neq 2\mathrm{BQ}$, each of which needs a special attention not encountered in the polynomial-time setting. For the proof of Theorem 2(a), we need to consider the following decision problem. Let $\#_b(x)$ indicate the total number of symbol b in a string x. Fix a number $n \in \mathbb{N}$ arbitrarily. Let $\mathcal{A}_{even} = \{1x \in \{0,1\}^{n+1} \mid \#_1(x) = 0 \pmod 2\}$ and $\mathcal{A}_{odd} = \{1x \in \{0,1\}^{n+1} \mid \#_1(x) = 1 \pmod 2\}$. For each Boolean function $f_n : 1\{0,1\}^n \to \{0,1\}$, by letting $\ell = |\mathcal{A}_{even} \cap f_n^{-1}(1)| - |\mathcal{A}_{odd} \cap f_n^{-1}(1)|$, it always follows that either $\ell = 0$ or $\ell \neq 0$. The notation $f_n^{-1}(i)$ expresses the inverse image $\{x \in 1\{0,1\}^n \mid f_n(x) = i\}$ for each index $i \in \{0,1\}$. As a decision problem, we are asked to determine whether or not $\ell \neq 0$. As shown in the following proof sketch, this problem is solved quantumly with accesses to f_n.

Proof Sketch of Theorem 2. We will verify only (a) by constructing a recursive oracle A for which (i) $2\mathrm{D}^A = 2\mathrm{BP}^A$ and (ii) $2\mathrm{BP}^A \neq 2\mathrm{BQ}^A$. We first enumerate all 2qfa's $\{M_i\}_{i\in\mathbb{N}}$ (with repetitions) and error bounds $\{\varepsilon_i\}_{i\in\mathbb{N}}$ (with repetitions) so that M_i is ε_i-error bounded.

We define an example problem family $\{(L_n^A, \bar{L}_n^A)\}_{n\in\mathbb{N}}$ relative to an oracle A for $2\mathrm{BQ}^A$. We split an oracle into two sections $0\Sigma^* \cap A$ and $1\Sigma^* \cap A$. For each index $n \in \mathbb{N}$, we take a function $f_n : 1\{0,1\}^n \to \{0,1\}$ and define $A = \{1x \in \{0,1\}^* \mid f_{|x|}(1x) = 1\}$. We further define $L_n^A = \{1^n \mid |\mathcal{A}_{even} \cap f_n^{-1}(1)| \neq |\mathcal{A}_{odd} \cap f_n^{-1}(1)|\}$ and $\bar{L}_n^A = \{1^n\} - L_n^A$. We set $\mathcal{L}^A = \{(L_n, \bar{L}_n^A)\}_{n\in\mathbb{N}}$.

We prove only (ii). First, we wish to claim that $\mathcal{L}^A \in 2\mathrm{BQ}^A$ holds for any A defined above. For the proof of this claim, we consider the following quantum algorithm. On input 1^n, we generate $2^{-n/2}\sum_{|x|=n}|1x\rangle|0\rangle$ on a query tape. We make a query and obtain $\sum_x |1x\rangle|0 \oplus A(1x)\rangle$ from $\sum_{x:|x|=n}|1x\rangle|0\rangle$. Since the vector $|0 \oplus A(1x)\rangle$ equals $|f_n(1x)\rangle$, we encode $f_n(1x)$ into its associated inner state, say, $r_{f_n(1x)}$. And then delete $f_n(1x)$. We further erase the query tape for the next query. For this purpose, we sequentially apply $H^{\oplus n}$ to $|x\rangle$. If $H^{\oplus n}$ produces 1^n, then we obtain $\sum_{x:|x|=n}(-1)^{\#_1(x)}|1^{n+1}\rangle|r_{f_n(1x)}\rangle$. In the case where $|\mathcal{A}_{even} \cap f^{-1}(1)| = |\mathcal{A}_{odd} \cap f_n^{-1}(1)|$, $|1^n\rangle$ is not observed. If $|\mathcal{A}_{even} \cap f_n^{-1}(1)| \neq |\mathcal{A}_{odd} \cap f_n^{-1}(1)|$, we enter an accepting state with probability at least $1/2^n$. By contrast, whenever H produces 0 at a certain step, we remember this fact in its inner state, write 1 instead, and remove 0 to a garbage tape. After making the query tape blank, we generate a computation path with probability $1/2^{n+2}$ by scanning the input 1^n, enter a rejecting state along this path, and restart the entire computation from scratch.

The proof of $\mathcal{L}^A \notin 2\mathrm{BP}^A$ is quite different from the proofs of Theorem 1 because 2pfa's can run for more than exponentially many steps. Here, we employ an idea that, for a certain polynomial p and for each oracle 2pfa M_n using at most $p(n)$ inner states, in order to compute the correct value of $|A \cap 1\Sigma^n|$ for every oracle A, a 2pfa must query every string in $1\Sigma^n$ exactly once (otherwise, the 2pfa cannot count the number $|A \cap 1\Sigma^n|$ correctly because it can neither remember all the query words nor hold the number more than $p(n)$). Thus, the 2pfa cannot solve $(L_n, \overline{L}_n)$ with bounded-error probability. □

3.2 Separations by Quantum Oracles

Unlike deterministic oracles that we have discussed so far, quantum oracles behave quite differently for underlying oracle quantum finite automata. In what follows, we are focused on constructing such quantum oracles that empower the computation of super quantum finite automata, compared to the standard quantum finite automata. Let us recall from Sect. 2.3 that a quantum oracle is a family of unitary operators, which are applied to quantum states of query words generated on query tapes by underlying oracle quantum finite automata. Earlier, in the polynomial-time setting, Aaronson and Kuperberg [1] constructed the first quantum oracle to separate QMA from QCMA as well as BQP/Qpoly from BQP/poly. In contrast, our goal here is to separate 2sBQ from 2BQ by a quantum oracle.

The proof of [1] heavily relies on the fact that underlying oracle quantum Turing machines run in polynomial time. Unfortunately, similarly to the case of deterministic oracles, since 2qfa's can run for more than exponentially many steps, the proof of [1] does not seem to work in its original form. We thus need to seek a different argument for the proof of Theorem 3.

Proof Sketch of Theorem 3. We wish to construct a quantum oracle U that guarantees that $2\mathrm{sBQ}^U \nsubseteq 2\mathrm{BQ}^U$.

Let $\Sigma = \{0,1\}$. We fix an arbitrary quantum state $|\phi_n\rangle = \sum_{i=1}^{n} \alpha_i |\bar{q}_i\rangle$, where $\alpha_i \in \mathbb{C}$ and $\bar{q}_i \in Q_n$. Take an arbitrary promise problem $(L_n, \overline{L}_n)$ for each $n \in \mathbb{N}$ with $L_n \cup \overline{L}_n \subseteq \Sigma^n$ and set $\mathcal{L} = \{(L_n, \overline{L}_n)\}_{n\in\mathbb{N}}$. The desired quantum oracle $U = \{U_n\}_{n\in\mathbb{N}}$ is defined as

$$U_n|b\rangle|\psi\rangle|x\rangle = \begin{cases} (-1)^{L_n(x)\oplus b'}|b\rangle|\psi\rangle|x\rangle & \text{if } |\psi\rangle = |\phi_n\rangle, \\ |b\rangle|\psi\rangle|x\rangle & \text{if } \langle\psi|\phi_n\rangle = 0, \end{cases}$$

where $b \in \{q'_0, q'_1\}$ and $b' = 0$ (resp., $= 1$) if $b = q'_0$ (resp., $= q'_1$). For clarity, we write U_{ϕ_n} to denote the transform obtained from the above transform U_n by setting $L_n(x) = 1$.

We define an oracle 1qfa M_n to solve $(L_n, \overline{L}_n)$ as follows. Let x be an input. At reading ¢, we generate $\frac{1}{\sqrt{2}}(|q'_0\rangle + |q'_1\rangle)$ from $|q'_0\rangle$. Following a transition table T, we copy x into a query tape. When we reach \$, we write $\bar{q}_i$ onto the query tape with amplitude α_i. We enter a query state and then obtain $(-1)^{L_n(x)\oplus b'}|b\rangle|\phi_n\rangle|x\rangle$ by the quantum oracle. By changing $|b\rangle$ to $\frac{1}{\sqrt{2}}(|q'_0\rangle + (-1)^{b'}|q'_1\rangle)$, we obtain $|L_n(x)\rangle|\phi_n\rangle|x\rangle$. We then measure $|b\rangle$ and accept (resp., reject) if $b = q'_1$ (resp., $b = q'_0$).

If $\mathcal{L} \in 2\mathrm{BQ}^U$, then we can distinguish between U_{ϕ_n} and I. However, we can show that this is basically impossible for 2pfa's with bounded-error probability. Therefore, $\mathcal{L}$ does not belong to $2\mathrm{BQ}^U$. □

4 Turing Reducibility: Proofs of Theorems 4–5

4.1 Closure Properties

We look into *Turing reducibility*, which extends the relativization of oracle finite automata families relative to nonuniform families of promise decision problems used as oracles. Hence, the Turing reducibility seems to be a powerful tool in expanding machine-based complexity classes. In certain cases, however, complexity classes are closed under Turing reductions. Theorem 4 exhibits some of those cases.

Proof Sketch of Theorem 4. (a) Hereafter, we show that $2\mathrm{D}^{2\mathrm{D}} = 2\mathrm{D}$. Since $2\mathrm{D} \subseteq 2\mathrm{D}^{2\mathrm{D}}$ is obvious, it suffices to show the converse inclusion. Let p and r denote two arbitrary polynomials. Take a family $\mathcal{L} = \{(L_n, \overline{L}_n)\}_{n\in\mathbb{N}}$ from $2\mathrm{D}^{2\mathrm{D}}$ and take another family $\mathcal{K} = \{(K_n, \overline{K}_n)\}_{n\in\mathbb{N}}$ in 2D such that $\mathcal{L} \leq_T^{2\mathrm{D}} \mathcal{K}$ via a family $\{M_n\}_{n\in\mathbb{N}}$ of oracle 2dfa's having at most $p(n)$ inner states. Let $\{N_n\}_{n\in\mathbb{N}}$ be a family of 2dfa's of at most $r(n)$ inner states that solves $\mathcal{K}$.

A basic idea of our simulation is that, whenever a machine writes a query word, we run a machine computing an oracle to process the query word, symbol by symbol. This simulation can be properly implemented by a certain 2dfa's with polynomially many inner states. This fact implies that $\mathcal{L}$ belongs to 2D.

(b) To show that $1\mathrm{BP}^{1\mathrm{BP}[1]} = 1\mathrm{BP}$, we can employ the proof, which is similar in essence to (1) except that an underlying oracle 1pfa makes at most one query along any computation path. The case of $1\mathrm{P}^{1\mathrm{P}[1]} = 1\mathrm{P}$ is similarly handled. □

4.2 The Second Level of a Hierarchy

Let us verify Theorem 5, which asserts the separation of 2BQ from $2\Sigma_2$ in an appropriate relativized world. A basic idea of the proof of the theorem comes from [17, Theorem 4.21, arXiv version], in which an appropriately chosen recursive oracle separates the bounded-error probabilistic CFL (denoted BPCFL) from the second level Σ_2^{CFL} of the so-called *CFL hierarchy* (see [17] for their formal definitions).

Proof Sketch of Theorem 5. We consider $(L_n, \overline{L}_n)$ defined in the proof sketch of Theorem 2. Next, we want to construct a recursive oracle A that separates 2BQ from $2\Sigma_2$. We utilize [17, Theorem 4.21, arXiv version], in which we represent an oracle computation of a 2nfa relative to oracle A as an $\ell(n)$-input leveled Boolean circuit of depth 3 with the top OR gate, alternating levels of OR and AND, with the bottom gate of fan-in at most cn, and any other gate of fan-in at most 3^{an} for certain constants $a, b > 0$, where $\ell(n)$ is a polynomial indicating the number of oracle queries (since 2nfa's can be thought to terminate within polynomially many steps). Similarly to [17, Theorem 4.21, arXiv version], we can construct an oracle A that separates $(L_n^A, \overline{L}_n^A)$ from such Boolean circuits that take A as inputs. □

5 A Brief Discussion of Future Research

This exposition has formally formulated the notion of relativizations by equipping various finite automata having polynomially many inner states with the oracle mechanisms of [14] and [17, arXiv version]. In the previous sections, we have studied the behaviors of various relativized finite automata and demonstrated inclusions and separations among nonuniform state complexity classes by constructing oracles that provide circumstantial evidences to the difficulty of proving such inclusions and separations in the unrelativized world. The relativization turns out to be a useful tool in providing such evidences to the structural properties of target nonuniform state complexity classes.

In the polynomial-time setting, relativization methodology has already paved a natural way to a comprehensive study of Turing reductions. As for probabilistic and quantum finite automata, the behaviors of Turing reducibility seems more complex than those of time-bounded Turing machines. For example, in the polynomial-time setting, we obtain $\mathrm{BQP}^{\mathrm{BQP}} = \mathrm{BQP}$ but we conjecture that $2\mathrm{BQ}^{2\mathrm{BQ}} \neq 2\mathrm{BQ}$ in the case of nonuniform state complexity.

Relativizations of finite automata [14,17], together with this exposition, are a relatively new challenge in automata theory and they require a further intensive investigation on the methodology of oracle constructions. We strongly hope that relativization will enrich the field of finite automata.

There remain numerous questions left unsolved in this exposition. For example, we do not know whether quantum oracle separations of Theorem 3 can be converted into deterministic oracle separations. Solving those questions surely enriches the automata research.

References

1. Aaronson, S., Kuperberg, G.: Quantum versus classical proofs and advice. Theory Comput. **3**, 129–157 (2007)
2. Ambainis, A., Beaudry, M., Golovkins, M., Kikusts, A., Mercer, M., Thérien, D.: Algebraic results on quantum automata. Theory Comput. Syst. **39**, 165–188 (2006)
3. Baker, T., Gill, J., Solovay, R.: Relativizations of the P=?NP question. SIAM J. Comput. **4**, 431–442 (1975)
4. Berman, P., Lingas, A.: On complexity of regular languages in terms of finite automata. Technical report 304, Institute of Computer Science, Polish Academy of Science, Warsaw (1977)
5. Dwork, C., Stockmeyer, L.: A time-complexity gap for two-way probabilistic finite state automata. SIAM J. Comput. **19**, 1011–1023 (1990)
6. Freivalds, R., Ozols, M., Mančinska, L.: Improved constructions of mixed state quantum automata. Theoret. Comput. Sci. **410**, 1923–1931 (2009)
7. Gruska, J.: Quantum Computing. McGraw-Hill, London (2000)
8. Kapoutsis, C.A.: Size complexity of two-way finite automata. In: Diekert, V., Nowotka, D. (eds.) DLT 2009. LNCS, vol. 5583, pp. 47–66. Springer, Heidelberg (2009). https://doi.org/10.1007/978-3-642-02737-6_4
9. Kapoutsis, C.A.: Minicomplexity. J. Autom. Lang. Comb. **17**, 205–224 (2012)
10. Kapoutsis, C.A.: Two-way automata versus logarithmic space. Theory Comput. Syst. **55**, 421–447 (2014)
11. Kapoutsis, C.A., Pighizzini, G.: Two-way automata characterizations of L/poly versus NL. Theory Comput. Syst. **56**, 662–685 (2015)
12. Kitaev, A., Shen, A., Vyalyi, M.: Classical and Quantum Computation. AMS, Providence (2002)
13. Sakoda, W.J., Sipser, M.: Nondeterminism and the size of two-way finite automata. In: Proceedings of STOC 1978, pp. 275–286 (1978)
14. Tadaki, K., Yamakami, T., Lin, J.C.H.: Theory of one-tape linear-time Turing machines. Theor. Comput. Sci. **411**, 22–43 (2010)
15. Villagra, M., Yamakami, T.: Quantum state complexity of formal languages. In: Shallit, J., Okhotin, A. (eds.) DCFS 2015. LNCS, vol. 9118, pp. 280–291. Springer, Cham (2015). https://doi.org/10.1007/978-3-319-19225-3_24
16. Yakaryilmaz, A., Say, A.C.C.: Unbounded-error quantum computation with small space bounds. Inf. Comput. **279**, 873–892 (2011)
17. Yamakami, T.: Oracle pushdown automata, nondeterministic reducibilities, and the hierarchy over the family of context-free languages. In: Geffert, V., Preneel, B., Rovan, B., Štuller, J., Tjoa, A.M. (eds.) SOFSEM 2014. LNCS, vol. 8327, pp. 514–525. Springer, Cham (2014). https://doi.org/10.1007/978-3-319-04298-5_45. A complete version is available at arXiv:1303.1717

18. Yamakami, T.: Nonuniform families of polynomial-size quantum finite automata and quantum logarithmic-space computation with polynomial-size advice. In: Martín-Vide, C., Okhotin, A., Shapira, D. (eds.) LATA 2019. LNCS, vol. 11417, pp. 134–145. Springer, Cham (2018). https://doi.org/10.1007/978-3-030-13435-8_10
19. Yamakami, T.: State complexity characterizations of parameterized degree-bounded graph connectivity, sub-linear space computation, and the linear space hypothesis. In: Konstantinidis, S., Pighizzini, G. (eds.) DCFS 2018. LNCS, vol. 10952, pp. 237–249. Springer, Cham (2018). https://doi.org/10.1007/978-3-319-94631-3_20. A complete and corrected version is found at arXiv:1811.06336

Self-stabilizing Gellular Automata

Tatsuya Yamashita, Akira Yagawa, and Masami Hagiya(✉)

The University of Tokyo, Tokyo, Japan
hagiya@is.s.u-tokyo.ac.jp

Abstract. Gellular automata are a class of cellular automata having the features: asynchrony, Boolean totality, and non-camouflage. Gellular automata have been introduced as models of smart materials made of porous gels and chemical solutions, which are expected to have abilities such as self-repair. Therefore investigating gellular automata that are self-stable is an important research topic as self-stability implies convergence to a target configuration even under external disturbances. In this paper, we present gellular automata which solve maze problems self-stably. We also briefly describe gellular automata that solve the leader election problem. We thus discuss the possibility of implementing self-stable distributed algorithms by gellular automata.

Keywords: Gellular automata · Maze problem · Self-stabilizing

1 Introduction

In recent years, research for making computers and robots using DNA molecules has been actively conducted [1,8,10]. Along the line of research is the study on gellular automata [4,11], which have been introduced as a class of cellular automata (e.g., refer to [10,12]) for modeling and designing smart materials made of porous gels [5,7]. They are partitioned into compartments containing chemical solutions with DNA molecules.

In this and previous papers, gellular automata are defined as cellular automata that are asynchronous, Boolean-totalistic, and non-camouflage as explained in the next section. In the previous papers, we studied the computational universality of gellular automata [6,14]. We also examined the potential of gellular automata in distributed computation through simulation of population protocols [13].

In this paper, we focus on self-stability (refer to [2]) of gellular automata because self-stability implies self-organization and self-repair, which are typical features of smart materials. Gellular automata are called self-stable if they always reach a target configuration from any initial configuration and continue to stay in target configurations. This means that even though they are damaged after a target configuration is reached, they repair themselves and reach a new target configuration. If the underlying models of smart materials are self-stable, they

I. McQuillan and S. Seki (Eds.): UCNC 2019, LNCS 11493, pp. 272–285, 2019.
https://doi.org/10.1007/978-3-030-19311-9_21

can survive in a harsh environment by repairing themselves even though they are damaged by external disturbances.

As a concrete example, we mainly deal with gellular automata that solve a maze problem self-stably because self-stability is evident in the context of solving a maze. Even though a maze is changed after a path from an entrance to an egress is found, a new path is eventually constructed if self-stability holds. Maze solving and maze generation have been studied from various perspectives in research on cellular automata (e.g., refer to [3,9]). In this paper, we focus on self-stability of a restricted class of cellular automata called gellular automata.

In the next section, we explain the conditions for cellular automata to become gellular automata. We then define gellular automata that solve a maze problem, and show that they are self-stable. We finally touch on leader election, which is another typical problem in distributed computing.

The reader is invited to visit https://cell-sim.firebaseapp.com/, which shows the two examples of gellular automata.

2 Constraints on Gellular Automata

We assume that in porous gel materials modeled by gellular automata, each chemical solution in a partitioned compartment consists of two types of molecules: those that pass through porous gel partition and those that keep staying inside a compartment. The molecules of the former type, called *signals*, are used for communication between compartments. Those of the latter have some stable states, and produce a signal depending on their current state. The chemical reaction between the molecules having states and the signals produced in the neighboring compartments is modeled as a state transition of cellular automata. The above assumptions require that cellular automata modeling such materials be asynchronous, Boolean-totalistic, and non-camouflage as defined below.

Cells in *asynchronous* cellular automata make transitions asynchronously in the sense that in one instance of time, each cell may either make a transition following a transition rule or remain in its current state. Asynchrony is required because chemical solutions in a compartment may not make transitions simultaneously.

Boolean-totalistic cellular automata are a variant of totalistic cellular automata. Each of their transition rules decides the next state of a cell depending only on the current state of the cell and the set of states of its neighbors. Unlike general cellular automata, Boolean-totalistic cellular automata ignore the direction and the number of neighboring cells in each state. They only recognize the existence or non-existence of a neighboring cell in each state.

A cell in state s of *non-camouflage* cellular automata makes a transition independently of the existence of a neighboring cell in state s. Gellular automata are non-camouflage because it is not possible to distinguish between a signal produced in a compartment from the same signal that has traveled from a neighboring compartment.

In this paper, we call asynchronous Boolean-totalistic non-camouflage cellular automata *gellular automata*. We assume that the cellular space in which gellular automata are defined is a two-dimensional square lattice with von Neumann neighborhood.

A transition rule of Boolean-totalistic cellular automata is written in the following form:

$$r\ (s_1 \wedge \ldots \wedge s_M \wedge \neg t_1 \wedge \ldots \wedge \neg t_N)\ \rightarrow\ u.$$

This rule means that if a cell is in state r and there are neighboring cells in states $s_1, \ldots, s_M$ and if there are no neighboring cells in states $t_1, \ldots, t_N$, then the cell can make a transition to state u. In a transition rule of non-camouflage cellular automata, all of $s_1, \ldots, s_M, t_1, \ldots, t_N$ should differ from r.

A *configuration* is defined as a mapping from cells to states, and a *run* is defined as an infinite sequence of configurations in which each configuration except the first is obtained from its predecessor by a transition. During a transition of a configuration, one transition rule or no rule is applied to each cell. If multiple rules are applicable to a cell, it makes a non-deterministic choice among the rules. It also has a choice not to make a transition. If no rule is applicable, it does not make a transition. A transition is called *completely asynchronous* if only one cell changes its state. A run is also called completely asynchronous if all transitions are completely asynchronous.

3 Gellular Automata Solving Maze Problems

In this section, we discuss how gellular automata solve maze problems and how they are said to be self-stabilizing.

3.1 Representation of Mazes and Solutions

First of all, we should define how to represent maze problems and their solutions in gellular automata. To represent maze problems, we introduce four states B, W, S, and T. They correspond to a blank, a wall, a starting point, and a terminal point, respectively.

A maze problem is represented by a configuration in which one cell is in state S, another cell is in state T, and other cells are all in state B or W. We call the cell in state S the *starting point*, and the cell in state T the *terminal point*, respectively. Cells in state W are called *walls* while other cells, i.e., cells in state B are called *passages*. We later introduce more and more states to which cells in state B may change. Cells in those states will also be called passages. When we show a part of a configuration in a figure, we express a cell in state B by a white square, a cell in state W by a black square, and a cell in another state by a square with the corresponding letter, respectively.

The goal of a maze problem is to find a path between the starting point and the terminal point. Let us discuss how to define a path and a solution of a maze problem.

Assume $n \geq 3$. We take $n = 3$ in some of the following examples, but we eventually take $n = 5$. We introduce n new states $P_0, \ldots, P_{n-1}$ and consider a sequence of neighboring cells in states $P_0, P_1, \ldots, P_{n-1}, P_0, \ldots$, in which the indices are increased by one (in $\mathbb{Z}/n\mathbb{Z}$). We call such a sequence a *path*. If a path ends in a cell in state P_i which is not adjacent to a cell in state P_{i+1} ($i \in \mathbb{Z}/n\mathbb{Z}$), we say that the path is *maximal*. Infinite paths (along a loop) are also considered as maximal.

We then define a solution of a maze.

Definition 1 (solution). *A solution of a maze is a configuration in which there exists a path from a cell adjacent to the starting point to a cell adjacent to the terminal point, and any maximal path from a cell adjacent to the starting point passes a cell adjacent to the terminal point.*

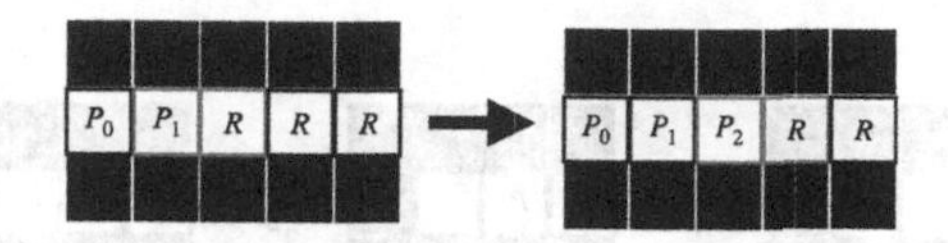

Fig. 1. Progress of a head

3.2 Constructing Paths

From now on, we discuss how gellular automata solve mazes. We take the following two-step approach similar to [3]. In the first step, a region of cells in a new state R is spread from the terminal point. In the second step, when the region reaches the starting point, paths are constructed backward to the terminal point.

If a passage has a neighboring cell in state T or R, it makes a transition to state R. We introduce the following transition rules.

$$B\ (T) \rightarrow R \tag{1}$$

$$B\ (R) \rightarrow R \tag{2}$$

After a passage adjacent to the starting point changes to state R, it further changes to state P_0 by the following transition rule.

$$R\ (S) \rightarrow P_0 \tag{3}$$

Of course, if a cell in state R has a neighboring cell in state P_0, it should make a transition to state P_1, and the next cell should again make a transition to state P_2. These transitions are derived from the following n rules

$$R\ (P_i) \rightarrow P_{i+1} \tag{4}$$

for any $i \in \mathbb{Z}/n\mathbb{Z}$. If cells in states R and P_i are adjacent, we say that they form a *head*. As shown in Fig. 1, if the cell in state R of a head is adjacent to another cell in state R, the application of the rule (4) makes the head move forward. We say that the head *progresses.*

Can the gellular automata defined only by the above transition rules solve mazes? The answer is "No." The gellular automata cannot cope with either of dead ends or loops, which are the main difficulties in solving a maze problem.

3.3 Coping with Dead Ends

Since the rule (4) may generate a path from the starting point not only to the terminal point but also to a dead end, we need another rule for deleting a path to a dead end.

$$P_i \ (\neg T \wedge \neg R \wedge \neg P_{i+1}) \ \rightarrow \ B \tag{5}$$

Figure 2 shows how this rule makes the head *regress* and deletes a path.

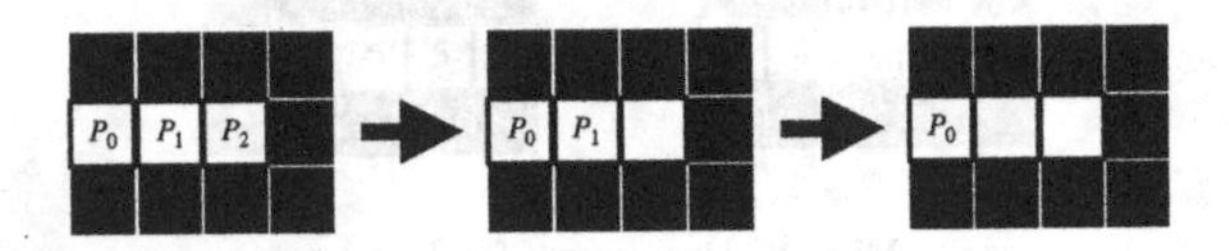

Fig. 2. Regress of a path

Fig. 3. A loop

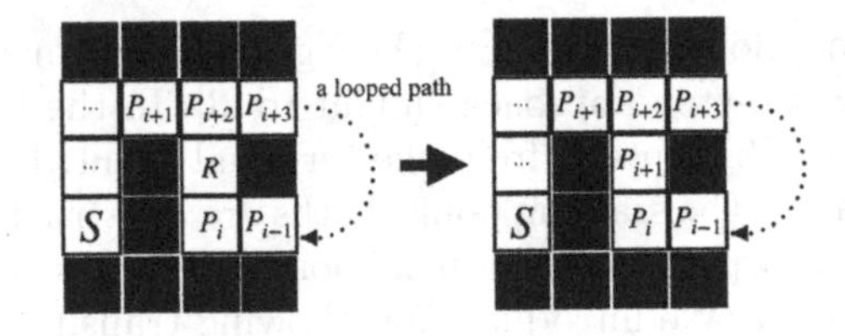

Fig. 4. Consistent collision at the entrance of a loop

3.4 Coping with Loops

A *passage* is now redefined as a cell whose state is not among W, S and T, i.e., cells in state P_i or R are also passages. Two distinct cells a and b are said to be *connected* iff there is a sequence of neighboring cells $a = c_0, c_1, \ldots, c_L = b \, (L \in \mathbb{N}^+)$ in which all intermediate cells $c_1, \ldots, c_{L-1}$ are passages.

A sequence of neighboring passages is called a *loop* if its first and last cells are identical. Consider a loop that is connected to both the starting point and the terminal point, as shown in Fig. 3. In such a loop, there is at least one cell that can be reached from the starting (terminal) point without passing the other cells in the loop. We call such a cell the *entrance* (*egress,* respectively) of the loop. The cell marked $\star$ in Fig. 3 is the entrance.

As shown in Figs. 5, 6 and 7, if multiple heads are adjacent to a cell in state R and one of them progresses, then the number of heads is reduced. We call such a transition a *collision* of heads. In particular, if the indices of states in the sequence generated by the colliding heads are i, $i+1$, i, then we say that the collision is *consistent* and call the cell in state P_{i+1} the *collision point* because the collision point is shared by the two paths. Note that in a consistent collision as in Fig. 6, the cell which makes the last transition is not necessarily the collision point.

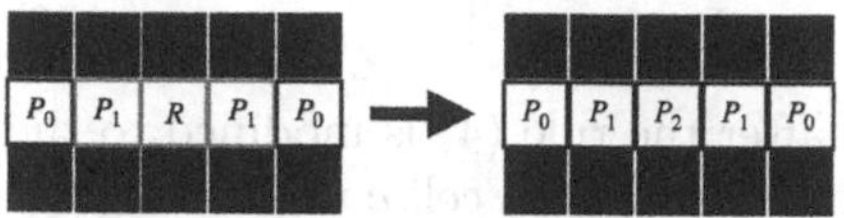

Fig. 5. Consistent collision of heads ($n = 3$)

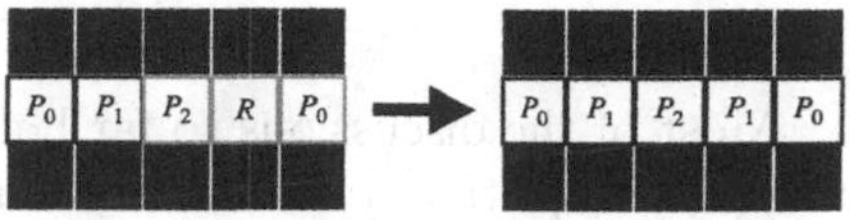

Fig. 6. Consistent collision of heads (another situation)

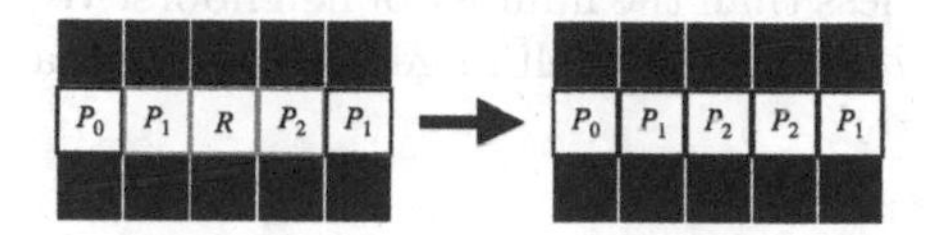

Fig. 7. Inconsistent collision of heads

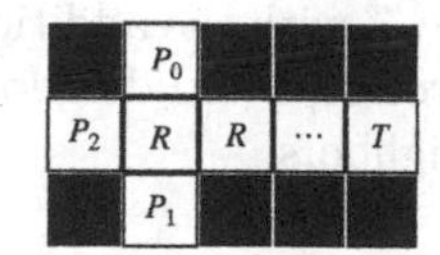

Fig. 8. Situation causing a deadlock (the case n = 3)

Assume that a head has progressed to the entrance from the starting point. The head will diverge at the entrance, and then the two heads will progress until their collision. The collision may be consistent or not.

If it is not consistent, the paths generated by the two heads will regress to the entrance or egress, and only the path from the entrance to the egress will remain.

If the collision is consistent, there are three detailed cases depending on the position of the collision point.

If the collision point is neither at the entrance nor egress of the loop, the rule (5) can be applied to the collision point even if the collision is consistent. Therefore, only the path from the entrance to the egress will remain as in the case of an inconsistent collision.

If the collision point is at the egress, both the paths generated by the two heads connect the entrance and egress. Since these paths may be contained in paths solving the maze, they need not be deleted.

The most difficult case occurs if the collision point is at an entrance. This case occurs, for instance, if only one of the two heads progresses until it circles the loop as shown in Fig. 4. Assume the entrance is in state P_{i+2}. There are at least two cells in state P_{i+1} and a cell in state P_{i+3} in the neighborhood of the entrance. Contrary to the case that the collision point is at the egress, an infinite path is obtained by tracing the loop.

We therefore need to avoid a consistent collision at an entrance. The configuration just before such a collision must be as in the left panel of Fig. 4, and the collision must have occurred by the transition of the cell in state R to state P_{i+1}. In this situation, the cell in state R can make a transition to state P_{i+3} instead of state P_{i+1}. This transition induces an inconsistent collision, and the redundant path will regress as we have already confirmed. To make a transition to state P_{i+3} constantly, we modify the rule (4) to the following one (6).

$$R\ (P_i \wedge \neg P_{i+2}) \ \rightarrow \ P_{i+1} \tag{6}$$

Most of the discussions so far hold even after the rule (4) is modified to the rule (6) except that a deadlock of multiple heads sharing a cell c in state R may occur as shown in Fig. 8. Assume that all cells between c and the terminal point are in state R. Then, in the case $n = 3$, $P_{2+2} = P_1$ and $P_{2+2+2} = P_0$ hold, so the configuration shown in Fig. 8 cannot make any transition. In general, if k heads with cells in states P_i, $P_{i+2}, \ldots, P_{i+2(k-1)}$ share a cell in state R, a deadlock occurs iff one of the heads has a cell in state P_{i+2k}, i.e., the order of 2 in the group $\mathbb{Z}/n\mathbb{Z}$ with the addition operation is less than the number of neighbors. We can, therefore, avoid deadlocks by taking $n = 5$ since a cell in gellular automata has 4 neighbors.

3.5 Self-stability

In general, a distributed system is *self-stable* if any fair run from any initial configuration reaches a step in the run from which only target configurations appear.

A (*completely asynchronous*) run $\mathcal{R}$ is called *fair* (or *globally fair*) if for any configuration $\mathcal{C}$ that appears infinitely often in $\mathcal{R}$, and for any configuration $\mathcal{D}$ obtained from $\mathcal{C}$ by a (*completely asynchronous*) transition, $\mathcal{D}$ also appears infinitely often in $\mathcal{R}$.

We define *initial configurations* of the gellular automata solving maze problems. We also define *ideal initial configurations.*

Definition 2 (initial configuration). *An initial configuration is a configuration that satisfies the following conditions (◇1–2).*

(◇1) *There are just one starting point and one terminal point, and they are not adjacent.*

(◇2) *The number of cells that are not in state W is finite.*

The condition (◇1) is required by the definition of a maze problem. The second one (◇2) means that the maze is not infinitely large but bounded. Since walls remain as walls in any transition, the conditions (◇1–2) are preserved by transitions.

Definition 3 (ideal initial condition). *An ideal initial configuration is an initial configuration in which all cells are in state W, B, S, or T.*

As we defined solutions of a maze, we define target configurations as follows.

Definition 4 (target configuration). *A target configuration is a configuration that satisfies the following conditions (⋆1–2).*

(⋆1) *If the starting point and the terminal point are connected, there is a path between them.*

(⋆2) *Any maximal path from a cell adjacent to the starting point passes a cell adjacent to the terminal point.*

These conditions are satisfied by solutions of the maze. If the starting point and the terminal point are not connected, the second condition (⋆2) implies that there is no path from a cell adjacent to the starting point.

For showing self-stability, we introduce the following additional conditions.

(⋆3) No cell connected to the starting point is in state R, and no cell adjacent to the terminal point is in state B.

(⋆4) For any cell in state P_i, there exists a path to the cell from a cell adjacent to the starting point.

The condition (⋆3) prevents a new path from being generated after a target configuration is obtained. If a configuration satisfies the conditions (⋆1–3), then any configuration obtained by a transition also does. The condition (⋆4) always holds from an ideal initial configuration. We can then mention a weak form of self-stability. Its proof is relatively easy as we assume fairness and boundedness, but we should also assume complete asynchrony.

Theorem 1. *Under the rules (1, 2, 3, 5, 6), for any completely asynchronous fair run $\mathcal{R} = \mathcal{C}_0, \mathcal{C}_1, \ldots$ starting from an ideal initial configuration $\mathcal{C}_0$, there exists $i \geq 0$ such that $\mathcal{C}_i, \mathcal{C}_{i+1}, \ldots$ all satisfy the conditions (⋆1–3).*

Assume that the starting point and the terminal point are connected. If the condition (⋆3) was violated infinitely often in $\mathcal{R}$, there would exist a configuration that does not satisfy the condition (⋆3) and appears infinitely often in $\mathcal{R}$ because the maze is bounded. If a configuration appears infinitely often in $\mathcal{R}$, any configuration that can be obtained from it by transitions also appears infinitely often due to the fairness of $\mathcal{R}$. Therefore we can obtain a configuration that satisfies the condition (⋆3) and also appears infinitely often. (Actually, the condition (⋆3) continues to hold.) Since we start from an ideal initial configuration, the condition (⋆4) holds. Therefore the condition (⋆1) is satisfied. The configuration does not contain looped paths (paths containing a loop) by the rule (6). Since we can delete paths to dead ends, we obtain a configuration that also satisfies (⋆2). (The following stronger condition holds: any maximal path from a cell adjacent to the starting point ends in a cell adjacent to the terminal point.)

If the starting point and the terminal point are not connected, the conditions (⋆1) and (⋆3) obviously hold. The condition (⋆2) also holds since there is no path from a cell adjacent to the starting point.

The gellular automata defined so far are not self-stable. We can extend them by adding the following rule.

$$P_i\ (\neg S \wedge \neg P_{i-1})\ \rightarrow\ B. \tag{7}$$

By this rule, it is possible to reach a target configuration even if a cell is dynamically changed to state W or B. Although self-stability does not hold in general. we have the following form of self-stability.

Theorem 2. *Let $\mathcal{C}_0$ be a configuration obtained from an ideal initial condition under the rules (1, 2, 3, 5, 6, 7) and the rules $W() \rightarrow B$, $B() \rightarrow W$, $P_i() \rightarrow W$ and $R() \rightarrow W$ under complete asynchrony. For any completely asynchronous fair run $\mathcal{R} = \mathcal{C}_0, \mathcal{C}_1, \ldots$ under (1, 2, 3, 5, 6, 7), there exists $i \geq 0$ such that $\mathcal{C}_i, \mathcal{C}_{i+1}, \ldots$ all satisfy the conditions ($\star$1–4).*

We first obtain a configuration that satisfies ($\star$4) by the rule (7). We can then obtain a configuration that satisfies ($\star$3) as in the previous theorem.

From now on, we will show a stronger form of self-stability by adding some new states and new transition rules. We introduce yet another condition.

($\star$5) A cell in state P_{i-1} that is not adjacent to the terminal point should not be adjacent to a cell in state P_i that is adjacent to another cell in state P_{i-1} or the starting point if $i = 0$.

If this condition holds, there exists no T-shaped junction as in the left panel of Fig. 4 except near the terminal point. The stronger form of self-stability is stated as follows. The rules (8–17) will be added later. (The rule (7) is excluded here.) We do not need complete asynchrony.

Theorem 3. *Under the rules (1, 2, 3, 5, 6, 8–17), for any fair run $\mathcal{R} = \mathcal{C}_0, \mathcal{C}_1, \ldots$ starting from an initial configuration $\mathcal{C}_0$, there exists $i \geq 0$ such that $\mathcal{C}_i, \mathcal{C}_{i+1}, \ldots$ all satisfy the conditions ($\star$1–3,5).*

In order to satisfy the condition ($\star$5), we add transition rules to detect a T-shaped junction of paths and make one of the paths regress. In order to detect a junction, a cell in state P_1 must change its transition depending on whether there are only one or more neighboring cells in state P_0. Such transitions are unfeasible since gellular automata are Boolean-totalistic. However, by employing the asynchrony of gellular automata, it is possible to create a mechanism that counts the number of neighboring cells. We add two variants P'_i and P^*_i of state P_i for each i. They are recognized as P_i by their neighboring cells except when the following rules are applied. (The definition of paths should also be extended.)

$$P_i\ ()\ \rightarrow\ P'_i \tag{8}$$
$$P'_i\ ()\ \rightarrow\ P_i \tag{9}$$
$$P_i\ (P_{i-1} \wedge P'_{i-1})\ \rightarrow\ P^*_i \tag{10}$$
$$P_0\ (S \wedge P'_{n-1})\ \rightarrow\ P^*_0 \tag{11}$$
$$P'_i\ (P^*_{i+1} \wedge \neg T)\ \rightarrow\ B \tag{12}$$
$$P^*_i\ (\neg P'_{i-1})\ \rightarrow\ P_i \tag{13}$$

According to the first two rules (8) and (9), cells in state P_i or P'_i can alternately and unconditionally make a transition to switch their states from P_i to P'_i or from P'_i to P_i. Since gellular automata are asynchronous, two cells in state P_0, for instance, make transitions to P'_0 and back to P_0 independently. We eventually obtain a configuration in which the rule (10) can be applied. The cell in state P_1^*, which was in state P_1 before the rule was applied, urges the neighboring cell in state P'_0 to make a transition to B by the rule (12) unless it is adjacent to the terminal point, and then it changes back to state P_1 by the rule (13). A path without a junction will not disappear because the rule (10) requires two cells in different variants of state P_{i-1}. We need the rule (11) for detecting a junction adjacent to the starting point.

If there exists a configuration that satisfies the condition ($\star$3) and appears infinitely often, we can remove infinite loops by the above rules, which preserve the condition ($\star$3). We can also remove dead ends as before. So we obtain a configuration that satisfies the conditions ($\star$1–3, 5).

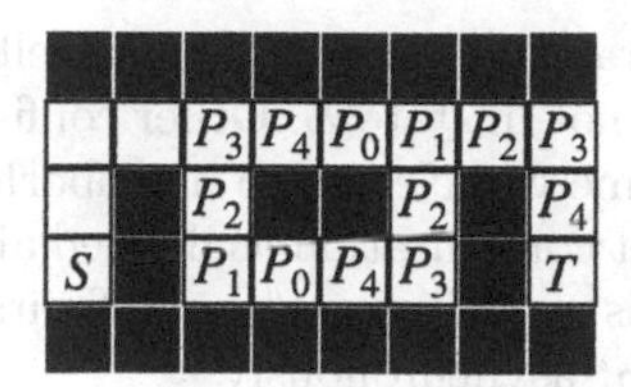

Fig. 9. No path from the starting point

s_2	s_1	s_0	s_1	s_2	s_3	s_4	s_0	W	W	W	W
s_1	s_0	s	s_0	s_1	s_2	s_3	s_4	s_0	W	W	W
s_0	s	S	s	s_0	s_1	s_2	s_3	s_4	s_0	W	W
s_1	s_0	s	s_0	s_1	s_2	s_3	s_4	s_0	W	W	W
s_2	s_1	s_0	s_1	s_2	s_3	s_4	s_0	W	W	W	W

Fig. 10. Propagation of state changes

However, since we start from an arbitrary initial configuration, not necessarily ideal, we can have configurations as in Fig. 9, where no cell adjacent to the starting point is in state P_i. (This kind of configuration can also result by the above rules.) Therefore, we add the following transition rules so that the starting point can broadcast error states, which will delete all paths already generated. In order to concisely present the rules, we use letters such as X and Y to denote states of passages, e.g., B, R, or P_0. Those states superscripted with E are the error states.

$$S\ (\bigwedge_i \neg P_i) \ \rightarrow\ S^E \tag{14}$$

$$X\ (Y^E) \ \rightarrow\ X^E \tag{15}$$

$$S^E\ () \ \rightarrow\ S \tag{16}$$

$$X^E\ () \ \rightarrow\ B \tag{17}$$

According to these transition rules, an ideal initial configuration is eventually obtained from which configurations that satisfy the conditions ($\star$1–5) can be reached by completely asynchronous transitions. On the other hand, the conditions ($\star$1–5) are preserved by the above rules since no error state appears.

4 Leader Election

Leader election is a process to elect a single node as the leader of all nodes in a distributed computing system. The elected node must remain unchanged unless there are some errors. In the context of gellular automata, this means that the leader cell needs to be unique among all cells.

4.1 Outline

All cells can take five types of states W, E, S^i, s^i, s^i_j ($i \in \mathbb{Z}/3\mathbb{Z}$, $j \in \mathbb{Z}/5\mathbb{Z}$) such that

- W is the initial state,
- E denotes an error,
- S^i is a state of a leader node,
- s^i is a state of a node adjacent to a leader in state S^i,
- s^i_j is a state of other nodes corresponding to S^i.

We assume that the cellular space is finite.

A *target configuration* is defined as a configuration in which only one cell is in state S^i and other cells are in state s^i or s^i_j and not in state E. Other configurations are called *bad configurations*. We also assume von Neumann neighborhood.

The goal of leader election is to reach a target configuration as defined above. The outline of our leader election algorithm is as follows. If a configuration contains cells in state W, they are changed to S^i asynchronously.

$$W\ () \ \rightarrow \ S^i \tag{18}$$

If a cell in state W is adjacent to the leader in state S^i, it is changed to s^i.

$$W\ (S^i) \ \rightarrow \ s^i \tag{19}$$

If a cell in state W is adjacent to a cell in state s^i, it is changed to s^i_0.

$$W\ (s^i) \rightarrow s^i_0 \tag{20}$$

If a cell in state W is adjacent to a cell in state s^i_j, it is changed to s^i_{j+1}.

$$W\ (s^i_j) \ \rightarrow \ s^i_{j+1} \tag{21}$$

If some cells detect an error (described in detail later), they are changed to E, and state E is propagated to other cells.

$$S^i\ (E) \ \rightarrow \ E \tag{22}$$

$$s^i\ (E) \ \rightarrow \ E \tag{23}$$

$$s^i_j\ (E) \ \rightarrow \ E, \tag{24}$$

Cells in state E are eventually changed to W.

$$E\ () \ \rightarrow \ W \tag{25}$$

Figure 10 shows how cells in state W are changed to S^i, s^i, or s^i_j.

4.2 Detecting Multiple Leaders

There may be multiple leaders whose states are S^i and S^{i+1}. By introducing the following rules, states of the same type as that of the leader in state S^{i+1} are propagated.

$$s^i\ (S^{i+1} \wedge \neg S^{i+2} \wedge \neg s^{i+2} \wedge \neg s_0^{i+2} \wedge \neg s_1^{i+2} \wedge \neg s_2^{i+2} \wedge \neg s_3^{i+2} \wedge \neg s_4^{i+2} \wedge \neg E) \rightarrow s^{i+1} \quad (26)$$

$$s^i\ (s_j^{i+1} \wedge \neg E) \rightarrow s_{j+1}^{i+1} \quad (27)$$

$$s_j^i\ (s^{i+1} \wedge \neg E) \rightarrow s_0^{i+1} \quad (28)$$

$$s_j^i\ (s_j^{i+1} \wedge \neg E) \rightarrow s_{j+1}^{i+1} \quad (29)$$

The leader in state S^i is changed to E by s^{i+1} and s_j^{i+1}.

$$S^i\ (s^{i+1}) \rightarrow E \quad (30)$$

$$S^i\ (s_j^{i+1}) \rightarrow E. \quad (31)$$

There may be multiple leaders whose states are S^i. In order to handle this situation in the same way as previously described, the following rule is introduced.

$$S^i\ (\neg s^{i-1} \wedge \neg E) \rightarrow S^{i+1} \quad (32)$$

4.3 Detecting Mismatch

In some configurations, there may be some cells in state s^i whose neighbors are neither in state S^i nor in S^{i+1}, or some cells in S^i whose neighbors are neither in state s^i nor in s^{i+2}. The cells in S^i and s^i must be adjacent to each other. In order to cope with this situation, the following rules are introduced.

$$s^i\ (\neg S^i \wedge \neg S^{i+1}) \rightarrow E \quad (33)$$

$$S^i\ (\neg s^i \wedge \neg s^{i+2} \wedge \neg W) \rightarrow E \quad (34)$$

The cells in states S^i and s_j^i must not be adjacent to each other.

$$s_j^i\ (S^i) \rightarrow E \quad (35)$$

In some configurations, there may be some cells in state s_j^i whose neighbors are not appropriate ones.

$$s_j^i\ (\neg s^i \wedge \neg s_{j-1}^i \wedge \neg S^{i+1} \wedge \neg s^{i+1} \wedge \neg s_0^{i+1} \wedge \neg s_1^{i+1} \wedge \neg s_2^{i+1} \wedge \neg s_3^{i+1} \wedge \neg s_4^{i+1} \wedge \neg W) \rightarrow E\ (i=0) \quad (36)$$

$$s_j^i\ (\neg s_{j-1}^i \wedge \neg S^{i+1} \wedge \neg s^{i+1} \wedge \neg s_0^{i+1} \wedge \neg s_1^{i+1} \wedge \neg s_2^{i+1} \wedge \neg s_3^{i+1} \wedge \neg s_4^{i+1} \wedge \neg W) \rightarrow E\ (i \neq 0) \quad (37)$$

$$s_j^i\ (s_{j-2}^i) \rightarrow E \quad (38)$$

$$s_j^i\ (s_{j+2}^i) \rightarrow E \quad (39)$$

4.4 Self-stability

All bad configurations can be handled by error propagation mentioned in the previous sections. After resetting a bad configuration, we can obtain a configuration in which all cells are in state W. We choose one cell as the leader and change it to state S^i. We then propagate states s^i and s^i_j through the cellular space with the rules (19–21) in a synchronous manner, and obtain a regular pattern with one leader, which is a target configuration. It continues to be a target configuration even if the leader changes its state by the rule (32). The gellular automata are thus self-stabilizing.

What we must note is that the gellular automata are not practical under fully asynchronous state transitions. The rules (19–21) should be applied synchronously. In the above argument on self-stability, synchronous transitions are allowed due to fairness and boundedness of the gellular automata.

5 Concluding Remark

In this paper, we defined gellular automata solving maze problems and showed their self-stability. If smart materials were implemented according to this model of gellular automata, they would autonomously form structures like blood vessels or axons of neurons, which would repair themselves under external damage.

In order to make the gellular automata fully self-stable, we had to introduce transition rules that allow a cell to count the number of neighbors in a specific state. The complexity of the rules is evidence of the weakness of Boolean totality against the full totality of cellular automata. We also had to introduce rules for detecting isolation of the starting point.

In addition to maze problems, we are currently working on gellular automata solving 2-distance coloring and spanning tree, which are also typical problems in distributed computing. They will show the expressive power of gellular automata compared with other models of distributed systems.

In this paper, we also defined gellular automata solving leader election. The leader should change its state periodically in order to detect the existence of other leaders. Some transition rules should be applied synchronously for choosing a leader efficiently. We are also working on a method to simulate synchronous transitions by fully asynchronous ones although we need more states and rules.

Since gellular automata are models of gel materials and chemical solutions, it might be more promising to introduce features of those physical or chemical entities rather than staying in fully discrete mathematical models of cellular automata. We are currently planning to extend gellular automata by modeling physical or chemical phenomena, such as a concentration gradient. The extensions are expected to employ features of natural environment to achieve more efficient or powerful computation.

Acknowledgement. This work was supported by JSPS KAKENHI Grant Number 17K19961.

References

1. Adleman, L.M.: Molecular computation of solutions to combinatorial problems. Science **266**(5187), 1021–1024 (1994)
2. Dolev, S.: Self-Stabilization. MIT Press, Cambridge (2000)
3. Golzari, S., Meybodi, M.R.: A maze routing algorithm based on two dimensional cellular automata. In: El Yacoubi, S., Chopard, B., Bandini, S. (eds.) ACRI 2006. LNCS, vol. 4173, pp. 564–570. Springer, Heidelberg (2006). https://doi.org/10.1007/11861201_65
4. Hagiya, M., et al.: On DNA-based gellular automata. In: Ibarra, O.H., Kari, L., Kopecki, S. (eds.) UCNC 2014. LNCS, vol. 8553, pp. 177–189. Springer, Cham (2014). https://doi.org/10.1007/978-3-319-08123-6_15
5. Hosoya, T., Kawamata, I., Shin-ichiro, M.N., Murata, S.: Pattern formation on discrete gel matrix based on DNA computing. New Gener. Comput. **37**(1), 97–111 (2019)
6. Isokawa, T., Peper, F., Kawamata, I., Matsui, N., Murata, S., Hagiya, M.: Universal totalistic asynchonous cellular automaton and its possible implementation by DNA. In: Amos, M., Condon, A. (eds.) UCNC 2016. LNCS, vol. 9726, pp. 182–195. Springer, Cham (2016). https://doi.org/10.1007/978-3-319-41312-9_15
7. Kawamata, I., Yoshizawa, S., Takabatake, F., Sugawara, K., Murata, S.: Discrete DNA reaction-diffusion model for implementing simple cellular automaton. In: Amos, M., Condon, A. (eds.) UCNC 2016. LNCS, vol. 9726, pp. 168–181. Springer, Cham (2016). https://doi.org/10.1007/978-3-319-41312-9_14
8. Murata, S., Konagaya, A., Kobayashi, S., Saito, H., Hagiya, M.: Molecular robotics: a new paradigm for artifacts. New Gener. Comput. **31**(1), 27–45 (2013)
9. Pech, A., Hingston, P., Masek, M., Lam, C.P.: Evolving cellular automata for maze generation. In: Chalup, S.K., Blair, A.D., Randall, M. (eds.) ACALCI 2015. LNCS (LNAI), vol. 8955, pp. 112–124. Springer, Cham (2015). https://doi.org/10.1007/978-3-319-14803-8_9
10. Rozenberg, G., Bäck, T., Kok, J.N.: Handbook of Natural Computing. Springer, Berlin (2012). https://doi.org/10.1007/978-3-540-92910-9
11. Wang, S., Imai, K., Hagiya, M.: On the composition of signals in gellular automata. In: 2014 Second International Symposium on Computing and Networking (CANDAR), pp. 499–502. IEEE (2014)
12. Wolfram, S.: A New Kind of Science. Wolfram Media, Champaign (2002)
13. Yamashita, T., Hagiya, M.: Simulating population protocols by gellular automata. In: 2018 57th Annual Conference of the Society of Instrument and Control Engineers of Japan (SICE), pp. 1579–1585. IEEE (2018)
14. Yamashita, T., Isokawa, T., Peper, F., Kawamata, I., Hagiya, M.: Turing-completeness of asynchronous non-camouflage cellular automata. In: Dennunzio, A., Formenti, E., Manzoni, L., Porreca, A.E. (eds.) AUTOMATA 2017. LNCS, vol. 10248, pp. 187–199. Springer, Cham (2017). https://doi.org/10.1007/978-3-319-58631-1_15

Correction to: Geometric Tiles and Powers and Limitations of Geometric Hindrance in Self-assembly

Daniel Hader and Matthew J. Patitz

Correction to: Chapter "Geometric Tiles and Powers and Limitations of Geometric Hindrance in Self-assembly" in: I. McQuillan and S. Seki (Eds.): *Unconventional Computation and Natural Computation*, LNCS 11493, https://doi.org/10.1007/978-3-030-19311-9_16

The original version of this paper contained a mistake. Theorem 3 in "Section 4," which claimed that for every temperature-1 system in the Dupled abstract Tile Assembly Model (DaTAM) there exists a temperature-1 system in the Geometric Tile Assembly Model (GTAM) which simulates it at scale factor 1 and using only 2 glues, was incorrect. That theorem and the section that contained it have been removed. Please note that that result was independent of all other results, and the fact that it was incorrect does not impact them or any of the other remaining portions of the paper.

The updated version of this chapter can be found at
https://doi.org/10.1007/978-3-030-19311-9_16

I. McQuillan and S. Seki (Eds.): UCNC 2019, LNCS 11493, p. C1, 2019.
https://doi.org/10.1007/978-3-030-19311-9_22

Author Index

Printed in the United States
By Bookmasters